MW00609045

A Gateway to Higher Mathematics

Jason H. Goodfriend, Ph.D.
Bureau of Transportation Statistics
of the U.S. Department of Transportation
and The George Washington University

JONES AND BARTLETT PUBLISHERS
Sudbury, Massachusetts
BOSTON   TORONTO   LONDON   SINGAPORE

World Headquarters
Jones and Bartlett Publishers
40 Tall Pine Drive
Sudbury, MA 01776
978-443-5000
info@jbpub.com
www.jbpub.com

Jones and Bartlett Publishers Canada
2406 Nikanna Road
Mississauga, ON L5C 2W6
CANADA

Jones and Bartlett Publishers
International
Barb House, Barb Mews
London W6 7PA
UK

Jones and Bartlett's books and products are available through most bookstores and online booksellers. To contact Jones and Bartlett Publishers directly, call 800-832-0034, fax 978-443-8000, or visit our website www.jbpub.com.

Substantial discounts on bulk quantities of Jones and Bartlett's publications are available to corporations, professional associations, and other qualified organizations. For details and specific discount information, contact the special sales department at Jones and Bartlett via the above contact information or send an email to specialsales@jbpub.com.

Production Credits
Acquisitions Editor: Tim Anderson
Production Director: Amy Rose
Editorial Assistant: Leslie Chiller
Associate Production Editor: Tracey Chapman
Marketing Manager: Matthew Payne
Manufacturing and Inventory Coordinator: Amy Bacus
Composition: ATLIS Graphics, Inc.
Cover Design: Kristin E. Ohlin
Cover Image: © Photos.com
Printing and Binding: Malloy, Inc.
Cover Printing: Malloy, Inc.

Library of Congress Cataloging-in-Publication Data

Goodfriend, Jason.
   A gateway to higher mathematics / Jason Goodfriend.— 1st ed.
      p. cm.
   Includes bibliographical references and index.
   ISBN 0-7637-2733-4 (casebound)
   1. Mathematics.   I. Title.
   QA373.G66 2005
   510—dc22

                                                                    2004025532

Printed in the United States of America
09 08 07 06 05   10 9 8 7 6 5 4 3 2 1

IN MEMORY OF

My father, Dr. Paul Louis Goodfriend.
He was a scholar and a gentleman.

Acknowledgments

I would like to thank my mentor and mathematics professor, Dr. Kenneth Rosen, for all of his advice. I would also like to thank the staff of Jones and Bartlett Publishers, Inc. for all their help, in particular my acquisitions editors, Stephen P. Solomon and Tim Anderson, and my editors, Tracey Chapman, Deborah Arrand, and Caroline Senay.

I would like to thank my family for all of their love and support, in particular, my wife Kathryn Goodfriend; my mother, Beverly Goodfriend; my brother, Benedict Goodfriend; and my cousins, Michael, Tricia, and Kimberly Weiler.

I would like to thank my friends for all of their moral support. In particular, I would like to thank Charles Titus, Gigi and Michael Louden, Diane Foster, Linda Boan, Lisa Schaefer, Tess Brothersen, and John DiLeo.

Finally, I would like to thank the employees and contractors at the Bureau of Transportation Statisics for their moral support. In particular, I would like to thank my colleagues Donald Bright, Sami Parbhoo, Ivan Lopez, and Suganthi Ramakrishnan.

Preface

Purpose of This Text

The motivation for writing this book is my own experience as a mathematics student, making the jump from elementary calculus and linear algebra to the "higher mathematics" of abstract algebra and analysis. As I read and contemplated the proofs of results in these latter subjects, I frequently found what I considered to be gaps in the reasoning of the arguments. Although mathematics is supposed to be a deductive science, I felt that I was still forced to make "leaps of faith" because the stepping stones used to get from one result to another were often unclear.

Later, while spending much time filling in these gaps, I began to understand the cause of them. It appeared to me that many of the gaps could be filled by a proper construction of the real number system. In such a construction, one has to deal with many issues that keep rearing their heads in higher mathematics courses, including 1) a basic understanding of elementary logic and set theory, 2) the axiomatic system of being provided with a set and some properties of that set, 3) the methods of using equivalence classes, 4) an understanding of the meaning of isomorphisms, and 5) an understanding of defining objects recursively. Of course, such a construction should not be at the expense of failing to include the necessary introductions to abstract algebra, number theory, and analysis. Instead, this construction should blend in with these subjects and complement them.

Many colleges and universities offer classes for mathematics majors that focus on preparing students for higher mathematics classes. The textbooks used in these classes can be referred to as "bridging" texts because they provide a bridge to higher mathematics. Typically, these classes emphasize the craft of proof-writing. I have discovered that including the construction of the real number system helps to fulfill the objectives of these classes with some added benefits. Upon performing this construction, students are able to not only do proofs, but they also are equipped to deal with mathematical issues that will come up time and time again in later courses.

This book provides a careful development of the real number system that is integrated with the process of laying the foundation stones for subjects such as abstract algebra, number theory, and analysis. The methods of correct proof-writing are emphasized. The student is gently led along the glorious path of learning how to interpret and construct sophisticated proofs of the type encountered in higher mathematics courses. The exercises at the end of each section are designed to further the student's understanding of the material and to assist in the student's development of proof-writing.

This text also can be used as an accessory to a student already enrolled in a higher-level mathematics course. Although anyone with a high school education could theoretically follow the material in this book (because it is entirely self-contained), I feel that calculus provides the student with the very useful mathematical maturity that is needed to understand the motivation behind the concepts in this book.

Defining Characteristics of this Text

The student is advised not to read this subsection. Instead, the student is advised to go to the following subsection entitled "To the Student."

Here I describe what I believe should be the components of a textbook that is designed to help the student make the transition to higher mathematics courses. This textbook contains all of these components.

1) **Teaching students how to understand and construct proofs of theorems that possess important mathematical content.**

   Not only should the text make the student aware of various proof techniques and teach how to apply them, it should soon apply them to theorems that will provide them with strong foundations in mathematics. Many of the texts that teach the craft of proof-writing are overzealous to classify and demonstrate every type of proof technique, but they do this at the expense of showing students how to understand and write proofs of deep and important results. Instead, they demonstrate the many types of proof techniques on simple and shallow results. Advanced textbooks frequently skip steps in proofs, so a bridging class can provide students with the means to fill in the gaps by working through deep proofs slowly and carefully. Furthermore, it makes sense to introduce the demonstration of some methods of proof at a point in the student's mathematical development in which the true usefulness of the methods can be shown in a natural setting. This is in lieu of teaching all the methods of proof at once.

2) Presenting the material in correct logical order.
Mathematicians promise students that most of mathematics can be built upon the foundation stones of naive set theory. At least once in his or her career, the student should have the opportunity to see how this works. A bridging book can fill that need. Many of the existing bridging textbooks skip around, so it is difficult for the student to construct the logical progression of mathematical results. A failure to lay secure mathematical foundation stones (without going overboard, of course) often leads students to feel that mathematics is a game in which the rules are unknown.

For example, after telling students that most of mathematics can be built upon naive set theory, many of these bridging books define an ordered pair (a, b) to be an "ordered collection of objects." Any discerning student is going to see that this definition does not use prior concepts! Students deserve to see the ordered pair (a, b) to be defined as $\{\{a\}, \{a, b\}\}$, along with a full explanation of why this works, at least once in their career! (Although a bridging book does not have to develop full-blown axiomatic set theory, it should point out the logical consequences of failing to do so, and the student should be encouraged to at least look at a full-blown axiomatic development someday.)

As another example, consider a binary operation on a set. Almost all textbooks make a jump from the assumption of commutativity and associativity to the fact that when it comes to applying the binary operation to given n elements of a set, the order of the application of the operation does not matter. (That is, the *generalized* commutative and associate laws hold.) It actually takes some work to establish this, and the student is not likely to ever see this work unless it is presented in a bridging book.

As a final example, the Recursion Theorem and the Generalized Recursion Theorem are of major importance throughout mathematics. The logical justification for these theorems is rarely presented in bridging courses, perhaps because this justification is a little complicated. Nonetheless, there are concepts, such as uniqueness, that can be learned by going through the proofs of them, and the student deserves to have a chance to see this. Some of the more complicated proofs can be made optional, but they should be included in the text. Many students go on to graduate mathematics courses without ever learning the justifications for proof steps taken in higher-level mathematics books. One of the purposes of this book is to correct this unfortunate omission.

Now, there is no one universally recognized curriculum for bridging courses. This text is also amenable to professors who wish to skip certain topics or need to do so. In such situations, the student can be told that some topics will be taught out of logical order. Students also can be told that this text can be used as a reference should they ever desire to see the topics addressed in logical order.

3) **Developing the real number system.**
   A development of the real number system based on naive set theory (and the Axiom of Infinity) in correct logical order, along with many sprinklings of intuitive explanations, can be invaluable to a bridging student. Interspersed with this development should be incursions into abstract algebra, number theory, and analysis. Of course, examples can, and should be, introduced concerning the real numbers that are out of logical order in order to aid the student's intuition that already exists about the real numbers. However, such digressions should not interfere with the formal development (i.e., theorem and proof). Also, the student should be informed when such necessary digressions occur. In addition to developing the real number system, the student should be exposed to decimal expansions, a discussion of some of the ordinary algorithms for arithmetic (that the student has employed since grade school without knowledge of why they work) and their justifications, the proofs of uniqueness of structures such as a Peano space and the real number system, and a discussion of real exponentiation. Some of the advantages of such a construction of the real number system are as follows:
   - Many of the steps taken in the development of the real number system demonstrate techniques and strategies that will be used time and time again in the student's mathematical career. For example, a development of the real number system demonstrates the power and usefulness of the concept of equivalence relations.
   - A development of the real number system provides the student with a very powerful intuitive feel for the true meaning of isomorphisms. For example, showing that any two Peano spaces are isomorphic is in concert with the student's intuitive feel that the symbols used to represent integers are immaterial. The importance of developing an intuition about isomorphisms cannot be overstated.
   - The teaching of the foundations of abstract algebra, number theory, and analysis occur in a more "natural" setting.
   - This approach assists students in integrating their everyday experience of the real numbers with the theory of real numbers.

I believe that it is important to remember the immortal words of Edmund Landau in his classic text, *Foundations of Analysis*:

> "... my daughters have been studying (Chemistry) at the University for several semesters already and think that they have learned the differential and integral calculus in College; and yet they still don't know why $x \cdot y = y \cdot x$."

4) **Presenting the role of notation in mathematics.**
One problem that mathematics students sometimes develop is confusion between the notation being used to convey mathematical results and the results themselves. Therefore, it is important that a bridging book demonstrate that notation is simply a convenience device, albeit a very important one. A good selection of notation can aid the learning and discovering process, just as a poor selection can muddy the waters.

Some textbooks designed to introduce the student to the axiomatic method and correct proof-writing focus heavily on formal logic and the axiomitization of set theory. In this text, those aspects of logic and set theory that are important to allowing the student to properly learn algebra, analysis, and number theory, are emphasized. For the most part, the use of Cantor's "naive" set theory suffices. It is pointed out how the use of this has led to some contradictions in mathematics but that students will not encounter any such problems for quite some time (if ever) in their pursuit of an understanding of the basics of higher mathematics. The Axiom of Infinity is discussed, as it is used to prove that the integers exist. There is also a short discussion of the Axiom of Choice. This axiom is used (in one of the appendices) in the proofs of several theorems about infinite sets.

To the Student

It is amazing that the whole stupendous mountain of mathematical knowledge can be built up from a few relatively simple concepts. This fact was discovered in the latter part of the 19th century, when mathematicians such as Georg Cantor, Richard Dedekind, and Giuseppe Peano struggled successfully to provide a logical foundation for arithmetics. You are probably somewhat familiar with the concept of starting with several postulates and then proving results based on them. Perhaps you had a geometry class in which you started with Euclid's postulates. If so, you may have wondered (as I did) why algebra was not taught in a similar manner. The answer is that to do so requires a level of sophistication that is usually not attained by high school students. In this text, we start with a few basic principles and use them to construct elaborate mathematical structures that every mathematician needs to know. In particular, the real number system will be carefully developed from first concepts.

Why is this a useful endeavor? Because mathematics is a deductive science. This means that each mathematical result should be provable from first concepts and/or prior results. You have been working with real numbers for many years now. However, can you come up with a precise definition of what a real number actually is? If not, how do you know what methods of proofs of theorems about real numbers are valid? Hopefully, you will find it interesting to discover how such intricate structures can be formed from simpler concepts.

While studying this text, you will develop insights and skills that will serve you well in future mathematical courses. One very important mathematical skill is the ability to understand how to read and construct proofs. It is unfortunate that many advanced mathematics textbooks gloss over some steps in proofs. Upon reading this book, you should be well prepared to know what is required to fill in these gaps.

Mathematics is not a spectator sport. To really understand what is going on, you must be a participant. Only by practice can you learn how to combine intuition with deduction in order to be able to correctly interpret and construct proofs. I strongly suggest that you read this book critically. Every statement made in a proof should be justifiable. If you cannot justify a statement that is made in a proof in your own mind, then continue to struggle with it until you see the light (or else make a note to come back to it later and ask someone). I also strongly suggest that you do as many of the exercises as possible because they will assist you in the strengthening of your skills.

Mathematicians teach that mathematics is a deductive science. However, in the presentation of higher mathematics courses, steps are often missing from proofs. Therefore, it is very useful that at least once in your mathematical journey that you see a presentation of some proofs without missing steps. Using this text, you will gradually reach the point where such missing steps are not a problem because you will have the necessary tools to fill them in whenever you feel the need to do so.

Finally, you are encouraged to read this book sequentially (at least someday). However, there are certain portions of the text that can be skipped, if necessary, without loss of continuity. For example, some of the proofs with a high degree of complexity (that have been included for completeness sake) can be taken on faith for the time being by the time-constrained student. We shall indicate this by writing "[OPTIONAL]". Of course, you are encouraged to read these portions when you find the time. This text can be used as reference throughout your mathematical career.

Guide to Reading the Text

One of the tasks of this book is to present a development of the real number system. However, properties of real numbers that should be well known to the student are frequently used as examples of concepts before these properties are formally introduced. This is because they aid the student in the development of intuition. Strictly speaking, these examples could be ignored in terms of the logical development of the book because the theorems, proofs, and definitions do not depend on them; but, to do so would deprive the reader of a great deal of the motivation that lurks behind the formal development of the real numbers.

Theorems, definitions, tables, and figures are referenced throughout the book by the chapter number and the number assigned within the chapter. For example, Theorem 2.7 refers to the seventh theorem of Chapter 2. Chapters are divided into sections, and many sections contain exercises at the end. For some sections that introduce many concepts or complicated concepts, subsection headers are used. These subsection dividers can be used as "stopping points." As mentioned in the "To the Student" section, a section (or portion of a section) labeled "[OPTIONAL]" can be skipped at first reading without loss of continuity.

Contents

Chapter 1

Logic and Techniques for Proofs

Mathematics is concerned with making inferences. A mathematical theory begins with a collection of statements or rules known as **axioms**. The only formal requirement of these axioms is that they do not contradict each other; however, they may or may not conform to our ideas of the physical universe. The axioms are the "starting points." The mathematician then attempts to find out what statements are implied by these axioms. These statements are called **theorems**. Theorems are established by invoking the axioms and previously established theorems. The demonstration that a theorem does indeed follow from axioms and previously established theorems is known as a **proof**. The construction of a proof utilizes rules of inference. The collection of these rules is known as **logic**. Logic has applications not only to mathematical proofs but also to fields such as computer science and electrical engineering.

However, mathematicians often do not explicitly list the axioms with which they are working. Instead, they have a collection of informal statements that act as "quasi-axioms." Indeed, this is the approach that we shall take, although we do explicitly list a few axioms of that branch of mathematics known as set theory. Also, you often will find that the proofs in advanced mathematics texts often skip steps. The reader is expected to know how to fill them in. In this text, we will work through proofs with more detail than is found in most advanced mathematics texts. This will prepare you to recognize when steps are missing and to know how to fill in the missing steps in the proofs of your future mathematics courses.

One excellent way to gain an understanding of how mathematics works is to do what we refer to as "develop the real number system." By this we mean we start with a few initial concepts that essentially work as our quasi-axioms. Using them as our starting points, we prove the important properties of real numbers that you will need to know when you take higher-level mathematics courses. Parallel to this, we also lay the foundation for the important mathematical topics of abstract algebra, number theory, and analysis.

This first chapter is about logic. Because logic is the set of tools that the mathematician uses to make inferences, it is important to understand how these tools work. In addition to introducing these tools, we demonstrate some important techniques that are used to construct proofs. Logic and proof techniques go hand in hand. It is logic that allows one to determine when a proof is correct. Additional proof techniques will be introduced in subsequent chapters as needed.

Mathematical logic is a complex field with many subtleties. In fact, there are axioms for logic, but we shall not concern ourselves with them. For your purposes, the material in this chapter will allow you to travel a long way down the mathematical road of discovery. Later along your mathematical journey, you can pause to discover more of the nuances of logic. In this chapter, the basic principles of logic that will be of most use to you are identified, followed by some time-honored methods of proof. You will probably find, however, that proof construction is, to a large degree, an art. Studying the proofs in this text and writing proofs for the exercises will assist you in becoming more proficient at proof construction.

Section 1.1 Statements and Truth Tables

In this text, a **statement** is an assertion that has truth value (i.e., it is either true or false). A statement is also sometimes called a **proposition**. The following example demonstrates statements:

Example 1.1

1) Paris is the capital of France.
2) $1 + 1 = 2$.
3) $1 + 1 = 3$.

The first sentence is a statement, provided that we have a preexisting understanding of what the terms *Paris, capital,* and *France* mean. This is a true statement, assuming these terms are defined in the usual way. The second sentence is also a true statement. The third sentence is a statement because it has truth value. It is a false statement.

The sentences in the following example do not constitute statements:

Example 1.2

1) Please pass the salt.
2) $1 + 1$.
3) $x + 1 = 3$ (where x is a variable).

It is clear why the first two expressions in Example 1.2 fail to be statements. They do not have truth value. The third sentence fails to be a statement unless the variable x is assigned a specific value; that is, only when x has been assigned a value does this sentence have truth value. So, for example, if we assigned the number 2 to x, then we would have the true statement $2 + 1 = 3$. If we assigned the number -6 to x, then we would have the false statement $-6 + 1 = 3$.

For notational ease, statements are frequently represented by lowercase letters (although this is by no means a requirement). For example, we might represent the statement *Paris is the capital of France* by the letter p.

We call a statement that is constructed from other statements a **compound statement**. Logic is very much concerned with rules for determining the truth value of a compound statement from the truth value of the statements that are used to construct the compound statement. There are some common means of forming compound statements, which are identified below. This brings us to our first formal definition. By the way, a **definition** is simply an agreement to call certain things by a simpler name for convenience sake.

Definition 1.1

If p and q are any two statements, the **conjunction** of p and q, written $p \land q$, is that statement that is defined to be true precisely when both p and q are true. Otherwise, it is false. This statement is pronounced "p and q."

Another way of writing $p \land q$ is p *and* q. A useful way to display the truth value of a compound statement as a function of the truth value of its constituent statements is the **truth table**. The truth table for a conjunction is displayed below. The first and second columns indicate all of the possible truth values (T, true; F, false) that p and q can have, respectively. The last column indicates the corresponding value of $p \wedge q$. Note that this table simply indicates the information in Definition 1.1; that is, the table shows that the conjunction statement is true only when both p and q are true.

Table 1.1
The truth table
for conjunction

p	q	$p \wedge q$
T	T	T
T	F	F
F	T	F
F	F	F

Definition 1.2

If p and q are any two statements, the **disjunction** of p and q, written $p \vee q$ (or p *or* q), is that statement that is defined to be true precisely when one of the following holds: 1) p is true and q is false; 2) p is false and q is true; or 3) both p and q are true. Otherwise, it is false. This statement is pronounced "p or q."

Note that this definition conforms to our intuitive notion of *or*, with the exception that in normal English, when we say something like *The paper is on my desk or the paper is on the kitchen table*, we usually mean to preclude the possibility that both portions of this compound sentence are true. Another way to word Definition 1.2 is $p \vee q$ *is false only when p and q are both false*. The truth table for disjunction is presented below.

Table 1.2
The truth table
for disjunction

p	q	$p \vee q$
T	T	T
T	F	T
F	T	T
F	F	F

In general, when the word *or* is used in a mathematical statement, this refers to the definition given in Definition 1.2. Sometimes we want to use the English definition of *or* mentioned above, in which we wish to declare the statement *p* or *q* to be false when both *p* and *q* are true. This definition of *or* is called **exclusive or**. When we want to use this in a mathematical statement, it can be worded as *either p or else q, but not both*, or as something similar to this.

We need to define two more major tools in the production of compound statements.

Definition 1.3

If p is any statement, the **negation** of p, written $\neg p$, is the statement that is true precisely when p is false. Otherwise, it is false. This statement is pronounced "not p."

Hence, the negation of p is simply the statement with truth value that is the opposite of the truth value of p. The truth table for negation is presented below.

Table 1.3
The truth table
for negation

p	$\neg p$
T	F
F	T

Example 1.3

Evaluate the truth value of the statement $\neg((\textit{Paris is the capital of France}) \lor (1 + 1 = 3))$.

Solution: Let p denote the statement *Paris is the capital of France*, and let q denote the statement $1 + 1 = 3$. Then the compound statement is $\neg(p \lor q)$. Now, p is true and q is false. Hence, by the definition of disjunction, the statement $p \lor q$ is true. Finally, the invocation of the definition of negation shows that $\neg(p \lor q)$ is false. Thus, the compound statement is false.

Note that for the above example, the answer is probably intuitive, and you may not need to go through a formal argument. However, it is important that you understand how to perform this formal process. It should also be noted that in most advanced mathematics books the symbols for conjunction, disjunction, and negation are rarely used, and we shall use them only in this chapter. Instead, the words *and, or,* and *not*, respectively, are usually used. For instance, the compound statement in Example 1.3 might be written *It is not the case that: Paris is the capital of France or* $1 + 1 = 3$. Again, it is important that you be able to break this English statement down into its components as was done in the example.

It is also important that you understand the importance of parentheses. Parentheses indicate that the operations within them are to be performed before any operations acting on them from outside the parentheses (and we have already implicitly assumed this in the above discussions). Note that the statement $\neg(p \vee q)$ is not the same as the statement $(\neg p) \vee q$. The former statement indicates that the disjunction $p \vee q$ is obtained first, and then the negation operator is applied to this disjunction. The latter statement is the disjunction of the statement $\neg p$ with the statement q.

If there are nested parentheses, the corresponding expressions are evaluated from the inside first and then toward the outside. For example, consider the statement $((s \wedge p) \vee q) \wedge r$. The nested statement $s \wedge p$ is first evaluated. Let us call this nested statement a. Moving outward, we then evaluate the statement $a \vee q$. Let us call this statement b. We finally evaluate the statement $b \wedge r$.

To cut down on the need for so many parentheses without introducing ambiguity, a convention is normally applied to the negation operator $\neg$. The convention is that this operator applies immediately to whatever lies immediately to the right of it. For examples, $\neg p \vee q$ is interpreted as $(\neg p) \vee q$, and $\neg(p \vee q) \wedge s$ is interpreted as $(\neg(p \vee q)) \wedge s$.

The following definition is of the utmost importance in mathematics. It is the means of asserting that one statement follows logically from another.

Definition 1.4

If p and q are any two statements, then the **implication** $p \Rightarrow q$ is the statement that is false precisely when p is true and q is false. Otherwise, it is true. This statement is pronounced "p **implies** q."

First, the truth table for implication is presented, and then this statement is discussed at length.

Table 1.4

The truth table for implication

p	q	$p \Rightarrow q$
T	T	T
T	F	F
F	T	T
F	F	T

Frequently, for the im
sis, and q is referred to as the
statement, the conclusion must
some texts, $p \Rightarrow q$ is written $p \rightarrow q$.
other ways of stating $p \Rightarrow q$ are

1) If p, then q.
2) p only if q.
3) p is sufficient for q.
4) q is necessary for p.

While expressions 1 and 3 are easy to understand, expressio
quire a little thought. To understand p *only if q*, we observe fro
table that if $p \Rightarrow q$ is true and q is not true, then p is also not true.
the only way to have p true (given the truth of $p \Rightarrow q$) is to have q true.
is why we write p *only if q*. Another way to say this is that the truth of q is ne
essary for the truth of p, and this is what expression 4 (*q is necessary for p*)
says.

You may wonder about the last two rows of Table 1.4. Note that these
rows indicate that if p is false, then $p \Rightarrow q$ is true regardless of the truth value
of q. This is a convention. We have to assign *some* truth value to $p \Rightarrow q$ in the
cases where p is false, even if we are not particularly interested in these
cases. We let $p \Rightarrow q$ be true *by default*. It turns out that this convention works
nicely.

Perhaps the best way to see the wisdom of this convention is by an example. From algebra, you would probably agree that for any real number
x, we have $x > 1 \Rightarrow x + 1 > 2$ without giving this much thought. Upon analyzing this in more detail, we see that the statements $x > 1$ and $x + 1 > 2$
are not really statements at all unless the value of the variable x is specified.
Nonetheless, we "want" this implication to be true without worrying about
the value of x, and the definition of implication allows us to accomplish this
because the above implication will be true regardless of the value of x.
Clearly substituting 3 for x works. Also plugging in -5 for x results in a true
statement by Definition 1.4, even though the statement $-5 > 1$ is false.

In the literature, the symbols $\wedge$, $\vee$, $\neg$, and $\Rightarrow$ are sometimes referred
to as **connectives**. This allows us to define a compound statement to be a
statement that contains one or more connectives. Of course, many compound statements are more complicated than those we have considered so
far. For example if p, q, and r are statements, then the statement $p \wedge q \wedge r$
is the statement that is true precisely when p, q, and r are all true. In general, a complex statement can be analyzed by applying one connective at a
time.

Example 1.4   Let p, q, and r be statements. Construct the truth table for the statement
$p \Rightarrow (q \text{ and } r)$.

values that p, q, and r can take
and r. Finally, we apply the im-
rth columns. The final column
he compound statement based
arts.

	$(q$ and $r)$
	T
	F
	F
	F
	T
	T
	T
	T

plication $p \Rightarrow q$. p is referred to as the **hypothe**

conclusion. Therefore, for $p \Rightarrow q$ to be a true

e true whenever the hypothesis is true. In

In the mathematical literature, some

7

F	F	F	P	T

ns 2 and 4 re-
the truth
That is,
This

It is important to realize that a different result may be obtained if the parentheses in Example 1.4 are altered to change the compound statement of interest to $(p \Rightarrow q)$ and r.

EXERCISES

1. Show that the truth table for $(p \Rightarrow q)$ and r is not the same as the truth table for $p \Rightarrow (q$ and $r)$.

2. Give an example of statements p and q such that $p \Rightarrow q$, but $\neg(q \Rightarrow p)$.

3. a. Construct the truth table for $(\neg(p \wedge q)) \Rightarrow r$.

   b. Let p be the statement *New York is in the United States*, let q be the statement *Canada is in Europe*, and r be the statement *Plants are mammals*. Evaluate the truth value of the statement defined in Exercise 3a.

4. Show that $(p \wedge (p \Rightarrow q)) \Rightarrow q$ for any statement p and any statement q. Why is this intuitively obvious?

5. Compare the truth tables for the following statements: $\neg(p \vee q)$ and $(\neg p) \wedge (\neg q)$. What do you find? Why is this not surprising?

6. Here is a well-known brainteaser: You have come to a fork in the road where you must go right or left. One direction leads to a wonderful place where your every

desire (within reasonable bounds) will be fulfilled. The other direction leads to a miserable place where you are guaranteed to be unhappy forever. Unfortunately, you do not know which path leads to the wonderful place. However, there are two women at an information booth at the fork (whose salaries are paid by the highway department). You are allowed to ask exactly one question of your choice, and you may address your question to either of the women (but not both). One of the women is always 100% truthful, and she is guaranteed to give you a truthful answer to whatever question that you ask. The other woman lies 100% of the time, and she is guaranteed to give you a false answer to whatever question that you ask. Unfortunately, you do not know which of the two women is the truthful one. What question should you ask to guarantee that you will know which road leads to the wonderful place upon receiving an answer? Prove that your method always works by exploring all of the possible implication statements.

Section 1.2   Logical Equivalence

Many mathematical results use the concept of logical equivalence, which is defined below.

Definition 1.5

Statements p and q are said to be **logically equivalent** provided both the implication $p \Rightarrow q$ and the implication $q \Rightarrow p$ are true. The logical equivalence of the statements p and q is denoted by $p \Leftrightarrow q$.

From the definition of implication, it can be seen that $p \Leftrightarrow q$ is simply another way of saying that the statements p and q have the same truth value; that is, the two statements p and q must both be true, or else they must both be false. Another notation commonly used in the mathematics community for expressing $p \Leftrightarrow q$ is p *iff* q, and this is pronounced "p if and only if q." Another common way to express $p \Leftrightarrow q$ is p *is necessary and sufficient for* q. For any two statements p and q, the implication $q \Rightarrow p$ is called the **converse** of the implication $p \Rightarrow q$. The implication $(\neg q) \Rightarrow (\neg p)$ is called the **contrapositive** of the implication $p \Rightarrow q$. The following example is important:

Example 1.5

Show that $(p \Rightarrow q) \Leftrightarrow ((\neg q) \Rightarrow (\neg p))$ for all statements p and all statements q.

Solution: This can be verified by a quick look at the following truth table.

Table 1.6
The truth table
for Example 1.5

p	q	$p \Rightarrow q$	$\neg q$	$\neg p$	$(\neg q) \Rightarrow (\neg p)$
T	T	T	F	F	T
T	F	F	T	F	F
F	T	T	F	T	T
F	F	T	T	T	T

Note that the column under $p \Rightarrow q$ is identical to the column under $(\neg q) \Rightarrow (\neg p)$. Example 1.5 simply states that an implication is logically equivalent to its contrapositive. This result (which should be quite intuitive) will be useful to you in proof construction. Frequently, you will be faced with attempting to prove that a given implication is true, and you will find it simpler to prove the truth of the contrapositive. Example 1.5 assures us that a proof of the truth of the contrapositive proves the truth of the given implication.

Another important logical equivalence is $(\neg(p \vee q)) \Leftrightarrow ((\neg p) \wedge (\neg q))$. This says that negating both statements p and q is logically equivalent to negating the statement p or q.

EXERCISES

1. Prove Show that $(\neg(p \vee q)) \Leftrightarrow ((\neg p) \wedge (\neg q))$ for any statements p and q.
2. Is it always the case that $((p \Rightarrow q) \wedge (q \Rightarrow r)) \Leftrightarrow (p \Rightarrow r)$ for any statements p, q, and r? Why or why not?
3. Specify a real number so that substituting the variable x in the expression $x^2 > 4$ and the expression $x > 2$ with this real number results in statements that are not logically equivalent.

Section 1.3   Quantifiers

This section discusses some additional (and important) techniques that are used to construct complex statements from simpler statements. Hopefully, you will begin to recognize these techniques in mathematical definitions and theorems that you have seen in previous mathematics courses. Before discussing quantifiers, we must introduce the concept of predicate functions.

Frequently, in mathematics, we encounter assertions that are functions of variables. The truth value of the assertion may depend on the value of the variables. Such a function is referred to as a **predicate function**. Perhaps this is best explained via examples.

Example 1.6

For each real number x, let $p(x)$ denote the statement $x > 5$. Clearly, $p(2)$ is false and $p(6.23)$ is true.

To be technical, the expression $p(x)$ in Example 1.6 is only a statement if the value of the variable x is specified; that is, the expression $x > 5$ only has truth value if the value of x is specified. This is because the expression $x > 5$ is a predicate function (i.e., $p(x)$ is a predicate function).

Example 1.7

For each pair of integers n, m, let $q(n, m)$ denote the statement *$n + m$ is prime*. Then $q(2, 3)$ is true and $q(1, 3)$ is false.

In general, a predicate function can have multiple variables, not necessarily just one variable as in Example 1.6 and two variables as in Example 1.7. The use of **quantifiers**, as discussed below, allows us to make statements about predicate functions. Quantifiers are groupings of words similar to "for all" or "there exists." We will now discuss two major types of quantifiers.

Frequently, we will make a statement about a whole class of objects. For example, we may want to assert that *for all real numbers x, we have $x^2 + 1 \geq 2x$*. (By the way, this turns out to be true.) To do this, we use the **universal quantifier** sign, which is denoted by $\forall$. This symbol is to be interpreted as "for all" or "for every." When a quantifier is employed, the collection of values that the variable can take on is called the **universe of discourse**. It is either stated explicitly, or it is clear from the context. For the inequality assertion, *for all real numbers x, $x^2 + 1 \geq 2x$*, the universe of discourse is, of course, all real numbers.

Example 1.8

The inequality assertion made above can be expressed as $\forall x \; x^2 + 1 \geq 2x$. It can also be expressed as follows: For each x in the universe of discourse (which is all of the real numbers in this example), let the predicate function $p(x)$ denote the statement $x^2 + 1 \geq 2x$. Then we can express the above assertion as $\forall x \; p(x)$.

Example 1.9

Express the statement *For all real numbers x, if $x^2 > 1$ then $x > 1$* in terms of the universal quantifier and other logical symbols, and evaluate the truth of this statement.

Solution: We set the universe of discourse to be all of the real numbers. Then we can write this as $\forall x (x^2 > 1 \Rightarrow x > 1)$. Now this statement is false, as can be demonstrated by the failure of the implication $x^2 > 1 \Rightarrow x > 1$ to hold when $x = -2$.

Example 1.10

Evaluate the truth of the statement in Example 1.9 if the universe of discourse is changed to all positive real numbers.

Solution: Because the statement $x^2 > 1 \Rightarrow x > 1$ is true for any positive number x, the statement $\forall x(x^2 > 1 \Rightarrow x > 1)$ is true for this new universe of discourse.

Example 1.11

Express the statement *For all rational numbers x and y, $x + y$ is rational* using universal quantifiers.

Solution: Here, the universe of discourse for both variables x and y is all rational numbers. Let the predicate function $p(x, y)$ denote $x + y$ *is rational.* Then we can express the statement as $\forall x \forall y \, p(x, y)$. (Note that we could also express this as $\forall y \forall x \, p(x, y)$.)

Frequently in mathematics, it is important to assert the existence of some object. For example, one such assertion is *There exists a real number x such that $x^2 - 2x + 1 = 0$.* To do this, we sometimes use the **existential quantifier** sign, which is denoted by $\exists$. This symbol is to be interpreted as "there exists" or "there is." In general, if $p(x)$ is a predicate function, then the statement $\exists x \, p(x)$ is to be interpreted as *there exists an x such that $p(x)$* or *for some x, $p(x)$.* There may be more than one such x; the assertion states that there is at least one. As is the case for the universal quantifier, when an existential quantifier is employed, the collection of values that the variable can possibly take on is called the **universe of discourse**.

Example 1.12

Express the statement *There exists a line L in the plane that is perpendicular to the x-axis* using the existential quantifier.

Solution: Here, the universe of discourse is the collection of lines in the plane. Let the predicate function $r(L)$ denote L *is perpendicular to the x-axis.* Then we can express the statement as $\exists L \, r(L)$.

Now, the fun begins when we combine universal and existential quantifiers.

Example 1.13

Express the statement *For every integer n, there exists a real number x such that $x > n$* using quantifiers.

Solution: Here, the universe of discourse for the universal quantifier is the collection of all integers. The universe of discourse for the existential quantifier is the collection of real numbers. Let the predicate function $G(x, n)$ denote $x > n$. Then we can express the given statement as $\forall n \exists x G(x, n)$.

An important point must be made about Example 1.13. What if we reverse the order of the quantifiers in that example, so that we have $\exists x \forall n G(x, n)$. Does this say the same thing as $\forall n \exists x G(x, n)$? The answer is no. For the statement in Example 1.13, the value of x can depend on the value of n. In other words, suppose we have distinct integers, n_1 and n_2. Then the statement in Example 1.13 says that there is a real number x_1 and a real number x_2, such that $x_1 > n_1$ and $x_2 > n_2$. The statement does not say that we necessarily have $x_1 = x_2$. On the other hand, the statement $\exists x \forall n G(x, n)$ says that there is at least one number x such that $x > n$ where you can "plug" in any integer n that you like while keeping x fixed. (Notice that this is false). This kind of ambiguity can be avoided by listing strict axioms for logic (which by the way would annoy most mathematicians). For our purposes, the context should be clear enough to avoid this sort of ambiguity.

Example 1.14

Express the statement *For each integer n, there is a real number x such that if* $x > n$ *then* $x^2 > n^2$ using quantifiers.

Solution: The universe of discourse for each of the quantifiers is the same as in Example 1.13. It should not be hard to see that the answer is $\forall n \exists x (x > n \Rightarrow x^2 > n^2)$. Note the use of parentheses to avoid ambiguity.

Example 1.15

(For students who have had calculus; otherwise, disregard and do not worry.)

Let f be a real-valued function in an open interval I of the real line containing the point a. Use quantifiers to state that f is continuous at a.

Solution: You will recall (potentially) that the definition of the continuity of f at a is as follows: For each $\varepsilon > 0$ there is a $\delta > 0$ such that for each $x \in I$, if $|x - a| < \delta$, then $|f(x) - f(a)| < \varepsilon$. We now let the universe of discourse for the variables ε and δ be all positive real numbers. We let the universe of discourse for the variable x be the interval I. We can now write the statement as $\forall \varepsilon \exists \delta (\forall x (|x - a| < \delta \Rightarrow |f(x) - f(a)| < \varepsilon))$.

It is common in mathematical texts to not always explicitly mention the universe of discourse, especially when it is obvious what universe of discourse is meant. At times, we will adapt this policy in this text.

EXERCISES

1. Express the following sentence using quantifiers: *Every person in the room has a cousin in the room.*

2. Express the following sentence using quantifiers: *For every person in the room, either that person has a cousin in the room or that person does not have a brother in the room.*

3. Evaluate the truth or falsity of the following statement: $\forall x \forall y (x < y \Rightarrow x^2 < y^2)$, in which the universe of discourse is all real numbers.

4. Express the following statement using quantifiers: *For every real number x, there is a real number y, such that x = y + n for some integer n.* Make sure to specify the universes of discourse.

5. Express the following statement using quantifiers: *For all real numbers x and y, if x < y, then there exists a rational number r such that x < r < y.* Make sure to specify the universes of discourse.

6. Are the following two statements equivalent where P is a predicate function $\neg(\forall x P(x))$ and $\exists x(\neg P(x))$? Why or why not?

Section 1.4   Some Techniques for Proofs

In this section some time-honored techniques used to construct the proofs of mathematical theorems are presented. By no means does this section claim completeness; we shall learn many proof techniques as we go along in this text and develop the real number system.

To present these techniques in a useful manner, we assume some basic facts about integers and real numbers. This allows for some examples that you will be able to relate to. Later, of course, we will derive these facts about integers and real numbers. To construct a proof, it is important for the student to know what initial facts may be used as starting points. Starting points are initial results that we assume are true. We then base additional results that we want to derive on these initial results. After all, you have to start somewhere in the deductive method. Therefore, for the purposes of this section, we informally state some of these starting points to demonstrate these proof methods. In the fully developed proofs of the remaining sections of this book, it is very clear as to what the starting points are. To simplify our lives (for now), we will not explicitly write down all of the simple rules of algebra that we make use of, which every high school student should know (although, as is shown later in this text, the proofs of these rules are not always so simple).

Informal Starting Points

Informal Definition A

An integer n is said to be **even** if there exists an integer m such that $n = 2m$. Note that this can be written as *Given an integer n, n is defined to be even* *provided* $\exists m(n = 2m)$, where the universe of discourse for the variable m is all of the integers. For example, we know that the integer 6 is even. Why? Because we can write 6 as *twice some integer*, with that *some integer* being the integer 3 in this case; that is, $6 = 2 \times 3$.

Informal Definition B

An integer n is said to be **odd** if there exists an integer m such that $n = 2m + 1$. Note that this can be written as *Given an integer n, n is defined to be odd provided* $\exists m(n = 2m + 1)$, where the universe of discourse for the variable m is all of the integers. For example, we know that the integer 7 is odd. Why? Because we can write 7 as *twice some integer plus* 1, with that *some integer* being the integer 3 in this case; that is, $7 = (2 \times 3) + 1$.

I suspect that very few people will disagree with the following, although you will have to wait until a later chapter to see the proof.

Informal Fact I

For every integer n, n is either odd or even, but not both.

Informal Definition C

A **rational number** is any number r that can be written as a quotient of integers, with the denominator nonzero. That is, to say that r is a rational number means that there exists integers m and n with $n \neq 0$ such that $r = \frac{m}{n}$. Note that the integers m and n need not be unique. For example, the rational number $\frac{1}{3}$ can also be written as $\frac{2}{6}$ and as $\frac{-7}{-21}$.

Informal Fact II

Given any rational number r, there are integers m and n with $n \neq 0$ with $r = \frac{m}{n}$ such that m and n are not **both** even. For example, consider the rational number $r = \frac{24}{28}$. We can cancel from the numerator and denominator the highest power of 2 that divides both of them. In this case, the highest power is $2^2 = 4$. After dividing both the numerator and denominator by 4, we can write r as $r = \frac{6}{7}$. Note that this does the trick.

Proof Technique 1: Using the Contrapositive

Suppose that we are trying to prove the implication $p \Rightarrow q$. Recall from Section 1.2 that the implication $(\neg q) \Rightarrow (\neg p)$ is called the contrapositive of the implication $p \Rightarrow q$. From Example 1.5, we see that an implication and its contrapositive are logically equivalent. That is, we have $(p \Rightarrow q) \Leftrightarrow ((\neg q) \Rightarrow (\neg p))$. Hence, if we can prove that the contrapositive is true, we are done. Sometimes it is easier to do this than to directly prove $p \Rightarrow q$, as we illustrate in Example 1.16.

Example 1.16

Suppose n is an integer such that n^2 is an even integer. Show that n is even.

Solution: We note that for the integer n, we are trying to prove the implication n^2 *even* $\Rightarrow$ n *even*. Let's first see why it is difficult to prove this directly. Informal Definition A tells us that we want to show that there exists an integer m such that $n = 2m$ because that is what it means for n to be even. All we have to go on is that we know that n^2 is even. Well, that means that there must be an integer; let's call it k, such that $n^2 = 2k$. (Here, we have again used Informal Definition A.) How do we get n from n^2? By taking the square root, as we recall from algebra. However, if you try to do this, it will not appear to get you very far.

Let's try proving the contrapositive. This contrapositive is the implication $(\neg(n \; even)) \Rightarrow (\neg(n^2 \; even))$. From Informal Fact I, we see that the statement $\neg(n \; even)$ is equivalent to the statement $n \; odd$. Also from Informal Fact I, we see that the statement $\neg(n^2 \; even)$ is equivalent to the statement $n^2 \; odd$. Hence, to complete our proof it is sufficient to show that $n \; odd \Rightarrow n^2 \; odd$. Let's do that. We start with the assumption that n is odd, and we hope that this assumption forces the conclusion that n^2 is odd.

Now, Informal Definiton B tell us that n^2 being odd means that there is an integer m such that $n^2 = 2m + 1$, so this is the result we want to end up with. But also by Definition B, saying that n is odd means that there is an integer k such that $n = 2k + 1$. Now, how do we get n^2 from n? By squaring it! Hence we have $n^2 = (2k + 1)^2$. Expanding this out using a little algebra gives us

$$n^2 = (2k + 1)^2 = 4k^2 + 4k + 1.$$

Can we express $4k^2 + 4k + 1$ as *twice some integer plus* 1? Well, we see that

$$4k^2 + 4k + 1 = 2(2k^2 + 2k) + 1.$$

Hence, letting m denote the integer $2k^2 + 2k$, we now have $n^2 = 2m + 1$, so we end up with our desired conclusion, and we are done. That is, we were able to show that the contrapositive is true, and this shows that the original implication is true.

Proof Technique 2: Proof by Contradiction

Suppose that we are trying to prove a particular proposition p. An ancient and common method of proof is known as **proof by contradiction**. In this method, we assume that the proposition is false; that is, we assume that the statement $\neg p$ is true. We then show that this forces us to conclude a nonsense result; that is, it forces us to conclude something that is inconsistent with something that we have already concluded. The following example is famous in mathematical circles.

| Example 1.17 | Prove that $\sqrt{2}$ is not a rational number. |

Solution: First of all, this example implies that there is such a thing as $\sqrt{2}$. Later, when we develop the real number system methodically, we will show that there is. For the moment, assume that $\sqrt{2}$ does exist; it is that positive number whose square is the number 2. To employ the proof-by-contradiction method, we assume that the result is false and then show that this leads to a contradiction.

Hence, assume that $\sqrt{2}$ *is* a rational number. Now, invoking Informal Definition C and Informal Fact II, there exist integers m and n with $n \neq 0$ with $\sqrt{2} = \frac{m}{n}$ such that m and n are not **both** even. Now, squaring both sides and using a little bit of algebra gives us $2 = \frac{m^2}{n^2}$. Again employing a little algebra, we have $m^2 = 2n^2$. By taking a look at Informal Definition A, we are forced to conclude that m^2 is even. Now, we invoke Example 1.16 and see that m is even.

If we can also prove that n is even, then we have our contradiction. Why? Because we have already concluded that m and n are not *both* even! OK, because m is even, we invoke Informal Definition A again and we see that there must be some integer, let's call it r, such that m is twice that integer; that is, $m = 2r$.

Now, plug the equation $m = 2r$ into the equation $m^2 = 2n^2$, which we derived above, and this gives us the equation $(2r)^2 = 2n^2$. Using a little algebra, this results in $4r^2 = 2n^2$. Dividing both sides of this equation by 2 gives us $2r^2 = n^2$; that is, $n^2 = 2r^2$. By Informal Definition A, this says that n^2 is even. Because that is the case, again invoking Example 1.16 tells us that n is even. But, we have now shown that both m and n are even. This contradicts our earlier statement that m and n are not both even! Hence, proof by contradiction!

Note: Do not worry if you reach the realization that you never would have thought of this argument on your own. I might not have either. The important thing is that you are able to follow the logic. After doing a lot of proofs, you will begin to develop an instinct as to what to do and when to do it.

Proof Technique 3: Proof by the Forward–Backward Methods

Some popular common methods of proof are the so-called **forward–backward methods**. Suppose that we wish to prove an implication $p \Rightarrow q$. Frequently, starting off with the statement p, we cannot see directly how to get to q, but we can find intermediate statements $p_1, p_2, p_3, \ldots, p_n$, such that we are able show $p \Rightarrow p_1, p_1 \Rightarrow p_2, p_2 \Rightarrow p_3, \ldots, p_{n-1} \Rightarrow p_n$, and finally $p_n \Rightarrow q$. We call this the **forward method** because we start with p and move forward until we finally reach q.

Example 1.18   Prove that $((x$ a real number$)$ and $(x^2 - 2x + 8 = 7)) \Rightarrow x = 1$.

Solution: Starting off with $x^2 - 2x + 8 = 7$ for a real number x, we move forward by subtracting 7 from both sides of the equation. This results in the equation $x^2 - 2x + 1 = 0$. Noticing that we can factor $x^2 - 2x + 1$ as $(x - 1)^2$, we have the equation $(x - 1)^2 = 0$. Now, from algebra, the only way that the square of a real number is 0 is if the number itself is 0. Hence, $x - 1 = 0$. Adding 1 to both sides of the equation gives us $x = 1$.   ∎

Sometimes, when trying to prove the statement $p \Rightarrow q$, you may find that rather than starting with p and working forward to q, it is easier to start with the conclusion q and work backward to p. Hence, we find intermediate steps $p_1, p_2, p_3, \ldots, p_n$, such that we are able show $p_1 \Rightarrow q$, $p_2 \Rightarrow p_1$, $p_3 \Rightarrow p_2, \ldots, p_n \Rightarrow p_{n-1}$, and finally $p \Rightarrow p_n$. It should come as no surprise that this is referred to as the **backward method**.

Example 1.19   Prove that $(x$ a real number$) \Rightarrow (x^2 + 1 \geq 2x)$.

Solution: Starting with the statement $x^2 + 1 \geq 2x$, we move backward and see that $(x^2 - 2x + 1 \geq 0) \Rightarrow (x^2 + 1 \geq 2x)$. We have put all the variables on the left-hand side of the inequality, with the hope of factoring the quantity. Indeed, we note that $x^2 - 2x + 1 = (x - 1)^2$, so we now have $((x - 1)^2 \geq 0) \Rightarrow (x^2 - 2x + 1 \geq 0) \Rightarrow (x^2 + 1 \geq 2x)$. To complete the proof, all we have to do is show that $(x$ a real number$) \Rightarrow ((x - 1)^2 \geq 0)$. But this is no problem because our knowledge of algebra reminds us that the square of *every* real number is nonnegative; in particular, the square of the real number $x - 1$ must be nonnegative.   ∎

It should be noted that in many real-world proofs, the discoverer may use both the forward method and the backward method in the same proof. It may be best to move a little forward and then jump to the desired conclusion and move a little backward, and so on.

Proof Technique 4: The Negation of Quantifiers

Sometimes in your mathematical journeys it will be necessary to negate a statement containing quantifiers. Suppose that we are attempting to negate $\forall x\, p(x)$, where $p(x)$ is some predicate function; that is, we would like to demonstrate $\neg(\forall x\, p(x))$. Hence, we are attempting to show that the statement $p(x)$ *is true for all x in the universe of discourse* is false. Clearly, we have accomplished this if we can find at least one x such that $p(x)$ is false. This is the same thing as saying $\exists x(\neg p(x))$. Such an x is referred to as a **counterexample**. (See Exercise 6 of Section 1.3.)

Example 1.20

Prove that $\neg(\forall x((x^2 > 1) \Rightarrow (x > 1)))$ where the universe of discourse is all real numbers.

Solution: By virtue of the above discussion, we seek a counterexample. That is, we seek a real number x such that $x^2 > 1$, in which $x \leq 1$. From algebra, it is easy to see that the number -2 does the trick.

Note: Most authors do not use as many parentheses for logical statements as are used here. For instance, it would not be unusual to write the statement in Example 1.20 as $\neg\forall x(x^2 > 1 \Rightarrow x > 1)$. Extra use of parentheses cannot hurt and can add clarity, but it is not required.

How do we negate a statement such as $\exists x\, p(x)$; that is, how do we demonstrate $\neg(\exists x\, p(x))$? We want to show that there is no x in the universe of discourse for which $p(x)$ is true. One way to do this is to show that $p(x)$ is never true. This is equivalent to saying $\forall x(\neg p(x))$.

Example 1.21

Prove that $\neg\exists x(x^2 = -1)$, where the universe of discourse is all real numbers.

Solution: From the above discussion and algebra, we can accomplish this by showing $\forall x(x^2 \neq -1)$. Now, from algebra, we know that the square of a real number cannot be negative. In particular the square of a real number cannot be equal to -1.

Note: Students who have studied complex numbers will be able to see that the argument of Example 1.21 fails if the universe of discourse becomes all the complex numbers instead of the real numbers. This is because there *is* a complex number whose square is equal to -1.

Proof Technique 5: Proving Uniqueness

Let $p(x)$ be a predicate function. There will be times when we wish to show that $\exists x\, p(x)$, and there is one and only one such x. One way to say this is that there is a **unique** x, such that $p(x)$. The notation $\exists! x\, p(x)$ is sometimes used to express this uniqueness in mathematical texts. One common way to prove to uniqueness is to show that $\exists x\, p(x)$ and then show that if we have $p(y)$ for some y in the universe of discourse, then $y = x$.

Example 1.22

Prove that there is a unique real number solution to the equation $3x - 7 = 8$.

Solution: A little algebra shows that there is at least one solution: namely the number 5. Now, to prove uniqueness, suppose that x and y are both solutions to the equation, so that $3x - 7 = 8$ and $3y - 7 = 8$. Then we have $3y - 7 = 3x - 7$. Adding 7 to both sides of this equation gives $3y = 3x$. Dividing both sides of this new equation by 3 gives us the desired result that $y = x$. Hence, 5 is the unique real number solution to the equation $3x - 7 = 8$.

Some Remarks on Proofs

When reading the proofs in some advanced mathematics books, it is easy to get intimidated. This is because many of the proofs look very polished and slick. Reading the proof does not always give the reader a sense of how the proof was discovered, even if it is easy to follow the proof. Although it may look slick now, it is very likely that the discoverer had to use a whole host of techniques, some that worked and some that did not. Similarly, you will find developing some proofs to be easy, whereas you will find that some require a great deal of work replete with false starts and some frustration. Do not be dismayed. As we proceed, we will introduce additional techniques for proofs.

EXERCISES

1. Prove that if r and s are rational numbers, then $r + s$ and rs are also rational numbers.

2. Prove that if r is a rational number and s is an irrational number, then $r + s$ is an irrational number. (Hint: Use proof by contradiction and Exercise 1.)

3. Prove that if $r + s$ is a rational number, then either both r and s are rational or else both r and s are irrational. (Hint: Prove the contrapositive.)

4. Disprove the following: $\forall x \forall y (x + y \text{ is irrational})$, in which the universe of discourse for both variables x and y is all irrational numbers.

5. Prove that if n is an even integer, then so is n^2.

6. Prove that if n is even and m is odd, then $n + m$ is odd and nm is even.

7. Prove that $\forall x (x^2 + 4 \geq 4x)$, where the universe of discourse is all real numbers.

8. Disprove the following: $\exists x (25 + 10x < -(x^2))$, where the universe of discourse is all real numbers.

9. Show that the point with coordinates $(0, 1)$ is the unique solution to the intersection of the circle with equation $x^2 + y^2 = 1$ and the line with equation $y = 1$.

Chapter 2

Elementary Set Theory

Now that we have explored some of the rules of logic, the next step is to use them to develop the foundation that the rest of the book will depend upon. This foundation is the basics of **set theory**. It is widely accepted that all or most of mathematics can be constructed from set theory. We will not have done our job in this book if we fail to define real numbers in terms of sets. This is because one of the purposes of this book is to demonstrate that mathematics is a deductive science. Therefore, we demonstrate how results are based on initial concepts and prior results. Some of the concepts discussed in this chapter will not be used subsequently in this book, but they will be seen again and again in future mathematics courses.

Were this a completely "pure" math book, we would begin by explicitly stating the axioms of this theory. Someday, you may want to study this. There is a good discussion of axioms for set theory in Goldrei (1996). However, this is a somewhat complicated matter that is probably not worth the effort at this point. We can get away with discussing some very intuitive concepts instead of using formal axioms. However, once this is done, the next step is to formally derive all of our results from these starting concepts. Later in the chapter, it will be shown how this "intuitive" set theory can lead to paradoxes. Axioms for set theory were developed in the beginning of the 20th century to prevent these paradoxes. These paradoxes are not the least bit serious for our purposes. They are discussed here only to give the reader some perspective.

Even though we are not going to list all of the axioms for set theory, we do discuss two such axioms in this text. The first one, which is discussed in Section 2.7, will prove useful in the process of developing the real number system. The second one, which is discussed in Chapter 4, plays an important role in some portions of advanced mathematics.

Section 2.1   Introduction to Sets

Intuitively, a **set** is a collection of objects. What could be simpler than that? The objects that comprise the set are called the **elements** or **members** of the set. One way to express the set is to list its elements between braces.

Example 2.1

Consider the set $\{1, 2, -6\}$. This is the set containing the numbers 1, 2, and -6; that is, the elements of the set are these numbers. Note that the number 8 is *not* a member of this set.

Letters are often used to represent sets. For example, we could denote the set $\{1, 2, -6\}$ of Example 2.1 by A. We may express this by saying, "Let $A = \{1, 2, -6\}$."

Definition 2.1

The symbol $\in$ is used to express set membership. Let x denote a given object. For any set S, the statement $x \in S$ is read "x is a member of S," or "x is an element of S." The symbol $\notin$ is used to claim that an object is not a member of a set. The statement $x \notin S$ is read "x is not a member of S," or "x is not an element of S."

Note that from the above definition, the statements $x \notin S$ and $\neg(x \in S)$ are identical. Using the definition and the set A in Example 2.1, we have $1 \in A$, $2 \in A$, and $-6 \in A$. Note that we have $8 \notin A$ because 8 is not a member of the set.

Sets can also be elements of other sets.

| Example 2.2 | Suppose there is a specific math class that contains the students Gigi and Vijay. Also, consider the letter 'B' of the alphabet. Define the set S by $S = \{-8.1, \{Gigi, B\}, Vijay\}$. List the elements of S. |

Solution: We have $-8.1 \in S$, $\{Gigi, B\} \in S$, and Vijay $\in S$.

Admittedly, the set in Example 2.2 is a little odd. Its members are the number -8.1, the set consisting of Gigi and the letter 'B', and Vijay. Although it may be odd, it is still perfectly valid. In fact, there will be many cases in this book where we are concerned with sets that are members of other sets. It is important to understand that in the above example, we have Gigi $\notin S$ and B $\notin S$. Is $\{Vijay\} \in S$? No, because S does not contain the set $\{Vijay\}$ as an element.

There will be many sets in this book that contain so many elements that it would be unwieldy or downright impossible to list all the elements. Another way to specify the elements of the set is to use the ellipsis symbol, which consists of three dots. It means *and so on*. For example, one way to write the set of all positive integers is $\{1, 2, 3, \ldots\}$. The ellipsis indicates that the pattern started with the listing of the first three positive integers is to be continued. Suppose we want to write the set of all positive integers up to the number 50. We can write this as $\{1, 2, 3, \ldots, 50\}$. How about the set of all integers? The easiest way to write this set is $\{\ldots, -3, -2, -1, 0, 1, 2, 3, \ldots\}$. This notation indicates that the pattern continues in both the negative and positive directions.

The only rule needed to write sets in this way is that the pattern that is to be continued should be clear. For example, it is unclear as to what set is meant by $\{1, \sqrt{5}, \ldots\}$ because that pattern is not obvious. On the other hand, the pattern expressed in the set $\{-2, -4, -6, \ldots\}$ is very clear.

The order that the objects in a set are listed is immaterial. For example, the set $\{1, 2\}$ is the same set as is $\{2, 1\}$. As another example, the set of all integers $\{\ldots, -3, -2, -1, 0, 1, 2, 3, \ldots\}$ could also be written as $\{0, -1, 1, -2, 2, -3, 3, \ldots\}$. Also, if an element of a set is listed twice, this does not change anything. The sets $\{1, 2, 1\}$ and $\{1, 2\}$ are identical because they have the same elements. The former way of expressing the set is not wrong; it is merely redundant.

We might as well take this opportunity to mention some sets that will be used throughout this book. When we have completed the construction of the real number system, we shall have four important sets. The set of all positive integers, the set of all integers, the set of all rational numbers, and the set of all real numbers shall be denoted by **N**, **Z**, **Q**, and **R**, respectively. (These are the symbols typically used in many mathematics texts for these sets). **N** is also frequently referred to as the set of **natural numbers**. Of course, we will be using these sets in examples for illustrative purposes before they have been formally constructed. Hence, these bolded letters are reserved for the rest of the book.

There is another method for identifying the members of sets. To see why another method is needed, how would you express *the set of all real numbers that are greater than or equal to 5*? Clearly, $\{5, 6, 7, \ldots\}$ does not work; this is the set of all integers that are greater than or equal to 5. How about $\{5, 5.1, 5.2, \ldots\}$? No, that fails too because it misses all the real numbers between 5 and 5.1, between 5.1 and 5.2, and so on. What is needed is a new notation.

Sets such as *the set of all real numbers that are greater than or equal to 5* can be expressed using **set-builder notation**. This is best explained via an example. Using this notation, *the set of all real numbers that are greater than or equal to 5* can be expressed as $\{x \in \mathbf{R} \mid x \geq 5\}$. The vertical bar is read "such that." This set is read "the set of real numbers x such that $x \geq 5$." It should be noted that the vertical bar is replaced with a colon in some mathematics texts.

As another example, consider the set of rational numbers, $\mathbf{Q}$. Recall the definition of a rational number given in Informal Definition C of Section 1.4. Using this, we could express $\mathbf{Q}$ as $\mathbf{Q} = \{\frac{m}{n} \mid m, n \in \mathbf{Z}, n \neq 0\}$. That is, $\mathbf{Q}$ is the set of quotients such that both the numerator and denominator are integers in which the denominator is nonzero.

The language of sets allows us to specify the universe of discourse for quantifiers more easily. For example, instead of saying $\forall x(x^2 + 1 \geq 2x)$, in which the universe of discourse is all real numbers, we could write $\forall x \in \mathbf{R}(x^2 + 1 \geq 2x)$. Sometimes, especially when dealing with arbitrary sets, the universe of discourse is not specified. For example, suppose A and B are sets. We might write a statement such as $\forall x(x \in A \Rightarrow x \in B)$. (Indeed, this is the definition of A being a *subset* of B, as you will shortly see.) Here, it is assumed that the universe of discourse is this vague undefined set called the **universal set**. This is the set of all objects under consideration. We will not concern ourselves with this "vagueness" here; it is the minor price that we pay for not formally writing out the axioms of set theory. We could also write the statement $\forall x(x \in A \Rightarrow x \in B)$ as $\forall x \in A(x \in B)$, and we shall not worry about some of the logical nuances here that would be ironed out in a formally "axiomitized" course on logic and set theory. (It should be pointed out that there are many such books that present this formalized approach. We recommend that you study this some day. A surprisingly large number of mathematicians are not familiar with this material because they do not need to deal with it on a day-to-day basis.)

We are now ready to present some definitions.

Definition 2.2

The **null set** (also called the **empty set**) is declared to be the set that has no elements. This set is normally written as $\emptyset$. It is also acceptable to write the null set as $\{\ \}$. Any set that is not the null set is said to be **nonempty**.

It should now be obvious to you that a set A is nonempty iff $\exists x \in A$. (Recall the definition of *iff* from Section 1.2.) By the way, in an axiomatic presentation of set theory, the existence of the empty set is presented as an axiom.

Definition 2.3

The set S is said to be a **subset** of the set T, provided that $\forall x(x \in S \Rightarrow x \in T)$. The statement "$S$ is a subset of T" is written $S \subseteq T$. It can also be written $T \supseteq S$.

Intuitively, the statement $S \subseteq T$ means that every element of S is also an element of T. This is also expressed sometimes by saying "T **contains** S." This can also be expressed by saying "T is a **superset** of S." We are now ready for our first theorem. Theorem 2.1 below states that the null set is a subset of every set.

Theorem 2.1

Let S be any set. Then $\varnothing \subseteq S$.

Proof: By Definition 2.3, we must show that $\forall x(x \in \varnothing \Rightarrow x \in S)$. Now, by Definition 2.2, we see that the statement $x \in \varnothing$ is false for every x because the null set has no members. Now, recall from the truth table for *implication* (Table 1.4) that an implication $p \Rightarrow q$ is true if the hypothesis p is false. Our hypothesis (for any x) is the statement $x \in \varnothing$, and we just said that this is a false statement. Therefore, the implication $x \in \varnothing \Rightarrow x \in S$ is true for any x. Hence, we have established that $\forall x(x \in \varnothing \Rightarrow x \in S)$. ∎

Example 2.3

Consider the set $S = \{a, b, c\}$ consisting of the first three letters of the alphabet. List all of the subsets of this set.

Solution: It is easy to see that the list of subsets of S is

1) $\{a\}$
2) $\{b\}$
3) $\{c\}$
4) $\{a, b\}$
5) $\{a, c\}$
6) $\{b, c\}$

7) $\{a, b, c\}$

8) $\varnothing$

Number 7 on the list is due to the fact that every set is a subset of itself; number 8 is due to Theorem 2.1.

Definition 2.4

Sets S and T are said to be **equal** (written $S = T$), provided $S \subseteq T$ and $T \subseteq S$. The statement $S \neq T$ is defined to mean the negation of the statement $S = T$.

Definition 2.4 says that sets S and T are equal provided that each element of S is also a member of T, and each member of T is also a member of S. That is, S and T contain the same members. This is the justification for saying, for example, that $\{1, 2, 1\} = \{1, 2\}$.

The need to show that two sets are actually equal arises often in mathematics. It should also be noted that there are many ways to list the empty set. For example, it is clear that $\{x \in \mathbf{R} | x^2 = -1\} = \varnothing$, because both of these sets contain no elements. Sometimes a mathematician will be working with a set that has many interesting properties, only to find out later that the set is in fact empty.

Definition 2.5

The set S is said to be a **proper subset** of the set T, provided $S \subseteq T$ and $S \neq T$. The statement "S is a proper subset of T" is written $S \subset T$. It can also be written $T \supset S$.

For example, it is clear that $\{1, 2\} \subset \{1, 2, 3\}$. Of course, if $S \subset T$, it is not wrong to write $S \subseteq T$; the statement $S \subset T$ simply provides more information. The reader should now provide a proof of the obvious statement $S \subset T \Rightarrow (\exists x (x \in T \text{ and } x \notin S))$.

EXERCISES

1. List all of the subsets of the set $S = \{-8.1, \{\text{Gigi, B}\}, \text{Vijay}\}$ in Example 2.2.

2. List all of the subsets of the set $S = \{\varnothing\}$. Note that S has the interesting property that we have both $\varnothing \subseteq S$ and $\varnothing \in S$.

3. Suppose A, B, and C are sets such that $A \subseteq B$ and $B \subseteq C$. Prove that $A \subseteq C$.

4. Suppose A, B, and C are sets such that $A \subseteq B$ and $B \subset C$. Prove that $A \subset C$.

5. Suppose A, B, and C are sets such that $A \subset B$ and $B \subseteq C$. Prove that $A \subset C$.

6. Using set-builder notation, write the set consisting of all positive rational numbers that are strictly less than the number 5.138.

7. List all the members of the set $\{x \in \mathbf{R} | x^2 = 4\}$.

8. List all the members of the set $\{x \in \mathbf{Z} | x^2 = 5\}$.

Section 2.2   Some Operations on Sets

We will now present some operations on sets, which are ways of combining sets. Before doing this, we present a pictorial way to represent relationships between sets. One well-known method for doing this is using a **Venn diagram**. In this diagram, circle-like figures represent sets. Whatever is outside the circle is outside the set. For example, a Venn diagram that shows that $S \subseteq T$ is given in Figure 2.1. The point of the figure is simply to demonstrate that S resides inside of T. The figure should not be taken to necessarily mean that S is a proper subset of T.

Definition 2.6

Let S and T be sets. The **union** of S and T, written $S \cup T$, is defined to be as follows: $S \cup T = \{x | x \in S \text{ or } x \in T\}$.

Figure 2.1
A Venn diagram showing a subset of a set

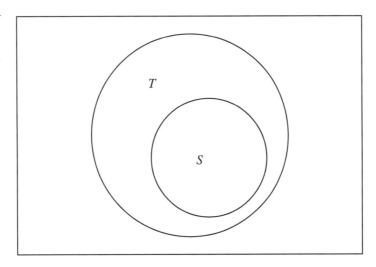

Please note that by *or* we mean the *disjunction*, which was defined in Definition 1.2. Hence, if an object x is in both S and T, it is still in $S \cup T$.

Example 2.4

The union of the sets $\{1, 6\}$ and $\{1,7\}$ is easily seen to be the set $\{1, 6, 7\}$; that is, $\{1, 6\} \cup \{1, 7\} = \{1, 6, 7\}$.

◾

Example 2.5

It is easy to see that the union of the sets $\{x \in \mathbf{R} | x \geq 5\}$ and $\{x \in \mathbf{R} | x < 5\}$ is $\mathbf{R}$; that is, $\{x \in \mathbf{R} | x \geq 5\} \cup \{x \in \mathbf{R} | x < 5\} = \mathbf{R}$. This is due to the fact that for every real number x, either $x \geq 5$ or else $x < 5$.

◾

Figure 2.2 displays the Venn diagram for the union of two sets. The union of the sets is the shaded region.

We will soon be encountering theorems, where we have to prove that two sets are equal. Sometimes, this can be done by inspection. Other times, a direct appeal will be made to Definition 2.4. Using this "subset" method of proof, we will demonstrate that sets S and T are equal by showing $S \subseteq T$ and $T \subseteq S$. Remember, to show that $A \subseteq B$ for given sets A and B, we show that $\forall x (x \in A \Rightarrow x \in B)$. By the definition of *implication*, it suffices to show that whenever the statement $x \in A$ is a true statement for any given x, then the statement $x \in B$ is also true. To do this, it is customary to say something in the proof like "let $x \in A$" or "consider any $x \in A$" or "suppose $x \in A$." Then some argument is employed to show that $x \in B$. For an example, see the proof of Part 4 of Theorem 2.2 that follows.

Figure 2.2
Venn diagram
showing the union
of S and T

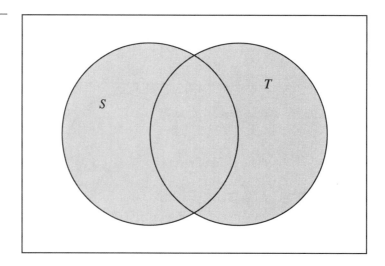

Theorem 2.2   Let S be any set. Then

1) $S \cup \varnothing = \varnothing \cup S = S$.

2) $S \cup S = S$.

3) $S \cup T = T \cup S$ for any set T. (commutative property of unions)

4) $(S \cup T) \cup U = S \cup (T \cup U)$ for any sets T and U. (associative property of unions)

5) If T is a set with $S \subseteq T$, then $S \cup T = T$.

Proof: Parts 1 and 2 are obvious. Part 3 is an immediate consequence of the fact that $(p \vee q) \Leftrightarrow (q \vee p)$ (because $p \vee q$ has the same truth table as does $q \vee p$) and the definition of the union of sets.

For Part 4, we prove the result by demonstrating that $(S \cup T) \cup U \subseteq S \cup (T \cup U)$ and $S \cup (T \cup U) \subseteq (S \cup T) \cup U$. To show the former, let $x \in (S \cup T) \cup U$. Then by the definition of *union*, we must have $x \in S \cup T$ or $x \in U$. We break this into cases (that are not necessarily mutually exclusive):

Case 1: $x \in S \cup T$. Then $x \in S$ or $x \in T$. We break this into subcases.
Subcase 1A: $x \in S$. Then $x \in S \cup (T \cup U)$.
Subcase 1B: $x \in T$. Then $x \in T \cup U$. Hence, $x \in S \cup (T \cup U)$.
Case 2: $x \in U$. Then $x \in T \cup U$. Hence, $x \in S \cup (T \cup U)$.

Note that in both cases we have $x \in S \cup (T \cup U)$. Therefore, we have established that $(S \cup T) \cup U \subseteq S \cup (T \cup U)$. The reader is asked to show $S \cup (T \cup U) \subseteq (S \cup T) \cup U$ by using the same type of argument. Once this is done, the result follows from Definition 2.4.

For Part 5 we can show that $S \cup T \subseteq T$ and $T \subseteq S \cup T$. Noticing that the latter is completely obvious, we demonstrate the former. So, let $x \in S \cup T$. Then $x \in S$ or $x \in T$. We break this into cases.

Case 1: $x \in S$. Now, we are given $S \subseteq T$, so $x \in T$ by the definition of *subset*.

Case 2: $x \in T$. Then there is nothing to prove.

Hence, in both cases we have shown that $x \in T$. Thus, $S \cup T \subseteq T$.   ∎

Note: Part 3 of Theorem 2.2 is referred to as the **commutative property** of unions, and Part 4 is referred to as the **associative property** of unions.

Example 2.6   Show that $\{x \in \mathbf{R} | x \geq 8\} \cup \{x \in \mathbf{R} | x > 5\} = \{x \in \mathbf{R} | x > 5\}$.

Solution: Let $S = \{x \in \mathbf{R} | x \geq 8\}$ and $T = \{x \in \mathbf{R} | x > 5\}$. Now, it is easy to see that $S \subseteq T$ because for any real number x, $x \geq 8 \Rightarrow x > 5$. By Part 5 of

Theorem 2.2, we have $S \cup T = T$. That is, $\{x \in \mathbf{R} | x \geq 8\} \cup \{x \in \mathbf{R} | x > 5\} = \{x \in \mathbf{R} | x > 5\}$.

Having discussed the union of sets, we next discuss another important operation on sets. This operation is the **intersection** of sets.

Definition 2.7

Let S and T be sets. The **intersection** of S and T, written $S \cap T$, is defined as follows: $S \cap T = \{x | x \in S \text{ and } x \in T\}$.

In other words, the intersection of S and T is the set consisting of those elements that have membership in both sets.

Example 2.7

The intersection of the sets $\{1, 6\}$ and $\{1, 7\}$ is easily seen to be the set $\{1\}$. That is, $\{1, 6\} \cap \{1, 7\} = \{1\}$.

Example 2.8

It is easy to see that the intersection of the sets $\{x \in \mathbf{R} | x \geq 5\}$ and $\{x \in \mathbf{R} | x < 5\}$ is $\varnothing$. That is, $\{x \in \mathbf{R} | x \geq 5\} \cap \{x \in \mathbf{R} | x < 5\} = \varnothing$. This is because for every real number x, if $x \geq 5$, it is not possible that we also have $x < 5$.

Figure 2.3 displays the Venn diagram for the intersection of two sets. The intersection of the sets in the figure is the shaded region that the sets share in common.

Figure 2.3
Venn diagram
showing the
intersection of S
and T

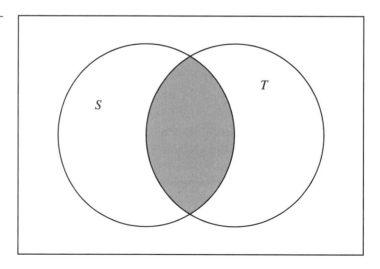

Theorem 2.3

Let S be any set. Then

1) $S \cap \varnothing = \varnothing \cap S = \varnothing$.
2) $S \cap S = S$.
3) $S \cap T = T \cap S$ for any set T. (commutative property of intersections)
4) $(S \cap T) \cap U = S \cap (T \cap U)$ for any sets T and U. (associative property of intersections)
5) If T is a set with $S \subseteq T$, then $S \cap T = S$.

Proof: We show only Part 5; the others are left to the reader. Hence, consider a set T such that $S \subseteq T$. To show that $S \cap T = S$, we show that $S \cap T \subseteq S$ and $S \subseteq S \cap T$. The former statement is clear because any element of both S and T is a member of S. To show the latter statement, suppose $x \in S$. Then, because we are given $S \subseteq T$, we have $x \in T$ by the definition of *subset*. Because we have $x \in S$ and $x \in T$, we have $x \in S \cap T$, by the definition of *intersection*. Hence, we have shown $S \subseteq S \cap T$. Because $S \cap T \subseteq S$ and $S \subseteq S \cap T$, we have $S \cap T = S$ by the definition of *set equality*. ∎

Part 3 of Theorem 2.3 is referred to as the commutative property of intersections, and Part 4 is referred to as the associative property of intersections. We now present two laws, known as the **distributive laws**, that tie unions and intersections together.

Theorem 2.4

Let A, B, and C be any sets. Then

1) $A \cap (B \cup C) = (A \cap B) \cup (A \cap C)$.
2) $A \cup (B \cap C) = (A \cup B) \cap (A \cap C)$.

Proof: We prove Part 1, with the proof of Part 2 being left to the reader. We first show that $A \cap (B \cup C) \subseteq (A \cap B) \cup (A \cap C)$. To that end, let $x \in A \cap (B \cup C)$; then $x \in A$ and $x \in B \cup C$. Hence, $x \in B$ or $x \in C$. If $x \in B$, then $x \in A \cap B$. If $x \in C$, then $x \in A \cap C$. Hence, $x \in A \cap B$ or $x \in A \cap C$. Thus, $x \in (A \cap B) \cup (A \cap C)$. We have shown that $A \cap (B \cup C) \subseteq (A \cap B) \cup (A \cap C)$.

We now show that $(A \cap B) \cup (A \cap C) \subseteq A \cap (B \cup C)$. Let $x \in (A \cap B) \cup (A \cap C)$; then $x \in A \cap B$ or $x \in A \cap C$. Note that $x \in A$ in either case. In the first case, we have $x \in B$. In the second case, we have $x \in C$. Hence, we have $x \in B \cup C$. Because $x \in A$ and $x \in B \cup C$, we have $x \in A \cap (B \cup C)$. We have shown that $(A \cap B) \cup (A \cap C) \subseteq A \cap (B \cup C)$.

Because $A \cap (B \cup C) \subseteq (A \cap B) \cup (A \cap C)$ and $(A \cap B) \cup (A \cap C) \subseteq A \cap (B \cup C)$, this proves that $A \cap (B \cup C) = (A \cap B) \cup (A \cap C)$ by Definition 2.4. ∎

We have one more set operation to consider in this section. It is defined in Definition 2.8 below.

Definition 2.8

Let S and T be sets. The **complement** of T with respect to S, also called the **difference** of S and T, is denoted by $S - T$ and defined by $S - T = \{x \in S | x \notin T\}$.

Hence, $S - T$ contains every member of S that is also not a member of T.

Example 2.9

The difference of the sets $\{1, 6\}$ and $\{1, 7\}$ is easily seen to be the set $\{6\}$. That is, $\{1, 6\} - \{1, 7\} = \{6\}$.

Example 2.10

It is not hard to see that the complement of the set $\{x \in \mathbf{R} | x \le 5\}$ with respect to the set $\{x \in \mathbf{R} | x < 8\}$ is the set $\{x \in \mathbf{R} | 5 < x < 8\}$. That is $\{x \in \mathbf{R} | x < 8\} - \{x \in \mathbf{R} | x \le 5\} = \{x \in \mathbf{R} | 5 < x < 8\}$.

Figure 2.4 displays the Venn diagram for $S - T$. $S - T$ in the figure is the striped region that lies in S but outside of T.

Figure 2.4
Venn diagram showing the complement of T with respect to S

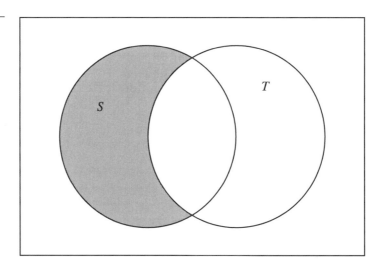

Definition 2.9

Sets S and T are said to be **disjoint** if $S \cap T = \emptyset$. A collection of sets (i.e., set of sets) is said to be **pairwise disjoint** if each pair of sets in the collection is disjoint.

In later mathematics courses, it will be important to write the union of sets as a union of sets in a pairwise disjoint collection of sets. For any sets S and T, it should be easy to see that the set of sets $\{S - T, T - S, S \cap T\}$ is pairwise disjoint. Furthermore, it is easy to see that $S \cup T = (S - T) \cup (T - S) \cup (S \cap T)$. This is illustrated in Figure 2.5. Hence, we have broken down $S \cup T$ into three sets. The first set is the set of elements that are in S but are not in T. This set is displayed in the figure by the region containing vertical lines. The second set is the set of elements that are in both S and T. It is displayed in the figure by the region of just white space. The third set is the set of elements that are in T, but are not in S. It is displayed in the figure by the region containing horizontal lines. (**Note:** the astute reader may have noticed that we have not officially defined the union of three sets, say $A \cup B \cup C$. We could define $A \cup B \cup C$ as $(A \cup B) \cup C$ or as $A \cup (B \cup C)$. Either definition will give us the same answer as shown by Part 4 of Theorem 2.2—the associative property of unions.)

We have one more theorem for this section. Theorem 2.5 states what is known as **De Morgan's laws**. These laws use all three of the set operations of this section: unions, intersections, and complements.

Figure 2.5
Venn diagram showing the breaking of a union into three sets

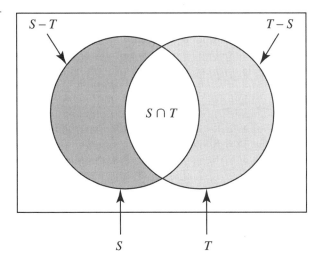

Theorem 2.5   Let A, B, and X be any sets. Then

1) $X - (A \cup B) = (X - A) \cap (X - B)$.
2) $X - (A \cap B) = (X - A) \cup (X - B)$.

Proof: We prove Part 2 with the proof of Part 1 being left to the reader. We first show that $X - (A \cap B) \subseteq (X - A) \cup (X - B)$. Let $x \in X - (A \cap B)$. Then, by definition, we have $x \in X$ and $x \notin A \cap B$. Therefore, $x \notin A$ or $x \notin B$. (Why?) If $x \notin A$, then $x \in X - A$. If $x \notin B$, then $x \in X - B$. Either way, we have $x \in (X - A) \cup (X - B)$. Hence, $X - (A \cap B) \subseteq (X - A) \cup (X - B)$.

Now we show that $(X - A) \cup (X - B) \subseteq X - (A \cap B)$. Let $x \in (X - A) \cup (X - B)$; then $x \in X - A$ or $x \in X - B$. If $x \in X - A$, then $x \notin A$. If $x \in X - B$, then $x \notin B$. In either case, we have $x \notin A \cap B$. Because $x \in X$ and $x \notin A \cap B$, we have $x \in X - (A \cap B)$. Hence, we have established $X - (A \cap B) \subseteq (X - A) \cup (X - B)$.

The result now follows the definition of the equality of sets. ∎

The reader is encouraged not only to follow the logic of the proof but also to attempt to understand the theorem intuitively. Drawing Venn diagrams can assist with this intuition. In Exercise 11, the student is asked to draw Venn diagrams that illustrate De Morgan's laws.

EXERCISES

1. Complete the proof of Part 4 of Theorem 2.2.
2. Write out the proof of Part 4 of Theorem 2.3.
3. Write out the proof of Part 2 of Theorem 2.4.
4. Write out the proof of Part 1 of Theorem 2.5.
5. Let $A = \{-2.5, 7, 9, 11\}$, $B = \{-2.5, 28, 90\}$, and $C = \{28, 59\}$.

   a. Find $A \cup B$, $A \cap B$, $A - B$, and $B - A$.

   b. Demonstrate that $A \cap (B \cup C) = (A \cap B) \cup (A \cap C)$ by listing $A \cap (B \cup C)$ and $(A \cap B) \cup (A \cap C)$.

6. Do the same as in Exercise 5 with the sets $A = \{x \in \mathbf{R} | x \geq -1\}$, $B = \{x \in \mathbf{R} | x < 50\}$, and $C = \{x \in \mathbf{R} | x \geq 0\}$.
7. Show that $A - \varnothing = A$ and $\varnothing - A = \varnothing$ for any set A.
8. For any sets A and B, show that A and $B - A$ are disjoint.
9. Suppose that there are sets A and B such that $A - B = B - A$. What can you conclude about A and B?
10. Suppose A and B are sets with $A \subseteq B$. Show $A - B = \varnothing$.

11. Draw Venn diagrams to "pictorially" convince yourself that the De Morgan's laws are true.

12. For any sets A and B, show that $A - (A - B) = A \cap B$.

Section 2.3   Relations

When we develop the real number system, we will need to define inequalities such as $a \leq b$ for real numbers a and b. We have an intuitive notion of what this means; it tells us that a is not to the right of b on the number line. This is the intuitive sense in which a and b are *related*. To lay the groundwork to be able to define inequalities, it will prove useful if we generalize this concept of an inequality to arbitrary sets. As you will see in time, this generalization has many applications in mathematics. At first glance, it may not appear that the work we are doing in this section has anything to do with inequalities of real numbers. The reader is advised to be patient. We are laying very strong and reliable foundations. Of course, we shall use the foundations of set theory discussed in the last section to define what is meant by two elements being *related*.

The Cartesian Product

Let A and B be sets, and suppose that we have elements $a \in A$ and $b \in B$. We will begin by defining the **ordered pair** (a, b). You are, no doubt, familiar with the concept of an ordered pair in the case where both A and B are the set **R** of real numbers. In this case, the ordered pair (a, b) specifies the coordinates of a point P in the plane, as shown in Figure 2.6.

Figure 2.6
The Cartesian coordinates of a point in the plane

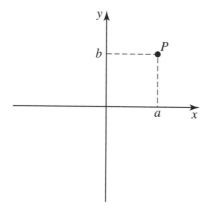

For the example in Figure 2.6, the first number indicates the position of the point with respect to the *x*-axis, and the second number indicates the position of the point with respect to the *y*-axis. Hopefully, you will recall from your algebra studies that the numbers *a* and *b* are called the *Cartesian coordinates* of the point.

Intuitively, however we choose to define an ordered pair, we want the order that the elements are listed to matter, so that we have a first element and a second element. Therefore, we want to define the equality of the ordered pair (a, b) with the ordered pair (c, d) to mean that $a = c$ and $b = d$. That is, we want to declare two ordered pairs to be equal provided the pairs' first element is the same and the pairs' second element is the same.

Why all the fuss about this? Why not just **define** (a, b) to be an ordered collection of elements? Because we want to be able to define an ordered pair in terms of sets. This book has already made the claim that most or all of mathematics can be built upon the foundations of set theory. Here, we begin to make good on that claim. We therefore want to define (a, b) in terms of set concepts that have already been defined. Besides, declaring an ordered pair to be "an ordered collection of elements" is cheating because we have not rigorously defined what is meant by this.

At this point, the reader may be thinking, "Why not just define (a, b) to be $\{a, b\}$? After all, $\{a, b\}$ is a valid set." The answer is that this attempted definition does not meet our objectives. Clearly it is possible to have $\{a, b\} = \{c, d\}$ without having that $a = c$ and $b = d$. This is due to the fact that order does not matter in the listing of the elements of a set.

Definition 2.10 below does the trick. The mathematician Norbert Wiener first gave this method of defining an order pair. Theorem 2.6 proves that this definition meets our objectives.

Definition 2.10

Let A and B be sets, and suppose that we have elements $a \in A$ and $b \in B$. The **ordered pair** (a, b) is defined to be the set $\{\{a\}, \{a, b\}\}$.

We now demonstrate that this definition works. First, a little notation is in order. For any set X, the statement $x, y \in X$ means $x \in X$ and $y \in X$. The statement $x, y, z \in X$ means $x \in X$ and $y \in X$ and $z \in X$. In general, if we present a list of elements separated by comma followed by the element sign and a set, it means all the elements are in the given set.

Theorem 2.6   Let A and B be sets with $a_1, a_2 \in A$ and $b_1, b_2 \in B$. Then $(a_1, b_1) = (a_2, b_2)$ iff $(a_1 = a_2$ and $b_1 = b_2)$.

Proof: It is clear that $(a_1 = a_2$ and $b_1 = b_2) \Rightarrow (a_1, b_1) = \{\{a_1\}, \{a_1, b_1\}\} = \{\{a_2\}, \{a_2, b_2\}\} = (a_2, b_2)$. We must show that $(a_1, b_1) = (a_2, b_2) \Rightarrow (a_1 = a_2$ and $b_1 = b_2)$. To do this, assume that $(a_1, b_1) = (a_2, b_2)$. Hence, this means $\{\{a_1\}, \{a_1, b_1\}\} = \{\{a_2\}, \{a_2, b_2\}\}$. We now break the problem down into cases.

Case 1: $a_1 = b_1$ Then $(a_1, b_1) = \{\{a_1\}, \{a_1, b_1\}\} = \{\{a_1\}, \{a_1, a_1\}\} = \{\{a_1\}, \{a_1\}\} = \{\{a_1\}\}$, so we have $\{\{a_1\}\} = \{\{a_2\}, \{a_1, b_2\}\}$. Now, the set $\{\{a_1\}\}$ has only one element, so the set $\{\{a_2\}, \{a_2, b_2\}\}$ must have only one element. Hence, we must have $\{a_2\} = \{a_2, b_2\}$. Hence, we must have $a_2 = b_2$. Hence, $\{\{a_2\}, \{a_2, b_2\}\} = \{\{a_2\}\}$. Because we have $\{\{a_1\}\} = \{\{a_2\}, \{a_2, b_2\}\} = \{\{a_2\}\}$, we must have $a_1 = a_2$. Thus, $b_1 = a_1 = a_2 = b_2$. Having shown that $a_1 = a_2$ and $b_1 = b_2$, we are done with this case.

Case 2: $a_1 \neq b_1$. Then the set $\{\{a_1\}, \{a_1, b_1\}\}$ consists of two distinct elements. One of these elements is a set containing one element, and the other element is a set consisting of two distinct elements. Therefore, the set $\{\{a_2\}, \{a_2, b_2\}\}$ must also consist of two distinct elements, with one element being a set consisting of one element and the other element being a set consisting of two distinct elements. Therefore, we must have $\{a_1\} = \{a_2\}$ and $\{a_1, b_1\} = \{a_2, b_2\}$. These equalities give us $a_1 = a_2$ and $\{a_1, b_1\} = \{a_1, b_2\}$. This last equality implies that $b_1 \in \{a_1, b_2\}$. Because $a_1 \neq b_1$ (as this is the case we are considering), we must have $b_1 = b_2$. Having shown $a_1 = a_2$ and $b_1 = b_2$, we are done. ∎

There is a useful point to be made about Theorem 2.6. The proof of a theorem often may look fairly complicated (as does the proof of Theorem 2.6), even though the basic idea is really quite simple. The idea behind defining an order pair as $\{\{a\}, \{a, b\}\}$ is very simple. Intuitively, if a and b are distinct (i.e., not equal to each other), then the first component of the ordered pair is taken from the member of $\{\{a\}, \{a, b\}\}$ that has only one member. Because $\{a, b\}$ has more than one member, and $\{a\}$ has only one member, this first component of the ordered pair must be a. Armed with the proper intuition, you will have a much easier time making it through the twists and turns of proofs such as the proof of Theorem 2.6.

Definition 2.11

Let A, B, and C be sets, and suppose that we have elements $a \in A$, $b \in B$, and $c \in C$. The **ordered triplet** (a, b, c) is defined to be the ordered pair $((a, b), c)$.

Note that by this definition, (a, b, c) is defined to be an ordered pair in which the first component is itself an ordered pair. Exercise 1 asks the reader to show that for $a_1, a_2 \in A$, $b_1, b_2 \in B$, and $c_1, c_2 \in C$, we have $(a_1, b_1, c_1) = (a_2, b_2, c_2) \Leftrightarrow (a_1 = a_2 \text{ and } b_1 = b_2 \text{ and } c_1 = c_2)$.

The reader may be interested (and probably relieved) to know that having defined ordered pairs and triplets in terms of sets, it is exceedingly rare to need to resort to the original definition of these terms. All we really care about is the fact that the order that the elements are listed makes a difference. The above work was performed to demonstrate that we can achieve the properties we want without having to define new concepts; the set theory that we have already developed suffices. Now we are ready to define the Cartesian product of two sets.

Definition 2.12

Let A and B be sets. The **Cartesian product** of A and B, written $A \times B$, is defined to be the set $\{(a, b) | a \in A \text{ and } b \in B\}$.

Example 2.11

Let $A = \{a, b, c\}$, and let $B = \{4, 9\}$. It is easy to see that $A \times B = \{(a, 4), (a, 9), (b, 4), (b, 9), (c, 4), (c, 9)\}$, and $B \times A = \{(4, a), (9, a), (4, b), (9, b), (4, c), (9, c)\}$.

Of course, the order of the listing of the ordered pairs does not matter. For example, in Example 2.11, it is just as legitimate to write $A \times B = \{(a, 4), (b, 4), (c, 4), (a, 9), (b, 9), (c, 9)\}$. This is because the order of the listing of the elements of any set does not matter, and $A \times B$ is a set.

It is common to write the Cartesian product of the set of real numbers with itself as $\mathbf{R}^2$. That is, $\mathbf{R}^2$ is frequently used to denote $\mathbf{R} \times \mathbf{R}$. $\mathbf{R}^2$ can be thought of as the set of Cartesian coordinates for all points in the Euclidean plane.

Definition and Examples of a Relation

Let A and B be sets. Most textbooks define a **relation between A and B** to be a subset of $A \times B$. (Actually, many textbooks call this a *binary relation*, but we will stick with *relation*.) This is the essence of our definition also, but we modify it slightly for a reason that is explained next.

Definition 2.13

Let A and B be sets. A **relation between A and B** is defined to be an ordered triplet (A, B, R), where $R \subseteq A \times B$.

The reason why an ordered triplet is used (which is the modification) is to ensure that two relations (A_1, B_1, R_1) and (A_2, B_2, R_2) are equal iff $(A_1 = A_2$ and $B_1 = B_2$ and $R_1 = R_2)$. Consider a relation (A, B, R) and suppose we have $a \in A$ and $b \in B$. Then, if $(a, b) \in R$, we can denote this by saying "a and b are **related** with respect to R." This can also be denoted by writing aRb.

Often, instead of using capital letters (such as R) to denote the given subset of $A \times B$, symbols are used. (There is no rule that states that only capital letters can denote sets). For example, the symbol $\leq$ could be used. So, a relation between sets A and B could be denoted $(A, B, \leq)$, and writing $a \leq b$ for an element $(a, b) \in A \times B$ would indicate that $(a, b) \in \leq$. In fact, it is rare to write $(a, b) \in \leq$; expressing the fact that a and b are **related** with respect to $\leq$ is much more commonly written as $a \leq b$.

Example 2.12

Let P denote the set of nonnegative real numbers, and consider the relation $(\mathbf{R}, \mathbf{R}, \leq)$ with $\leq$ defined as follows: $\forall a, b \ni \mathbf{R}$, $a \leq b$ iff $b - a \in P$. In this case, it is not hard to see that the relation is the *less than or equal to* inequality for real numbers that we are all familiar with! (Of course, this makes sense only after the set of real numbers $\mathbf{R}$ and the set of nonnegative real numbers P have been defined and after the subtraction of real numbers has been defined.)

It is important for the reader to get used to the idea that many symbols, such as $\leq$, that have a preexisting meaning with respect to real numbers are used in a context that may have nothing to do with the preexisting meaning. The *context* of the discussion should always eliminate any ambiguity. Of course, the *motivation* for using the symbol often comes from the real numbers.

Example 2.13

Suppose S is a set of given people, with $S = \{$Tess, Bob, Maxine, Vijay$\}$. Suppose T is a set of given jobs, with $T = \{$Engineer, Accountant, Clerk, Mayor, Chef$\}$. We can define a relation $(S, T, >)$ between S and T as follows: Tess $>$ Mayor, Bob $>$ Engineer, Vijay $>$ Accountant, and Tess $>$ Engineer. Hence, $>$ is a set of ordered pairs with $> = \{$(Tess, Mayor), (Bob, Engineer), (Vijay, Accountant), (Tess, Engineer)$\}$. Now, because $< \subseteq S \times T$, this is a valid relation. If we give this relation the intuitive meaning that $a > b$ means that person a has job b, then we can note the following: 1) Maxine does not have any of the jobs listed in T;

2) Nobody in S has the job of Clerk or Chef; and 3) Poor Tess has been assigned two jobs in T.

■

A special name is often used in the mathematics literature when we have a relation (A, A, R) between A and itself. In this case, we frequently refer to this situation by calling R a **relation on** A. This is a slight abuse of our notation because we have previously defined the relation to be the ordered triple (A, A, R), and now we are calling R the relation. This is all right as long as the context is clear. For (A, A, R), unless otherwise stated, we mean the set R when we refer to the relation.

There is no problem here as long as we understand the following: Consider the relation (A, A, R), and suppose we have a set B with $A \subset B$. Then we also have the relation (B, B, R) because $R \subseteq A \times A \subset B \times B$. When referring to (A, A, R), we call R a relation on A, and when referring to (B, B, R), we call R a relation on B. By the way, note that with respect to the relation $(\mathbf{R}, \mathbf{R}, \leq)$ in Example 2.12, we can now say that $\leq$ is a relation on $\mathbf{R}$.

EXERCISES

1. Let A, B, and C be sets. Show that for $a_1, a_2 \in A$, $b_1, b_2 \in B$, and $c_1, c_2 \in C$, we have $(a_1, b_1, c_1) = (a_2, b_2, c_2) \Leftrightarrow (a_1 = a_2$ and $b_1 = b_2$ and $c_1 = c_2)$.

2. Show that the result of Exercise 1 also would hold were we to have defined the ordered triple (a, b, c) as $(a, (b, c))$ instead of as $((a, b), c)$.

3. Come up with a "reasonable" way to define the Cartesian product of three sets A, B, and C.

4. Suppose A and B are sets. Show that $A \times B = \emptyset \Leftrightarrow (A = \emptyset$ or $B = \emptyset)$.

5. Give an example of nonempty sets A and B such that $A \times B = B \times A$.

6. Let W be the set $W = \{a, b, c\}$ consisting of the first three letters of the English alphabet. Define a relation R on W as follows: $\forall w_1, w_2 \in W$ ($w_1 R w_2$ iff w_1 occurs before w_2 in the listing of the alphabet). List the ordered pairs that comprise the set R.

7. Let S be the set defined by $S = \{\{1\}, \{1, 2\}, \{3\}, \emptyset\}$. Define a relation $\leq$ on S as follows: $\forall x, y \in S$ ($x \leq y$ iff $x \subseteq y$). (Note here that we are using small letters to denote sets.)

    a. Do we have $\{1\} \leq \{3\}$?

    b. List all the ordered pairs in $\leq$.

8. For the set S in Exercise 7, define a relation $<$ on S as follows: $\forall x, y \in S$ ($x < y$ iff $x \subset y$). List all of the ordered pairs in $<$.

Section 2.4   Some Important Types of Relations on a Set

In this section we consider a relation on a set. There are a number of important types of relations on a set that will be discussed. We will make much use of them in the chapters to come.

Definition 2.14

A relation R on a set S is said to be **reflexive** if aRa for every $a \in S$.

Of course, another way to write this is $(a, a) \in R$ for every $a \in S$. Hence, a relation R on a set S is called reflexive provided every element of S is related to itself.

Definition 2.15

A relation R on a set S is said to be **symmetric** if $\forall a, b \in S(aRb \Rightarrow bRa)$.

Hence, if a relation R on a set S is symmetric, this means that for any elements $a, b \in S$, if a is related to b, then b is related to a. That is, $(a, b) \in R \Rightarrow (b, a) \in R$ for all $a, b \in S$.

Definition 2.16

A relation R on a set S is said to be **antisymmetric** if $\forall a, b \in S\,((aRb \text{ and } bRa) \Rightarrow a = b)$.

Definition 2.16 says that we declare a relation R on a set S to be antisymmetric if has the following property: The only way that we can have both $(a, b) \in R$ and $(b, a) \in R$ is if $a = b$.

Definition 2.17

A relation R on a set S is said to be **transitive** if $\forall a, b, c \in S\,((aRb \text{ and } bRc) \Rightarrow aRc)$.

Definition 2.17 indicates that a relation R on a set S is called transitive if the following condition holds: Whenever we have $(a, b) \in R$ and $(b, c) \in R$, then we must also have $(a, c) \in R$.

Example 2.14 　　Let S be any set. We define the relation R on S as follows: $\forall a, b \in S$, we define aRb iff $a = b$. It will now be shown that R is reflexive, symmetric, antisymmetric, and transitive.

Reflexive: For any $a \in S$, we have $a = a$, so $aRa \; \forall a \in S$.

Symmetric: Suppose we have aRb for some $a, b \in S$. Then $a = b$, so $b = a$. Hence, bRa.

Antisymmetric: Suppose we have aRb and bRa for some $a, b \in S$. Then, of course, $a = b$.

Transitive: Suppose we have aRb and bRc for some $a, b, c \in S$. Then $a = b$ and $b = c$. Hence, $a = c$, so aRc.

Example 2.14 is, in fact, quite trivial. It is easy to see that $R = \{(a, a) | a \in S\}$. Example 2.15 below is more interesting.

Example 2.15

Let S be any nonempty set. Let $\mathbf{P}(S)$ denote the set of all subsets of S. Define a relation $\leq$ on $\mathbf{P}(S)$ as follows: $\forall A, B \in \mathbf{P}(S)$, we define $A \leq B$ iff $A \subseteq B$. Which of the four properties defined in Definitions 2.14 to 2.17 does $\leq$ possess?

Solution: We check each of the four properties.

Reflexive: For any $A \in \mathbf{P}(S)$, it is clear that $A \subseteq A$, so $A \leq A$. Hence, the relation is reflexive.

Symmetric: Suppose $A, B \in \mathbf{P}(S)$ with $A \leq B$. Hence, $A \subseteq B$. Does this imply that $B \subseteq A$? It is not enough just to think "not necessarily." We have to come up with a counterexample to show that it is not necessarily the case that $B \subseteq A$. Here is where we use the given condition that S is nonempty. Because S is nonempty, there must exist an object $s \in S$. Thus, $\{s\} \subseteq S$, so $\{s\} \in \mathbf{P}(S)$. Note that we also have $\varnothing \subseteq S$, so $\varnothing \in \mathbf{P}(S)$. Clearly, $\varnothing \leq \{s\}$ because $\varnothing \subseteq \{s\}$. Hence, for the relation to be symmetric, we must also have $\{s\} \leq \varnothing$. But if this were true, it would indicate that $\{s\} \subseteq \varnothing$, which is clearly false. We therefore conclude that the relation is not symmetric.

Antisymmetric: Suppose $A, B \in \mathbf{P}(S)$ with $A \leq B$ and $B \leq A$. Then, $A \subseteq B$ and $B \subseteq A$. Hence, $A = B$ by Definition 2.4, so the relation is antisymmetric.

Transitive: Suppose $A, B, C \in \mathbf{P}(S)$ with $A \leq B$ and $B \leq C$. Then $A \subseteq B$ and $B \subseteq C$. By Exercise 3 of Section 2.1, we have $A \subseteq C$, so $A \leq C$. Hence, the relation is transitive.

By the way, for any set S, the set $\mathbf{P}(S)$ of all subsets of S is called the **power set** of S. We will be working with power sets later in the text.

Definition 2.18

A relation R on a set S is said to be an **equivalence relation** if it is reflexive, symmetric, and transitive.

If R is an equivalence relation on a set S, and $a, b \in S$ with aRb, then we say a is **equivalent** to b. By the symmetric property of equivalence relations, if a is equivalent to b, then b is equivalent to a. Note that the relation in Example 2.14 is an equivalence relation. The relation in Example 2.15 is not an equivalence relation because it is not symmetric.

Example 2.16

We define a relation S on **R** as follows: $\forall a, b \in \mathbf{R}$, we define aSb iff $a - b \in \mathbf{Z}$. (Recall that **Z** is the set of integers.) We will show that S is an equivalence relation.

 Reflexive: For any $a \in \mathbf{R}$, $a - a = 0 \in \mathbf{Z}$. Hence, aSa.

 Symmetric: Suppose $a, b \in \mathbf{R}$ with aSb. Now, because $a - b \in \mathbf{Z}$, so is $-(a - b)$. But $-(a - b) = b - a$. Hence, $b - a \in \mathbf{Z}$, so bSa.

 Transitive: Suppose $a, b, c \in \mathbf{R}$ with aSb and bSc. Then $a - b \in \mathbf{Z}$ and $b - c \in \mathbf{Z}$. Now, the sum of two integers is an integer. Also, $a - c = (a - b) + (b - c)$. Thus, $a - c \in \mathbf{Z}$, so aSc.

Definition 2.19

Suppose R is an equivalence relation on a set S. For each $a \in S$, **the equivalence class** of a with respect to R is denoted by $[a]_R$ and defined by $[a]_R = \{x \in S | xRa\}$.

That is, $[a]_R$ is the set of all elements in S that are equivalent to a. Often, as long as we are clear about which relation R is under consideration, we will simply write $[a]$ instead of $[a]_R$ for the equivalence class of a.

Example 2.17

Consider the equivalence relation of S in Example 2.16. What is $[0.5]$?

Solution: By definition, $[0.5] = \{x \in \mathbf{R} | x - 0.5 \in \mathbf{Z}\}$. A little thought indicates that we can write this set as $[0.5] = \{n + 0.5 | n \in \mathbf{Z}\}$. Why? Because if $x - 0.5 \in \mathbf{Z}$, then there must exist an $n \in \mathbf{Z}$ such that $x - 0.5 = n$, so $x = n + 0.5$. It is easy to see that $[0.5] = \{\pm 0.5, \pm 1.5, \pm 2.5, \ldots\}$.

With a little thought (do it!) the reader will see that $[-0.5] = [0.5]$, so that the equivalence class of -0.5 is the same set as is the equivalence class of 0.5. Noting that -0.5 is related to 0.5 (because $-0.5 - 0.5 = -1 \in \mathbf{Z}$), the next theorem proves that $[-0.5] = [0.5]$.

Theorem 2.7

Suppose R is an equivalence relation on a set S. Then for $\forall a, b \in S$, we have aRb iff $[a] = [b]$.

Proof: We first show that $\forall a, b \in S(aRb \Rightarrow [a] = [b])$. So, suppose aRb for a given $a, b \in S$. Recalling that $[a]$ and $[b]$ are sets, we shall show $[a] = [b]$ by showing $[a] \subseteq [b]$ and $[b] \subseteq [a]$. To show that $[a] \subseteq [b]$, suppose $x \in [a]$. We would like to conclude that $x \in [b]$. Starting with $x \in [a]$, we have xRa. To show that $x \in [b]$, we need only show that xRb. Well, we have xRa and aRb. Hence, because R is transitive, we must have xRb. Thus, we have shown $[a] \subseteq [b]$. Using the same type of argument, the reader is now asked to show that $[b] \subseteq [a]$.

Next, we must show that $\forall a, b \in S([a] = [b] \Rightarrow aRb)$, so suppose $[a] = [b]$. Now, because R is reflexive, aRa, so $a \in [a]$. Thus, we must have $a \in [b]$, so aRb. Having shown that $\forall a, b \in S([a] = [b] \Leftrightarrow aRb)$, we are done. ■

Theorem 2.8

Suppose R is an equivalence relation on a set S. Then for $\forall a, b \in S$, we have aRb or $[a] \cap [b] = \varnothing$.

Proof: First convince yourself (using truth tables if necessary) of the following: Given statements p, q, we have $(p \vee q) \Leftrightarrow ((\neg q) \Rightarrow p)$. Keeping this in mind, it suffices to show that $\forall a, b \in S$, we have $[a] \cap [b] \neq \varnothing \Rightarrow aRb$. Now, if $[a] \cap [b] \neq \varnothing$, then $\exists x \in [a] \cap [b]$. Hence, xRa and xRb. Because R is symmetric, we must have aRx. Finally, because R is transitive, we have aRb. ■

Definition 2.20

A relation $\leq$ on a set S is called a **linear order** if it has the following properties:

1) The relation is reflexive, antisymmetric, and transitive.
2) $\forall a, b \in S$, we have $a \leq b$ or $b \leq a$.

A linear order is sometimes referred to also as a **total order** in the mathematics literature. With respect to a given linear order $\leq$, the set S is often referred to as a **linearly ordered** (or **totally ordered**) set. Part 2 of Definition 2.20 is often referred to by saying any two elements $a, b \in S$ are **comparable**. The reader is now asked to confirm that the relation in Example 2.12 is a total order.

Indeed, Example 2.12 is the motivation behind calling the relation an *order*. Definition 2.20 is simply the generalization to arbitrary sets of the well-known *less than or equal to inequality* for real numbers. In fact, because of this generalization, the symbol '≤' is usually reserved for linear orders, and from now on in this book we shall only denote a relation by '≤' if it turns out to be a linear order. We will have more to say about linear orders later in the book.

A few final notes would be useful. You may have noticed that we have become a little more relaxed with our notation involving quantifiers. For example, Definition 2.20, Part 2 was expressed by "$\forall a,\ b \in S$, we have $a \le b$ or $b \le a$." Previously, we would have written this as "$\forall a, b \in S\ (a \le b\ \text{or}\ b \le a)$." The reader should get used to a more relaxed style of writing this because this style is employed in most mathematics textbooks. We could also express the condition by saying "$\forall a,\ b \in S,\ a \le b$ or $b \le a$." You might also see this written as "$a \le b$ or $b \le a$ for every $a, b \in S$." The important thing is that the meaning is clear and that the reader is able to write the condition 'formally' if necessary.

Also, there has been a lot of "hand-holding" in the proofs presented so far. This was our intent because one of the major goals of the book is to show students how to do proofs. The student should require less "hand-holding" as more experience is gained. We will not always be as verbose as we were, for example, in the proof of Theorem 2.7.

EXERCISES

1. With respect to the relation in Example 2.15, find an example of a set S such that Part 2 of Definition 2.20 fails to hold. (By the way, a relation on a set for which Part 1 of Definition 2.20 holds is called a **partial order**. Hence, a total order (i.e., a linear order) is a partial order that also satisfies the condition of Definition 2.20, Part 2).

2. Let Ω denote the set of people attending a particular family reunion. Define a relation R on Ω as follows: For a given $u, v \in \Omega$, define uRv to mean that u is a first cousin of v. Which of the conditions specified in Definitions 2.14 to 2.17 must hold? (**Note**: This is the first time in this book that Greek letters have been used to denote sets. Doing so is a very common occurrence in mathematics textbooks).

3. Let L denote the set of lines in the Euclidean plane. Define a relation R on L as follows: For lines $l, m \in L$, define lRm to mean that l is parallel to m. (We declare a line to always be parallel to itself.) Show that R is an equivalence relation on L. What is the equivalence class of the line given by the equation $y = 3x + 7$? (Hint: Recall the *slope* of a line from high school analytic geometry.)

4. Let $A = \{a, b, c\}$ (i.e., A is the set consisting of the first three elements of the alphabet).

   a. Construct an equivalence relation on A.

   b. Construct a linear order on A.

5. Define a relation R on $\mathbf{N} \times \mathbf{N}$ as follows: For each (a, b), $(c, d) \in \mathbf{N} \times \mathbf{N}$, define $(a, b)R(c, d)$ to mean $a + d = b + c$. Show that R is an equivalence relation on

$\mathbf{N} \times \mathbf{N}$. (By the way, you will be seeing this equivalence relation again when we construct the set of integers from the set of natural numbers).

Functions

There is a special type of relation that is pervasive throughout mathematics: the function. No doubt you have had some exposure to the concept of a function. Before we launch into the formal definition of a function in terms of a relation, we review the "intuitive idea" behind it that most students should be familiar with. Of course, this intuitive idea is insufficient for our purposes because we must base everything that we do squarely on the concepts of set theory.

The Function Concept

Let X and Y be sets. Many elementary mathematics books define a function f from X to Y to be a rule that assigns exactly one element of Y to each element of X. Consider an $x \in X$, and suppose y is the member of Y that has been assigned to x. Then we would indicate this by writing $y = f(x)$. It is very likely that so far in your mathematical training, the sets X and Y have always been the set $\mathbf{R}$ of real numbers, or subsets of $\mathbf{R}$ (such as open or closed intervals of real numbers). For example, one well-known function is the one given by $f(x) = x^2 \; \forall x \in \mathbf{R}$. This is the mapping (another word for function) that assigns the number x^2 to each real number x. So, the number 2 is assigned the number 4, the number 3 is assigned the number 9, the number -2 is assigned the number 4, and so on. Note that the number 4 has been assigned twice; it was assigned both to 2 and to -2.

From an engineering perspective, a function is often viewed as an input–output process. This is displayed in Figure 2.7. The input, x, is fed into the function "machine." The machine processes the input and outputs the value $f(x)$.

A key point about functions is that there is only one value assigned to each $x \in X$. This does not mean that the assigned value cannot be used again. Indeed, with the function $f(x) = x^2$, we have already seen that the value 4 was assigned for both -2 and 2. However, we cannot assign two different output values to the input value 2. Many elementary mathematics texts display this point pictorially by introducing the *vertical line test*. A collection of points in the Euclidean plane is a function (of real numbers) if any vertical line intersects the collection only once. Hence, the collection of points in Figure 2.8 constitutes a function because it passes this test. On the other hand, the collection of points graphed in Figure 2.9 fails this test because there is at least one vertical line that intersects the collection more than once.

Figure 2.7
A function *f* viewed
as an input–output
process

Figure 2.8
A collection of
points passing the
vertical line test

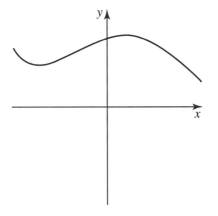

Figure 2.9
A collection of
points failing the
vertical line test

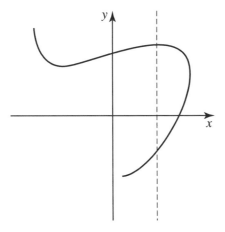

The previous intuitive definition of a function is not sufficiently rigorous for our purposes. We do not have a definition of what is meant by an
"assignment rule." We now are ready to define functions in terms of set
theory.

Definition and Examples of a Function and Some Initial Concepts

---

Definition 2.21

Let X and Y be sets. A **function**, also called a **mapping**, from X to Y is a relation (X, Y, f) with the following property:

For each $x \in X$, there is a unique $y \in Y$ such that $(x, y) \in f$.

First of all, Definition 2.21 says that given an $x \in X$, $\exists y \in Y$ such that $(x, y) \in f$. The declaration of its *uniqueness* means the following: $\forall z \in Y$, if $z \neq y$, then $(x, z) \notin f$. Another way to word this is $\forall z \in Y$, $(x, z) \in f \Rightarrow z = y$. Hence, of all the ordered pairs in f, exactly one of them has x as its first component. This is why the collection of points displayed in Figure 2.9 fails to be a function. The vertical line test reveals that at least two of the points have the same x-coordinate.

Instead of writing a function as (X, Y, f), it is usually expressed by writing $f: X \to Y$. Also, although f is actually a set of ordered pairs, the notation is usually abused, and we refer to the function as f. There is no problem here, as long as we understand that the function is actually the ordered triple (X, Y, f). The set X is called the **domain** of f, and the set Y is called the **range** of f. Instead of writing $(x, y) \in f$, we normally express this by writing $y = f(x)$. We also say that y is the **image** of x with respect to f. Finally, we also say that f **maps** x to y. Thus, for a given $x \in X$, $f(x)$ is defined to be that unique $y \in Y$ such that $(x, y) \in f$. Keep in mind that other symbols besides small English letters can be used to denote function. In fact, capital English letters and Greek letters are also often used to represent functions.

Example 2.18

Consider the function $f: \mathbf{R} \to \mathbf{R}$ specified by $f(x) = x^2 \; \forall x \in \mathbf{R}$. What this *really* means is that we have the function $(\mathbf{R}, \mathbf{R}, f)$, in which $f = \{(x, x^2) \,|\, x \in \mathbf{R}\}$. Now, we have $f(2) = 4$ because $(2, 4) \in f$.

Example 2.19

Let $A = \{1, 2, 3\}$, $B = \{7, 8, 50\}$, $f = \{(1, 7), (1, 8), (2, 8), (3, 50)\}$, $g = \{(2, 8), (3, 50)\}$, $h = \{(1, 8), (2, 8), (3, 8)\}$, and $\phi = \{(1, 50), (2, 8), (3, 7)\}$. Which of the relations f, g, h, and ϕ are functions from A to B?

Solution: f fails to be a function from A to B because 1 has been "assigned" two elements of B. g is not a function from A to B because 1 has not been "assigned" an element of B. h and ϕ are valid functions from A to B because they meet the criteria of Definition 2.21.

Note that the function $h\colon A \to B$ of Example 2.19 can be specified by $h(x) = 8 \forall x \in A$. That is, h "assigns" the value 8 to each element of A (which is a perfectly valid "assignment"). The function $\phi\colon A \to B$ in Example 2.19 can be specified by $\phi(1) = 50$, $\phi(2) = 8$, and $\phi(3) = 7$. It should also be noted that although relation g of Example 2.19 fails to be a function from A to B, it is a valid function from $A - \{1\}$ to B (i.e., we have $g\colon \{2, 3\} \to B$).

How do we show that two functions (X, Y, f) and (A, B, g) are equal? Because we have defined functions as ordered triplets, we must have $X = A$, $Y = B$, and $f = g$. (See Exercise 1 in Section 2.3.) Indeed, the desire to force this condition is the reason why we have defined a function as an ordered triplet. Note that the condition $f = g$ can be written as $f(x) = g(x) \forall x \in X$.

Definition 2.22

Suppose that $f\colon X \to Y$, and suppose $A \subseteq X$ and $B \subseteq Y$. The **image** of A with respect to f is the set denoted by $f(A)$ and is defined by $f(A) = \{f(x) \mid x \in A\}$. The **inverse image** of B with respect to f is the set denoted by $f^{-1}(B)$ and is defined by $f^{-1}(B) = \{x \in X \mid f(x) \in B\}$.

Note that if $f\colon X \to Y$, and $A \subseteq X$ and $B \subseteq Y$, then $f(A) \subseteq Y$ and $f^{-1}(B) \subseteq X$. Intuitively, the image of A is the set of elements in Y that have been assigned to elements in A. The inverse image of B is the set of objects in X that have been assigned an element in B. Do not let the notation f^{-1} confuse you. It has nothing to do with exponents.

Example 2.20

Consider the function $f\colon \mathbf{R} \to \mathbf{R}$ specified by $f(x) = x^2 \; \forall x \in \mathbf{R}$. Let $A = \{x \in \mathbf{R} \mid 1 \leq x \leq 2\}$, and let $B = \{x \in \mathbf{R} \mid 4 \leq x \leq 9\}$. Find $f(A)$ and $f^{-1}(B)$.

Solution: The reader should have an easy time using simple algebra to determine that $f(A) = \{x \in \mathbf{R} \mid 1 \leq x \leq 4\}$ and $f^{-1}(B) = \{x \in \mathbf{R} \mid -3 \leq x \leq -2\} \cup \{x \in \mathbf{R} \mid 2 \leq x \leq 3\}$. This is illustrated in Figure 2.10.

Figure 2.10
Illustrating Example
2.20

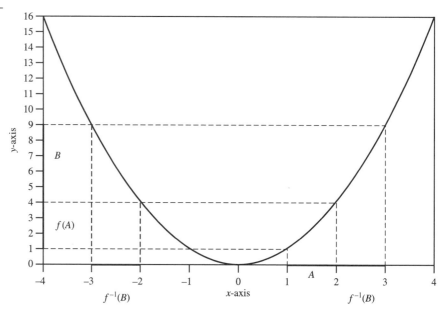

Example 2.21   Consider the functions h and ϕ in Example 2.19. It should be easy to see that $h(\{1, 2\}) = \{8\}$, and $\phi^{-1}(\{7, 8\}) = \{2, 3\}$.

Theorem 2.9   Suppose $f: X \to Y$ with $E \subseteq Y$ and $F \subseteq Y$, then

1) $f^{-1}(E \cup F) = f^{-1}(E) \cup f^{-1}(F)$.
2) $f^{-1}(E \cap F) = f^{-1}(E) \cap f^{-1}(F)$.

Proof: We prove Part 1 and leave Part 2 to the reader. To prove Part 1, we show that sets on both sides of the equality sign are subsets of one another. To show that $f^{-1}(E \cup F) \subseteq f^{-1}(E) \cup f^{-1}(F)$, let $x \in f^{-1}(E \cup F)$. By definition, there must exist a $y \in E \cup F$ such that $y = f(x)$. Hence, $y \in E$ or $y \in F$. If $y \in E$, then $x \in f^{-1}(E)$. If $y \in F$, then $x \in f^{-1}(F)$. Either way, $x \in f^{-1}(E) \cup f^{-1}(F)$.

To show that $f^{-1}(E) \cup f^{-1}(F) \subseteq f^{-1}(E \cup F)$, suppose $x \in f^{-1}(E) \cup f^{-1}(F)$. Then $x \in f^{-1}(E)$ or $x \in f^{-1}(F)$. If $x \in f^{-1}(E)$, then $f(x) \in E$.

If $x \in f^{-1}(F)$, then $f(x) \in F$. Either way, $f(x) \in E \cup F$. Hence $x \in f^{-1}(E \cup F)$.   ∎

Theorem 2.10   Suppose $f: X \to Y$ with $A \subseteq X$ and $B \subseteq X$, then $f(A \cup B) = f(A) \cup f(B)$.

Proof: We show $f(A \cup B) \subseteq f(A) \cup f(B)$ and $f(A) \cup f(B) \subseteq f(A \cup B)$. To show the former, suppose $y \in f(A \cup B)$. Then $\exists x \in A \cup B$ such that $y = f(x)$. Hence, $x \in A$ or $x \in B$. If $x \in A$, then $y \in f(A)$. If $x \in B$, then $y \in f(B)$. Either way, $y \in f(A) \cup f(B)$.

To show $f(A) \cup f(B) \subseteq f(A \cup B)$, suppose $y \in f(A) \cup f(B)$. Hence, $y \in f(A)$ or $y \in f(B)$. If $y \in f(A)$, then $\exists x \in A$ such that $y = f(x)$. If $y \in f(B)$, then $\exists x \in B$ such that $y = f(x)$. Either way, $\exists x \in A \cup B$ such that $y = f(x)$. Hence, $y \in f(A \cup B)$.   ∎

The astute reader may notice that Theorem 2.10 does not claim that $f(A \cap B) = f(A) \cap f(B)$ always holds. There is good reason for this. In the Exercises, the student is asked to produce a counterexample to the claim that this always holds.

The following definition will prove useful on a number of occasions.

Definition 2.23

Suppose that $f: X \to Z$, and suppose $g: A \to Y$ such that $A \subseteq X$, $Y \subseteq Z$, and $g(x) = f(x)$ for each $x \in A$. Then we say that f is an **extension** of g to X and that g is a **restriction** of f to A. A restriction of f to A can be denoted by $f|A$.

One-to-One and Onto Functions

Definition 2.24

A function $f: X \to Y$ is said to be **one-to-one** if it has the following property: $\forall x_1, x_2 \in X, f(x_1) = f(x_2) \Rightarrow x_1 = x_2$. A one-to-one function is also called an **injection** and is said to be **injective**.

It is important that the reader understands the meaning of Definition 2.24. Looking at the contrapositive of the implication may be easier. The statement of the definition then reads $\forall x_1, x_2 \in X, x_1 \neq x_2 \Rightarrow f(x_1) \neq f(x_2)$. Yet another way to look at the definition is as follows: $f: X \to Y$ is one-to-one iff for each $y \in Y$ there is at most one $x \in X$ such that $y = f(x)$.

Example 2.22

Consider the function $f: \mathbf{R} \to \mathbf{R}$ specified by $f(x) = x^2 \; \forall x \in \mathbf{R}$. Then it is seen that f is not one-to-one. The statement $f(-2) = 4 = f(2)$ immediately shows this. Using the notation introduced in Definition 2.23, if we let $g = f|P$ where P is the set of *nonnegative* real numbers, then we see that g is one-to-one. Let us show this formally using algebra. Suppose $g(a) = g(b)$, in which a and b are nonnegative numbers. Then we have $a^2 = b^2$. Taking square roots gives us $|a| = |b|$. But $|a| = a$ and $|b| = b$ because a and b are nonnegative. Hence, $a = b$. Thus, we have shown that $g(a) = g(b) \Rightarrow a = b$ $\forall a, b \in P$, so we declare g is one-to-one by Definition 2.24.

Definition 2.25

A function $f: X \to Y$ is said to be **onto** if it has the following property: For each $y \in Y$, $\exists x \in X$ such that $y = f(x)$. An onto function is also called a **surjection** and is said to be **surjective**.

Hence, a function is surjective if each member of the range is the image of at least one element in the domain. The reader should become convinced that $f: X \to Y$ is onto iff $f(X) = Y$.

Example 2.23

Consider the function $f: \mathbf{R} \to \mathbf{R}$ specified by $f(x) = x^2 \; \forall x \in \mathbf{R}$. Then f is not onto. Why not? Well, for example, there is no $x \in \mathbf{R}$ such that $f(x) = -9$. Now, let P be the set of nonnegative real numbers and consider $g: \mathbf{R} \to P$ specified by $f(x) = x^2 \; \forall x \in \mathbf{R}$. To show that g is onto, let $y \in P$. Then $y \geq 0$, so y has a square root $\sqrt{y}$. Now, $f(\sqrt{y}) = (\sqrt{y})^2 = y$. (In fact, we also have $f(-\sqrt{y}) = (-\sqrt{y})^2 = y$.)

Definition 2.26

A function $f: X \to Y$ is said to be a **one-to-one correspondence** if it is both one-to-one and onto. A one-to-one correspondence is also called a **bijection** and is said to be **bijective**.

Note that $f: X \to Y$ is bijective iff for each $y \in Y$ there is an $x \in X$ such that $\{x\} = f^{-1}(\{y\})$.

The concept of a bijection is believed to predate counting. For example, ancient herdsmen would know if they were missing a sheep by setting up a one-to-one correspondence between the set of sheep in their herd and a collection of objects, say rocks. As each sheep returned home, the herdsman could move a rock from one pile to another. At the end of the day, any rocks remaining in the first pile would indicate missing sheep.

Example 2.24

Let X be a nonempty set, and define the function $i_X: X \to X$ by $i_X(x) = x$ $\forall x \in X$. Then i_X is called the **identity function** on X because it maps each element of X to itself. It is clear that this mapping is a bijection.

Inverse Functions and the Composition of Functions

Suppose $f: X \to Y$ is a one-to-one correspondence. Then, for each $y \in Y$, there is exactly one $x \in X$ whose image is y. Therefore, we can define a function $g: Y \to X$ by defining $g(y)$ to be that $x \in X$ such that $f(x) = y$. This is done in Definition 2.27. It is illustrated in Figure 2.11.

Definition 2.27

Suppose $f: X \to Y$ is a bijection. The **inverse function** of f is the function from Y to X whose image at a given $y \in Y$ is that $x \in X$ such that $f(x) = y$. The inverse function of f is denoted f^{-1}.

The astute reader may be concerned that this notation clashes slightly with some previous notation. Recall that if $f: X \to Y$ is any function, and $B \subseteq Y$,

Figure 2.11
The relationship between a bijection and its inverse

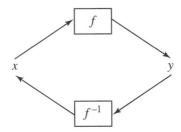

then we have the set $f^{-1}(B)$, which was defined in Definition 2.22 to be $f^{-1}(B) = \{x \in X | f(x) \in B\}$. Now, the *function* f^{-1} only exists if f is bijective. In this case, $f^{-1}(B)$ could also be taken to mean $f^{-1}(B) = \{f^{-1}(y) | y \in B\}$, by applying Definition 2.22 to the *function* f^{-1}. However, it is easily seen that in this case we have $\{x \in X | f(x) \in B\} = \{f^{-1}(y) | y \in B\}$. Hence, the ambiguity in the notation makes no difference! Finally, the reader should have no difficulty showing that $f^{-1}(f(x)) = x \ \forall x \in X$, and $f(f^{-1}(y)) = y \ \forall y \in Y$.

Example 2.25   Let $h: \mathbf{R} \to \mathbf{R}$ be defined by $h(x) = 3x + 7 \ \forall x \in \mathbf{R}$. Show that h is bijective, and find its inverse h^{-1}.

Solution: We first show that h is injective. To do this, suppose $h(x_1) = h(x_2)$ with $x_1, x_2 \in \mathbf{R}$. Then we have $3x_1 + 7 = 3x_2 + 7$. Using simple algebra, this gives us $3x_1 = 3x_2$, and dividing both sides by 3 gives us the desired result of $x_1 = x_2$.

To show that h is surjective, we must show that for each $y \in \mathbf{R}$, there exists an $x \in \mathbf{R}$ such that $h(x) = y$. To do this, suppose $y \in \mathbf{R}$. We must show that there exists an $x \in \mathbf{R}$ such that $h(x) = y$. Hence, we must solve the equation $3x + 7 = y$ for x. It is easily seen that doing so results in $x = \frac{1}{3}(y - 7)$. That is, $h(\frac{1}{3}(y - 7)) = y$.

The results for demonstrating that h is surjective give us what we need to find the inverse function. It is now clear that defining h^{-1} by $h^{-1}(y) = \frac{1}{3}(y - 7) \forall y \in \mathbf{R}$ does the trick.

By the way, it is important that the reader not get hung up on particular letters. We could just have well defined h^{-1} in the above example by saying $h^{-1}(x) = \frac{1}{3}(x - 7) \forall x \in \mathbf{R}$ (i.e., using the letter x instead of y).

To motivate the next definition, suppose we have functions $g: X \to Y$ and $f: Y \to Z$. Now, for any $x \in X$, note that $g(x) \in Y$. Hence, not only is $g(x)$ in the range of g, it is also in the domain of f. Now the image of $g(x)$ with respect to f is $f(g(x))$. This latter expression is pronounced "f of g of x."

Definition 2.28

Suppose we have functions $g: X \to Y$ and $f: Y \to Z$. We define the composition of f and g, denoted by $f \circ g$, to be the following function: $f \circ g: X \to Z$, with $f \circ g(x) = f(g(x)) \forall x \in X$.

Example 2.26   Define the sets X, Y, and Z by $X = \{4, 5, 6\}$, $Y = \{a, b\}$, and $Z = \{a, 60, 90\}$ (where by 'a' and 'b' we mean the first two letters of the alphabet). Define $g: X \to Y$ by $g(4) = a$, $g(5) = b$, and $g(6) = a$. Define $f: Y \to Z$ by $f(a) = 90$ and $f(b) = a$. Find $f \circ g$.

Solution: We first note that $f \circ g: X \to Z$ is well defined (i.e., makes sense) because the range of g is the domain of f. $f \circ g(4) = f(g(4)) = f(a) = 90$. $f \circ g(5) = f(g(5)) = f(b) = a$. Finally, $f \circ g(6) = f(g(6)) = f(a) = 90$. ■

By the way, for Example 2.26, note that the function $g \circ f$ is not defined because the range of f is not the domain of g.

The notation in Example 2.27 below has been chosen to be a little different on purpose, although it is perfectly "legal."

Example 2.27

Let $\Gamma = \{x \in \mathbf{R} | x \geq 0\}$. Let $\theta: \mathbf{R} \to \Gamma$ be defined by $\theta(x) = x^2 \ \forall x \in \mathbf{R}$. Let $h: \Gamma \to \Gamma$ be defined by $h(x) = x^{\frac{1}{4}} \ \forall x \in \Gamma$. Find $h \circ \theta$.

Solution: We first note that $h \circ \theta: \mathbf{R} \to \Gamma$ is well defined (i.e., makes sense) because the range of θ is the domain of h. Now, $\forall y \in \mathbf{R}$, we have $h \circ \theta(y) = h(\theta(y)) = h(y^2) = (y^2)^{\frac{1}{4}} = y^{\frac{1}{2}}$. That is, $h \circ \theta$ is that function $f: \mathbf{R} \to \Gamma$ defined by $f(x) = x^{\frac{1}{2}} \ \forall x \in \mathbf{R}$. ■

Theorem 2.11

Suppose we have functions $g: X \to Y$ and $f: Y \to Z$.

1) If f and g are both injective, then so is $f \circ g$.

2) If f and g are both surjective, then so is $f \circ g$.

3) If f and g are both bijective, then so is $f \circ g$.

Proof: We prove Part 2 and leave Part 1 to the reader. Note that Part 3 follows immediately from Parts 1 and 2. To prove Part 2, let $z \in Z$. We must show that there exists an $x \in X$ such that $f \circ g(x) = z$. Now, because the function f is onto, there must exist a $y \in Y$ such that $f(y) = z$. Because the function g is onto, there must exist a $x \in X$ such that $g(x) = y$. Hence, $f \circ g(x) = f(g(x)) = f(y) = z$. ■

Suppose $f: X \to Y$ is a bijection. Then we have the inverse mapping $f^{-1}: Y \to X$. Note that $f \circ f^{-1}: Y \to Y$ and $f^{-1} \circ f: X \to X$. Furthermore, the reader should have no trouble showing (do it!) that $f \circ f^{-1}(y) = y \ \forall y \in Y$, and $f^{-1} \circ f(x) = x \ \forall x \in X$. Hence, $f \circ f^{-1}$ is the identity function on Y, and $f^{-1} \circ f$ is the identity function on X. (See Example 2.24.)

Binary Operations

One important application of functions is to use them to express **binary operations**. Let X be a nonempty set. Intuitively, a binary operation is simply a rule that assigns a value in X to a pair of values from X. For example, for every pair of numbers $a, b \in \mathbf{R}$, the binary operation known as **addition** assigns a number $a + b$ (called the **sum**) in $\mathbf{R}$. Another well-known binary operation known as **multiplication** assigns a number ab (called the **product**) in $\mathbf{R}$.

A major branch of mathematics known as **abstract algebra** generalizes the well-known binary operations of addition and multiplication of numbers in $\mathbf{R}$ to arbitrary sets. This is the motivation behind the next definition.

Definition 2.29

Let X be a nonempty set. A binary operation on X is a mapping $\varphi \colon X \times X \to X$.

Example 2.28

Let X be the set of letters defined by $X = \{a, b, c\}$. Define $\varphi \colon X \times X \to X$ by $\varphi((a, a)) = b$, $\varphi((a, b)) = c$, $\varphi((a, c)) = c$, $\varphi((b, a)) = c$, $\varphi((b, b)) = a$, $\varphi((b, c)) = a$, $\varphi((c, a)) = c$, $\varphi((c, b)) = a$, and $\varphi((c, c)) = b$. Then φ is a binary operation on X.

Why have we used double parenthesis such as $\varphi((a, a))$ in Example 2.28? Because each element of the domain is an ordered pair, such as (a, a). In many math texts, this is simply written as $\varphi(a, a)$ instead of $\varphi((a, a))$. Speaking of notation, often, instead of using a letter (such as φ) to denote a binary operation, the symbols for addition or multiplication are reused. The context will always make it clear what is meant. For example, for Example 2.28, we might have expressed this binary relation as follows:

Let X be the set of letters defined by $X = \{a, b, c\}$, and define a binary relation on X as follows: $a + a = b$, $a + b = c$, $a + c = c$, $b + a = c$, $b + b = a$, $b + c = a$, $c + a = c$, $c + b = a$, and $c + c = b$.

Here the context makes it clear that by using the plus sign '+', we do not mean the addition of real numbers because X is not a set of real numbers, even though the binary relation may be referred to as a "sum" of elements of X. Frequently, the multiplication sign '·' is used to denote a binary operation. Also frequently, the notation of simply writing ab is utilized to denote the image of (a, b) under the binary operation. This is also referred to as multiplication and/or "juxtaposition." Using multiplication, we could have expressed the binary relation of Example 2.28 as follows:

Let X be the set of letters defined by $X = \{a, b, c\}$, and define a binary relation on X as follows: $aa = b$, $ab = c$, $ac = c$, $ba = c$, $bb = a$, $bc = a$, $ca = c$, $cb = a$, and $cc = b$.

Sometimes, symbols such as $\otimes$, $\oplus$, or $*$ may be used to express a binary operation. This is especially true in situations where many binary operations are being used simultaneously, and it is important to distinguish one operation from the other. So, for example, if we have a binary operation on a set X, we may use the notation $x \otimes y$ for a given $x, y \in X$ to indicate the image of the ordered pair (x, y) with respect to the mapping.

Definition 2.30

A binary operation $\oplus$ on a set X is said to be **commutative** if $x \oplus y = y \oplus x$ $\forall x, y \in X$. $\oplus$ is said to be **associative** if $x \oplus (y \oplus z) = (x \oplus y) \oplus z$ $\forall x, y, z \in X$.

Of course, the usual addition and multiplication on **R** are both commutative and associate (as we will formally show when we develop the real number system). It is easy to show that the binary operation in Example 2.28 is commutative.

We now discuss a very important associative binary operation. Let X be any nonempty set, and let Ω_X be the set of all functions from X to itself. That is, $\Omega_X = \{f \mid f\colon X \to X\}$. Note that for each $f, g \in \Omega_x$, the composition of f and g, $f \circ g$, is also a member of Ω_X. The composition of functions in Ω_X is a binary operation on Ω_X. Theorem 2.12 states that this operation is associative.

Theorem 2.12

Suppose we have functions f, g, and h from a set nonempty X to itself. Then $(f \circ g) \circ h = f \circ (g \circ h)$.

Proof: Noting that $(f \circ g) \circ h$ and $f \circ (g \circ h)$ have the same domain and range, all we have to do to complete the proof is to show that $((f \circ g) \circ h)(x) = (f \circ (g \circ h))(x) \forall x \in X$. Now, repeatedly using Definition 2.28, we have $((f \circ g) \circ h)(x) = f \circ g(h(x)) = f(g(h(x))) = f(g \circ h(x)) = (f \circ (g \circ h))(x) \forall x \in X$. ∎

Finally, an important class of binary operations analyzed at lengh in abstract algebra texts should be mentioned here, although it does not play a prominent role in this text.

Definition 2.31

Suppose that G is a nonempty set with $e \in G$ and that '$\cdot$' is a binary operation on G. Then the triple $(G, \cdot, e)$ is referred to as a **group** provided

1) The binary operation '$\cdot$' is associative.
2) $\forall a \in G, a \cdot e = e \cdot a = a$.

3) For each $a \in G$, there exists an element $a^{-1} \in G$ such that $a \cdot a^{-1} = a^{-1} \cdot a = e$.

Consider a group $(G, \cdot, e)$. The element e is referred to as the **identity element** of the group. In the exercises, the reader is asked to show that it is unique. For a given $a \in G$, the element $a^{-1} \in G$ is referred to as the **inverse** of a. In the exercises, the reader is asked to show that each element has a unique inverse.

An important notational note needs to be made. Although Definition 2.31 uses the multiplication symbol to define a group, any notation describing a binary operation suffices provided it satisfies the conditions of the definition. For example, in cases where the addition symbol '+' is employed instead of multiplication, it is customary to write $-a$ instead of a^{-1} to express the inverse of a. Also, we do not need to use the symbol e to express the identity element.

Example 2.29 Consider the set **Z** of integers with ordinary addition that you have known since grade school. Then it is clear that $(\mathbf{Z}, +, 0)$ is a group. In this case, the inverse of any $n \in \mathbf{Z}$ is $-n$. Of course, the number 0 is the identify element.

Example 2.30 Let X be any nonempty set, and let Γ_X be the set of all bijections from X to itself. That is, Γ_X is the set of all functions $f: X \to X$ such that f is both one-to-one and onto. Now consider the mapping $i_X: X \to X$ in Example 2.24. (Recall that this is called the identity function on X because it maps each element of X to itself.) Finally, consider the binary operation 'o' of Definition 2.28 that designates the composition of functions. Recall from Theorem 2.12 that this binary operation is associative. The reader should now have an easy time verifying that $(\Gamma_X, \mathrm{o}, i_X)$ is a group, in which the inverse of any $f \in \Gamma_X$ is the inverse function f^{-1} defined in Definition 2.27.

Indexed Sets and Arbitrary Unions and Intersections

Another important application of functions is to use them to create an **indexed family of sets**. Suppose we have a nonempty set S and a nonempty set of sets Ω. (A set of sets is frequently called a *family* of sets.) Furthermore, suppose we have a function $A: S \to \Omega$. Hence, for each $x \in S$, we have a set

$A(x) \in \Omega$. The set of sets $\{A(x) \mid x \in S\}$ is often called an indexed family of sets. The set S is called the **index set**.

Example 2.31

To illustrate the above, let S denote the set $\mathbf{R}$, and let Ω denote the set of all subsets of $\mathbf{R} \times \mathbf{R}$. Define $A: S \to \Omega$ as follows: For each $x \in S$, $A(x)$ is the set of points in the plane that lie on the line going through the origin of the coordinate system with the line's slope being x. Then the indexed family of sets $\{A(x) \mid x \in S\}$ is the set of nonvertical lines in the plane. (Why novertical? Because a vertical line has no slope.)

With respect to an indexed family of sets $\{A(x) \mid x \in S\}$, we often write A_x instead of $A(x)$. Hence, we can now write the indexed family of sets as $\{A_x \mid x \in S\}$. This is also frequently written as $\{A_x\}_{x \in S}$. What is the point of all of this? There are many reasons in mathematics. One answer is that by using an indexed family of sets, we can now define the union and intersection of an unlimited number of sets. The first application in this text of taking the intersection of an arbitrary number of sets occurs in the proof of Theorem 3.1 in Section 3.1.

Definition 2.32

Let $\{A_x\}_{x \in S}$ be an indexed nonempty family of sets. We denote the **union of the family** and the **intersection of the family** as $\bigcup_{x \in S} A_x$ and $\bigcap_{x \in S} A_x$, respectively, with these terms defined as

$\bigcup_{x \in S} A_x = \{y \mid y \in A_x \text{ for some } x \in S\}$.
$\bigcap_{x \in S} A_x = \{y \mid y \in A_x \text{ for every } x \in S\}$.

The first thing that the reader should do is become convinced that Definition 2.32 *generalizes* Definitions 2.6 and 2.7. That is, if the indexed set S consists of only two elements, then the union of the family coincides with the union of the two sets defined in Definition 2.6, and similarly for the intersection.

Now, a special notation is often employed in the case where the index set is the set $\mathbf{N}$ of natural numbers. Instead of writing $\bigcup_{n \in \mathbf{N}} A_n$ and $\bigcap_{n \in \mathbf{N}} A_n$, we can write this as $\bigcup_{n=1}^{\infty} A_n$ and $\bigcap_{n=1}^{\infty} A_n$, respectively.

Example 2.32

Let $\{A_n \mid n \in \mathbf{N}\}$ be the indexed family defined by $A_n = \{x \in \mathbf{R} \mid -n \le x \le n\}$ for each $n \in \mathbf{N}$. Find $\bigcup_{n=1}^{\infty} A_n$ and $\bigcap_{n=1}^{\infty} A_n$.

Solution: With a little thought, it should become obvious to the reader that $\bigcup_{n=1}^{\infty} A_n = \mathbf{R}$ and $\bigcap_{n=1}^{\infty} A_n = A_1 = \{x \in \mathbf{R} \mid -1 \le x \le 1\}$.

Theorem 2.13   Let $\{A_x\}_{x \in S}$ be an indexed family of sets and let Z be any set. Then,

1) $Z \cap (\cup_{x \in S} A_x) = \cup_{x \in S}(Z \cap A_x)$.
2) $Z \cup (\cap_{x \in S} A_x) = \cap_{x \in S}(Z \cup A_x)$.
3) $Z - (\cup_{x \in S} A_x) = \cap_{x \in S}(Z - A_x)$.
4) $Z - (\cap_{x \in S} A_x) = \cup_{x \in S}(Z - A_x)$.

Proof: We prove Part 1 and leave the rest to the reader. First of all, the notation of theorem "cheats" a little. For example, by $\cup_{x \in S}(Z \cap A_x)$, it is meant the union of the family $\{B_x | x \in S\}$ in which each B_x is defined to be the set $Z \cap A_x$. To prove Part 1, we will show that $Z \cap (\cup_{x \in S} A_x) \subseteq \cup_{x \in S}(Z \cap A_x)$ and $\cup_{x \in S}(Z \cap A_x) \subseteq Z \cap (\cup_{x \in S} A_x)$.

To show $Z \cap (\cup_{x \in S} A_x) \subseteq \cup_{x \in S}(Z \cap A_x)$, let $a \in Z \cap (\cup_{x \in S} A_x)$. Then $a \in Z$ and $a \in (\cup_{x \in S} A_x)$. Hence, there must exist a $y \in S$ such that $a \in A_y$. So, $a \in Z \cap A_y$. This shows that $a \in \cup_{x \in S}(Z \cap A_x)$.

To show $\cup_{x \in S}(Z \cap A_x) \subseteq Z \cap (\cup_{x \in S} A_x)$, suppose $a \in \cup_{x \in S}(Z \cap A_x)$. Then there must exist a $y \in S$ such that $a \in Z \cap A_y$. Hence, $a \in Z$ and $a \in (\cup_{x \in S} A_x)$. Hence, $a \in Z \cap (\cup_{x \in S} A_x)$. ∎

It should be noted that Theorem 2.13 is a generalization of Theorems 2.4 and 2.5.

Finally, it should be mentioned that there is another way to define the union and intersection of an arbitrary number of sets without using an index set. Suppose that Ω is a nonempty family of sets. We can define the union of the sets in Ω, denoted by $\cup_{S \in \Omega} S$, as $\cup_{S \in \Omega} S = \{x | x \in S \text{ for some } S \in \Omega\}$. Similarly, we can define the intersection of the sets in Ω, denoted by $\cap_{S \in \Omega} S$, as $\cap_{S \in \Omega} S = \{x | x \in S \text{ for each } S \in \Omega\}$. It is immediate that the results of Theorem 2.13 apply to this definition of union and intersection. For instance, without using index sets, Part 1 of Theorem 2.13 would read $Z \cap (\cup_{S \in \Omega} S) = \cup_{S \in \Omega}(Z \cap S)$.

A Note on Rigor and Notation

Most mathematics books do not define a function as an ordered triplet. I have partially explained why we did. A little more discussion on this point would be useful. An example will help. Consider the set Y defined by $Y = \{y \in \mathbf{R} | y \geq 0\}$, and define the functions $f: \mathbf{R} \to \mathbf{R}$ and $g: \mathbf{R} \to Y$, in which both $f(x) = x^2$ and $g(x) = x^2$ $\forall x \in \mathbf{R}$. Now, many mathematicians intuitively would regard the functions f and g as the same function because they have the same domain and the same image at each point in this common domain. By our definition of functions, they are not the same function because the range of f is not the same as the range of g. We make a distinction between these functions because one of the functions is onto and the other is not onto. Instead of considering f and g to be different functions, many mathematicians might say something like "f is onto Y, but is not onto $\mathbf{R}$."

It is important to know how to be rigorous, and this is why we have defined a function as we have. However, it is also important to know how to read advanced mathematics texts. Rigor frequently is relaxed in such advanced texts in situations where maintaining absolute rigor is not pertinent to the issue at hand. It is not that the authors of these texts are wrong. Rather, they relax the rigor for the sake of brevity, and they assume that the reader knows what is meant. That is, they assume that the student can mentally adjust the argument to obtain complete rigor should this be necessary. When you are reading an argument in which rigor has been somewhat relaxed, it is a good idea to convince yourself that you can make the argument rigorous if you have concerns that the relaxation of rigor could possibly result in an incorrect conclusion.

As a related example, consider Definition 2.23. Define the sets A and B by $A = \{x \in \mathbf{R} | -2 \le x \le 2\}$ and $B = \{y \in \mathbf{R} | 0 \le y \le 4\}$. Define functions $f: \mathbf{R} \to \mathbf{R}$, $g: A \to \mathbf{R}$, and $h: A \to B$, in which $f(x) = x^2 \ \forall x \in \mathbf{R}$, $g(x) = x^2 \ \forall x \in A$, and $h(x) = x^2 \ \forall x \in A$. Note that each of h and g is a restriction of f to A, in accordance with Definition 2.23. In an "abuse of notation," a mathematician might say that either of these functions is **the** restriction of f to A, depending on the point that is being made. So, for example, this mathematician might choose to call h the restriction of f to A if this mathematician has a need to work with an onto function. As mentioned in Definition 2.23, the notation often used in the mathematics literature to denote a restriction of f to A is $f|A$. Hence, the mathematician might write $h = f|A$. Although this is technically unrigorous, the student should realize that this means that the mathematician has "chosen" the function h to be the restriction of f to A that is of interest.

EXERCISES

1. Define the sets A and R as follows: $A = \{x \in \mathbf{R} | -1 \le x \le 1\}$ and $R = \{(x, y) \in A \times A | x^2 + y^2 \le 1\}$. Show that the relation (A, A, R) is not a function, and graph the set R.

2. Let $X = \{1, 2\}$ and $Y = \{a, b\}$ (i.e., the set consisting of the first two letters of the alphabet). List all the functions that exist from X to Y. Which of them are one-to-one? Which are onto? Which are bijective?

3. Define $f: \mathbf{R} \to \mathbf{R}$ specified by $f(x) = 3x - 6 \ \forall x \in \mathbf{R}$. Define sets A and B by $A = \{x \in \mathbf{R} | -4 \le x < 1\}$ and $B = \{x \in \mathbf{R} | 5 < x \le 100\}$. Find $f(A)$ and $f^{-1}(B)$.

4. Define $g: \mathbf{R} \to \mathbf{R}$ specified by $g(x) = x^2 + 3x - 18 \ \forall x \in \mathbf{R}$. Find $g(\mathbf{R}) g^{-1}(\{0\})$, and $g^{-1}(\{-1000, -500\})$.

5. Suppose $f: X \to Y$ with $A \subseteq X$ and $B \subseteq X$. Show $f(A \cap B) \subseteq f(A) \cap f(B)$. Show by an example that is not necessarily the case that $f(A) \cap f(B) \subseteq f(A \cap B)$. (Hint: Choose a function that is not one-to-one.)

6. Define $f: \mathbf{R} \to \mathbf{R}$ specified by $f(x) = x^3 - 7 \; \forall x \in \mathbf{R}$. Find the inverse function f^{-1}. Demonstrate that $f \circ f^{-1}(x) = x \; \forall x \in \mathbf{R}$ and $f^{-1} \circ f(x) = x \; \forall x \in \mathbf{R}$.

7. Suppose that $f: X \to Y$ is a bijection. Show that the inverse function f^{-1} is also a bijection, and show that $\left(f^{-1}\right)^{-1} = f$.

8. Suppose we have functions $g: \mathbf{N} \to \mathbf{R}$ and $f: \mathbf{R} \to \mathbf{R}$ defined by
$$g(n) = \begin{cases} 2n & \text{if } n \in \{1, 2, 3, \ldots 10\} \\ 3n + 1 & \text{otherwise} \end{cases}, \text{ and } f(x) = x^2 - 17 \; \forall x \in \mathbf{R}. \text{ Find}$$
$f \circ g(5), f \circ g(10),$ and $f \circ g(15)$.

9. Find an example of a set X and functions $f: X \to X$ and $g: X \to X$ such that $f \circ g \neq g \circ f$. This shows that the composition of functions is not necessarily a commutative operation on the set of functions from X to X.

10. Let $Y = \{a, b\}$ (i.e., the set consisting of the first two letters of the alphabet). Specify all of the binary operations on Y. Which of them are associative? Which are commutative?

11. Find the union and intersection of the family of sets defined in Example 2.31.

12. Suppose $(G, \cdot, e)$ is a group. Suppose $f \in G$ is also an identity element (i.e., $\forall a \in G, \; a \cdot f = f \cdot a = a$. Show $f = e$. This demonstrates that a group has a unique identity element.

13. Show that each element of a group has a unique inverse.

14. Prove Theorem 2.9, Part 2.

15. Prove Theorem 2.11, Part 1.

16. Prove Theorem 2.13, Part 2.

17. Prove Theorem 2.13, Part 3.

18. Prove Theorem 2.13, Part 4.

Section 2.6   Mathematical Induction

This section is out of logical order. The student could skip this section and proceed to Section 2.7 without any loss of logical continuity. However, the student is strongly advised not to do this for two reasons. The first reason is that the information in this section helps to provide the *intuition* necessary to appreciate the formal development of the integers that occurs in the next chapter. The second reason is that the method of **mathematical induction** provides the reader with a very important additional method for proof construction. The reason why we include it here rather than in Section 1.4 is that a proper discussion of it requires the use of sets.

The first thing to be said about mathematical induction is that the name is misleading. The word *induction* usually has connotations of determining a general rule by looking at specific cases. This is not the meaning that we attach to the word for our purposes. The method of mathematical induction is every bit a deductive method carrying the full force of mathematical proof.

Mathematical induction is a method of proof employed to prove a statement about the positive integers. Although it has several nuances (such as the *second principle of mathematical induction*), we shall focus on what many texts refer to as *the first principle of mathematical induction* in this section. We will discuss the second principle of mathematical induction in Section 3.3. Finally, there are no formal theorems in this section because this discussion is out of logical order. The student is asked to use intuitive knowledge about the positive integers and real numbers that hopefully has been built up over the years. Whereas some books on proof construction have a lot of material on how to do proofs using mathematical induction, we will merely have a strong introduction to the topic here. As the reader will see, there will be plenty of opportunities to use mathematical induction in the upcoming chapters.

Consider a statement $P(n)$ about the positive integer n. For example, $P(n)$ could be the statement $n < 2^n$. Hence, $P(1)$ is the statement $1 < 2^1$, $P(2)$ is the statement $2 < 2^2$, and so on. Many times a mathematician will have a need to prove the truth of $P(n)$ for each $n \in \mathbf{N}$. Hence, with respect to the above example, the mathematician may want to prove that $n < 2^n$ for each $n \in \mathbf{N}$. The use of mathematical induction usually occurs when the mathematician first has an intuitive "feel" of the truth of $P(n)$ for each $n \in \mathbf{N}$, and mathematical induction is employed to confirm the belief.

To proceed, we need to examine a property of the set $\mathbf{N}$ of positive integers to which our intuition should agree. (Keep in mind that we formally develop the properties of positive integers in Chapter 3.)

Consider a set $S \subseteq \mathbf{N}$ that has the following two properties:

Property 1: $1 \in S$.

Property 2: $\forall n \in \mathbf{N}, n \in S \Rightarrow n + 1 \in S$.

We will now show why intuitively the set S must in fact be the set $\mathbf{N}$. Because we already are given that $S \subseteq \mathbf{N}$, it suffices to show that $\mathbf{N} \subseteq S$. By Property 1, we have $1 \in S$. Now, we will show that Property 2 forces the conclusion that $2 \in S$. Why? Because Property 2 says that for *every* positive integer n, if $n \in S$, then so is $n + 1$. What does Property 2 say about the positive integer 1? It says that if $1 \in S$, then we also have $1 + 1 = 2 \in S$. But we have already said that $1 \in S$. Therefore, $2 \in S$. Now, what does Property 2 say about the positive integer 2? It says that if $2 \in S$, then we also have $2 + 1 = 3 \in S$. But we have just concluded that $2 \in S$. Therefore, $3 \in S$.

Intuitively, we can keep applying this argument to show that every positive integer is in S. To do this, we shall perform an intuitive "proof" by contradiction. Suppose (to reach a contradiction) that every positive integer is *not* a member of S. Then there must be a *smallest* positive integer that fails to be a member of S. Let us denote this positive integer by m. Because m is the smallest positive integer to fail to be a member of S, we have $m \notin S$.

Now, we know that $m > 2$, having already shown that $2 \in S$. Hence, $m - 1$ is also a positive integer (i.e., it is not possible to have $m - 1 = 0$). Now, we just declared m to be the smallest positive integer to fail to be a member of S. Because $m - 1$ is smaller than m, we must have $m - 1 \in S$. Now, what does Property 2 above say about $m - 1$? It says that if $m - 1 \in S$, then $(m - 1) + 1 \in S$. But $(m - 1) + 1 = m$. Hence, $m \in S$. But, wait a minute! We already said that $m \notin S$. Here is our contradiction! Therefore, we must reject the possibility that there is a positive integer failing to be a member of S. Hence, we conclude that $S = \mathbf{N}$. We write this up below. However we go about defining the set $\mathbf{N}$ of positive integers, we want to define it in such a way as to make the following true:

Property of Mathematical Induction

Consider a set $S \subseteq \mathbf{N}$ that has the following two properties:

Property 1: $1 \in S$.

Property 2: $\forall n \in \mathbf{N}, n \in S \Rightarrow n + 1 \in S$.

Then $S = \mathbf{N}$.   ∎

What does this have to do with proving $P(n)$ for each $n \in \mathbf{N}$, where $P(n)$ is a statement about the positive integer n? Let $S = \{n \in \mathbf{N} | P(n)\}$. That is, S is the set of positive integers for which the statement holds true.

Now suppose we can show the following:

Property 1: $P(1)$.

Property 2: $\forall n \in \mathbf{N}, P(n) \Rightarrow P(n + 1)$.

Then we can conclude $P(n)$ $\forall n \in \mathbf{N}$. Why? Because the set S defined above satisfies both properties of the *property of mathematical induction* presented above.

This gives rise to the following method of proof, which is normally called **proof by induction**:

Step 1:   Show $P(1)$.

Step 2:   Assume $P(n)$ for some arbitrary $n \in \mathbf{N}$. (This is often called the **inductive step**. It is also called making the **induction hypothesis**.) Show that this forces $P(n + 1)$.

| **Example 2.33** | Let $P(n)$ be the statement $1 + 2 + \cdots + n = \dfrac{n(n + 1)}{2}$. Hence, $P(1)$ states that $1 = \dfrac{1(1 + 1)}{2}$; $P(2)$ states that $1 + 2 = \dfrac{2(2 + 1)}{2}$; $P(3)$ states that $1 + 2 + 3 = \dfrac{3(3 + 1)}{2}$; $P(50)$ states that $1 + 2 + \cdots + 50 = \dfrac{50(50 + 1)}{2}$; and so on. Use mathematical induction to prove $P(n)$ for $\forall n \in \mathbf{N}$. |

Solution:

Step 1: We need to show $P(1)$. $P(1)$ states that $1 = \frac{1(1 + 1)}{2}$. But $\frac{1(1 + 1)}{2} = \frac{2}{2} = 1$, so $P(1)$ clearly holds.

Step 2: Assume that $P(n)$ holds for some $n \in \mathbf{N}$. We need to show that this forces $P(n + 1)$ to hold. Hence, by our assumption, we have $1 + 2 + \cdots + n = \frac{n(n + 1)}{2}$. We are trying to show this assumption forces $P(n + 1)$ to hold. That is, we are trying to show that $1 + 2 + \cdots + n + (n + 1) = \frac{(n + 1)(n + 2)}{2}$. At this point, we have to use some algebra to show that the statement $1 + 2 + \cdots + n = \frac{n(n + 1)}{2}$ implies the statement $1 + 2 + \cdots + n + (n + 1) = \frac{(n + 1)(n + 2)}{2}$. How do we do this? Well, we begin by noting that $1 + 2 + \cdots + n + (n + 1) = (1 + 2 + \cdots + n) + n + 1$. Now, we use our inductive step to replace with $1 + 2 + \cdots + n$ with $\frac{n(n + 1)}{2}$. Doing so gives us $1 + 2 + \cdots + n + (n + 1) = \frac{n(n + 1)}{2} + (n + 1) = \frac{n(n + 1) + 2(n + 1)}{2} = \frac{(n + 1)(n + 2)}{2}$. (The last equality was obtained by factoring out $n + 1$.) Having shown that $1 + 2 + \cdots + (n + 1) = \frac{(n + 1)(n + 2)}{2}$ provided we have $1 + 2 + \cdots + n = \frac{n(n + 1)}{2}$, we are done. ∎

Often, when students are learning mathematical induction for the first time, they object to the inductive step. The inductive step calls for assuming the truth of $P(n)$. Then the inductive step is used to show the truth of $P(n + 1)$. Some students will say something like, "What gives us the right to assume $P(n)$ in order to show $P(n + 1)$?" The answer is a proper understanding of implication. If we know that $P(1)$ is true and $P(1) \Rightarrow P(2)$, then we can conclude that $P(2)$ is true. If we know that $P(2)$ is true and $P(2) \Rightarrow P(3)$, then we can conclude that $P(3)$ is true. This process keeps going on.

Figure 2.12
Falling dominoes

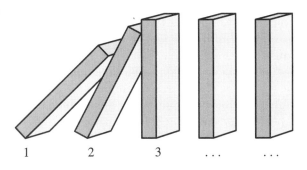

1 2 3 $\cdots$ $\cdots$

Another way to think about mathematical induction is to imagine dominoes falling down (Figure 2.12). Imagine that we have an infinite line of dominoes, one for each positive integer. Suppose we know that the first domino falls. Suppose we also know that if *any* domino falls, it causes the next domino to fall. Then we can conclude that all the dominoes will fall.

Let us do another proof by induction. For the next example, we make use of the obvious fact that $1 < 2^n$ for any positive integer n.

Example 2.34   Prove that $n < 2^n$ for every positive integer n.

Solution: For each $n \in \mathbf{N}$, let $P(n)$ be the statement $n < 2^n$. We will use mathematical induction to show the truth of $P(n)$ for each $n \in \mathbf{N}$.

Step 1.   We first need to show $P(1)$. $P(1)$ states that $1 < 2^1$ (i.e., that $1 < 2$, which is known to be true).

Step 2.   Assume the truth of $P(n)$ for some $n \in \mathbf{N}$, and show that this forces the truth of $P(n + 1)$. Hence, we assume that $n < 2^n$ for some $n \in \mathbf{N}$. We want to show that this forces the truth of $n + 1 < 2^{n+1}$. Starting with $n < 2^n$, we add 1 to both sides of the inequality to get $n + 1 < 2^n + 1$. Now $2^n + 1 < 2^n + 2^n = 2(2^n) = 2^{n+1}$. Hence, $n + 1 < 2^{n+1}$, and this is precisely what $P(n + 1)$ says.

When we are performing a proof via mathematical induction, we will frequently directly use the *property of mathematical induction* presented above instead of using the "$P(n)$" notation. For example, consider Example 2.34, where we were asked to prove that $n < 2^n$ for every positive integer n. We could have presented the proof in the following manner:

Let $S = \{n \in \mathbf{N} | n < 2^n\}$. Then $1 \in S$ because $1 < 2$. Suppose $n \in S$ so that $n < 2^n$. Then $n + 1 < 2^n + 1$. Now $2^n + 1 < 2^n + 2^n = 2(2^n) = 2^{n+1}$. Hence, $n + 1 < 2^{n+1}$, so $n + 1 \in S$. Hence, $S = \mathbf{N}$.

Mathematical induction can also be used to prove statements "starting" at numbers other than 1. For example, we may want to prove that some statement $P(n)$ is true for all positive integers n such that $n \geq 6$. In this case, we have to 1) show that that $P(6)$ is true and 2) show that $P(n) \Rightarrow P(n + 1) \forall n \in \{m \in \mathbf{N} | m \geq 6\}$. Hence, we are "starting" with the number 6 in this example instead of 1. We could also use this method to "start" at 0 or a negative integer. For instance, we may want to prove that some statement $P(n)$ is true for all integers n such that $n \geq -3$.

There is one very important application of mathematical induction that needs to be mentioned. In mathematics, there are many functions that will be used that have domain $\mathbf{N}$. Often it is difficult to explicitly state the value of the function for each positive integer. Instead, it may be easier to specify the value of the function at $n + 1$ in terms of the value of the function at n. In other words, we can specify a function f with domain $\mathbf{N}$ as follows: 1) Specify $f(1)$, and 2) Specify how to obtain $f(n + 1)$ from the value of $f(n)$. The process of specifying a function with domain $\mathbf{N}$ in this manner is called **recursion**. A function specified by recursion is said to be **defined recursively**.

Example 2.35   Consider the function $f: \mathbf{N} \to \mathbf{R}$, in which $f(n)$ is the exponential function 2^n. In other words, $f(1) = 2$, $f(2) = 2 \cdot 2 = 4$, $f(3) = 2 \cdot 2 \cdot 2 = 8$, and so on. This function can be defined recursively as follows: 1) $f(1) = 2$, and 2) $\forall n \in \mathbf{N}$, $f(n + 1) = 2f(n)$.

In the previous example, how would you find $f(4)$? To do this, we "plug" the value 3 in for n in our recursion equation $f(n + 1) = 2f(n)$ to get $f(4) = 2f(3)$. Hence, the problem of finding $f(4)$ has now been reduced to finding $f(3)$. To find $f(3)$, we "plug" the value 2 in for n in the recursion equation $f(n + 1) = 2f(n)$ to get $f(3) = 2f(2)$. Using the same process to find $f(2)$, we get $f(2) = 2f(1)$. Hey, but we were given that $f(1) = 2$. Now, replacing $f(1)$ with 2 in the equation $f(2) = 2f(1)$, we get $f(2) = 2f(1) = 2 \cdot 2 = 4$. Now, replacing $f(2)$ with 4 in the equation $f(3) = 2f(2)$, we get $f(3) = 2 \cdot 4 = 8$. Finally, replacing $f(3)$ with 8 in the equation $f(4) = 2f(3)$, we get $f(4) = 2 \cdot 8 = 16$.

Now, many mathematics texts "justify" defining a function via recursion as follows: Let S denote the set of positive integers for which f is defined. Then $1 \in S$ because $f(1)$ has been explicitly given. Suppose $f(n)$ has been specified for some $n \in \mathbf{N}$. Then because we have a "rule" that tells us how to obtain $f(n + 1)$ from $f(n)$, we have $n \in S \Rightarrow n + 1 \in S$. Thus, by mathematical induction, the function f is defined for all $n \in \mathbf{N}$.

Although the above argument is intuitively correct, it is not sufficiently rigorous. Recall that one of the goals of the book is to demonstrate that we can base all of our concepts on set theory. What does it mean to say that there is "a rule that tells us how to obtain $f(n + 1)$ from $f(n)$"? Also, what does it mean to say that S is "the set of positive integers for which f is defined"? What is needed is a rigorous argument that proves the existence and uniqueness of a function f having the desired properties. This justification, called *the Recursion Theorem*, is presented in Section 3.2. In fact, recursion will be used to *define* the binary operations of addition and multiplication of natural numbers.

EXERCISES

1. Show that $1^2 + 2^2 + \cdots + n^2 = \dfrac{n(n + 1)(2n + 1)}{6}$ for each $n \in \mathbf{N}$.

2. Let $a, r \in \mathbf{R}$ with $r \neq 1$. Show that $a + ar + ar^2 + \cdots + ar^n = \dfrac{ar^{n+1} - a}{r - 1}$ for each $n \in \mathbf{N}$.

3. Show that $2^n > n^2$ for each integer $n \geq 5$.

4. What function $f: \mathbf{N} \rightarrow \mathbf{R}$ is defined recursively by $f(1) = 3$ and $f(n + 1) = 3f(n)$ for each $n \in \mathbf{N}$?

5. The Tower of Hanoi is a well-known puzzle. You are provided with three pegs and a collection of rings of different sizes. All of the rings are initially on one of the pegs (which we shall call Peg 1), placed in order of size, with the largest on the bottom. Your goal is to transfer all of the rings from Peg 1 to Peg 2 by making a series of "moves." Each move consists of taking one ring from the top of any of the pegs and placing this ring on another peg, with the stipulation that you are not allowed to place any ring on top of a smaller ring. Of course, you will want to use Peg 3 as an auxilliary peg. Show via mathematical induction that the minimum number of moves needed to complete the task for n rings is $2^n - 1$.

Section 2.7   The Axiom of Infinity

We mentioned earlier in this chapter that if this were a completely "pure" math book, we would begin by explicitly stating the axioms of set theory. We chose not to take this course due to the additional complexity it would add without much benefit to us. However, one axiom that we shall explicitly mention is the **Axiom of Infinity**. This axiom is worth mentioning because it, in essence, "jump-starts" the process of developing the real number system. It also allows us to prove the existence of an *infinite set* (which we will do in Chapter 4).

Earlier in this chapter, we also mentioned that a failure to explicitly "axiomitize" set theory results in some interesting paradoxes. It was, in fact, the discovery of these paradoxes that led to the axiomitization of set theory. We briefly discuss one of these paradoxes.

The Axiom of Infinity

There exists a set Δ with the following properties:

Property 1: $\varnothing \in \Delta$.

Property 2: $\forall x, x \in \Delta \Rightarrow \{x\} \in \Delta$.   ■

It will now be shown intuitively how to construct such a set Δ. First, we simply put in $\varnothing$ as a member of Δ. (Of course, it is also a subset of Δ.) This satisfies Property 1. Now, because $\varnothing \in \Delta$, Property 2 requires that we also put $\{\varnothing\}$ in as a member of Δ. Because $\{\varnothing\} \in \Delta$, Property 2 requires that we also have $\{\{\varnothing\}\} \in \Delta$. Property 2 requires that this process continue *ad infinitum*.

What does this have to do with "jump-starting" the process of constructing the real number system? Intrinsic to the positive integers is the concept of a **successor**. For every positive integer n, we get the next one by adding 1. This gives us the successor to n, which is $n + 1$. Intuitively, we could define the number 1 to be $\varnothing$. We could define the number 2 to be $\{\varnothing\}$. We could define the number 3 to be $\{\{\varnothing\}\}$. The Axiom of Infinity allows us to keep getting successors *ad infinitum*. This is the basic idea, but there are many details to be worked out. You, the reader, are now ready to cross over to the "promised land" and begin constructing the real number system.

Before we begin this process, a paradox will be presented to remind the reader that studying a completely axiomatic approach to set theory is a worthwhile activity to be performed eventually. The following paradox is known as *Russell's Paradox*. It was discovered by the mathematician and philosopher Bertrand Russell. For a discussion of paradoxes arising from a failure to set proper axioms for set theory, see Goldrei (1996).

Let Ω denote the family of sets that are not members of themselves. That is, $\Omega = \{A | A$ is a set with $A \notin A\}$. Now, either $\Omega \in \Omega$, or else $\Omega \notin \Omega$. Clearly, we cannot have both $\Omega \in \Omega$ and $\Omega \notin \Omega$ because the statement $\Omega \notin \Omega$ is the negation of $\Omega \in \Omega$.

Suppose $\Omega \in \Omega$. Then, by the definition of Ω, Ω cannot be a member of itself because Ω is precisely the set of sets that are not members of themselves. Hence, $\Omega \notin \Omega$. Because the assumption that $\Omega \in \Omega$ leads to the conclusion that $\Omega \notin \Omega$, we must have $\Omega \notin \Omega$.

Ok, so $\Omega \notin \Omega$. By the definition of Ω, this means that $\neg(\Omega \notin \Omega)$, so we must have $\Omega \in \Omega$. There is the paradox: $\Omega \in \Omega \Rightarrow \Omega \notin \Omega$ and $\Omega \notin \Omega \Rightarrow \Omega \in \Omega$. The paradox was resolved by putting in axioms to forbid the existence of sets such as Ω that are too "bizarre." The method of doing this is beyond the scope of this book.

EXERCISES

1. Suppose sets Δ and Ψ both satisfy the two properties of the Axiom of Infinity. Show that $\Delta \cap \Psi$ also satisfies these two properties.

2. Suppose we define 1 to be $\varnothing$, 2 to be $\{\varnothing\}$, and 3 to be $\{\{\varnothing\}\}$. Show $1 \in 2$ and $2 \in 3$.

Chapter 3

The Development of the Integers

We have come a long way. We are ready to embark on a formal development of the system of integers based on set theory. First, we introduce the positive integers, define the binary operations of addition and multiplication of positive integers, and define a linear order on them. Then we use the positive integers to create the system of all integers, define addition and multiplication on them, and define a linear order on them. Along the way, we encounter a concept known as an **isomorphism** that it is critically important to higher mathematics. We also explore a concept known as a **ring**, which is also critically important.

We continue as always to make appeals to the student's intuition, but, of course, the logical flow of the development of the integers will not depend on this intuition. As before, some of the examples and end-of-section exercises will depend on some concepts of real numbers not yet formally developed. Now we have to be more careful to make a distinction between the formally developed concepts and those that the student should know about via previous experience with real numbers. Fortunately, as we continue to formally develop important properties, there will be less dependence on undeveloped material in the examples and problems.

Section 3.1   Peano's Postulates

We begin with an activity that the mathematics student will encounter again and again in higher mathematics. We start with a blueprint for a set by specifying properties of the set. Next, we show that there actually exists a set satisfying the blueprint. We then set out to discover the implications of these properties. Although the properties that we investigate appear to be quite simple, there is an amazing amount of deep structure associated with them that will eventually be discovered. There will also be subtleties that will be discovered along the way.

What we are presenting here is a slightly modified version of **Peano's postulates**. These postulates are specified conditions on a set. Such a set will turn out to have the properties that we "want" the positive integers to have. The mathematician Giuseppe Peano introduced Peano's postulates.

Definition 3.1

Let P be a nonempty set. Let $s: P \rightarrow P$, and consider an element e of P. The ordered triplet (P, s, e) is called a **Peano space** provided the following three properties hold:

1) s is one-to-one.
2) $\forall n \in P, s(n) \neq e$.
3) Suppose A is any subset of P such that
   a) $e \in A$, and
   b) $\forall n \in P$, we have $n \in A \Rightarrow s(n) \in A$.
   Then $A = P$.

The function s is called a **successor function**. The element e is called an **initial element** with respect to s. Properties 1 through 3 are called **Peano's postulates**.

Some comments and intuitive explanations are definitely in order here. It is going to turn out that what we have called an initial element is going to have all of the properties of what we have intuitively thought of as the number 1 since grade school. Intuitively, the image of the successor function at each $n \in P$ is the next member; that is, the reader should think of $s(n)$ as $n + 1$ (using the arithmetic we have all been familiar with since grade school). Keeping this intuition in mind will be of great value to the student.

Armed with this intuition, what is Property 1 saying? Well, because s is injective, if we have $s(n) = s(m)$ for some $n, m \in P$, then we must have $n = m$.

Hence, intuitively, this says that for every n, $m \in P$, the only way we can have $n + 1 = m + 1$ is if $n = m$. Intuitively, Property 2 says that 1 is not the successor of any member of P. That is, $\forall n \in P$, $n + 1 \neq 1$. Property 3 is the mandating of mathematical induction! Hopefully, the student has read Section 2.6 and has an intuitive understanding of the meaning of mathematical induction.

Given a Peano space (P, s, e), we cannot (yet) draw any conclusions just because we have called e an initial element with respect to s. Notation should be a matter of convenience. It does not, however, have any intrinsic inference-making powers. We still have to do all the work. While we are on this topic, why does Definition 3.1 refer to e as *an* initial element instead of *the* initial element? This has to do with the issue of uniqueness. (See the discussion on uniqueness in Section 1.4.) That is, in addition to having e for the Peano space (P, s, e), how do we know that there is not another element $f \in P$ with $f \neq e$ such that (P, s, f) is a Peano space? If there were to exist such another element, a Peano space could not conform to our ideas about how positive integers are supposed to behave. We will show that there does not exist another such element. After doing this, it will be legitimate to refer to e as *the* initial element with respect to s, and we shall begin to call this element 1. Before the uniqueness proof, we provide an example of a Peano space.

Normally when the definition of a set with specified properties is given (such as in Definition 3.1), one of the first things to set straight is that there actually exists a set with the specified properties. It would be a shame if we did a lot of work exploring the implications of the properties only to discover later that we cannot even provide an example of a set obeying the properties. It turns out that the Axiom of Infinity (Section 2.7) provides us with what we need to show that there does exist a Peano space. The proof of this is very instructive because higher mathematics uses many proofs of a similar flavor.

Theorem 3.1   There exists a Peano space.

Proof: Let Δ be set satisfying the Axiom of Infinity, so that it satisfies the following properties:

1) $\varnothing \in \Delta$.
2) $\forall x$, $x \in \Delta \Rightarrow \{x\} \in \Delta$.

Let us refer to *any* set that satisfies Properties 1 and 2 above as **inductive**. Then, by definition, Δ is inductive. Let Ω denote the family of all subsets of Δ that are inductive. That is, $\Omega = \{A | A \subseteq \Delta \text{ and } A \text{ is inductive}\}$. Ω is not empty because $\Delta \in \Omega$.

Define the set Φ by $\Phi = \cap_{A \in \Omega} A$. (See the last part of Section 2.5 that deals with the issue of taking the intersection of an arbitrary number of sets.) Hence, Φ is the intersection of all inductive subsets of Δ. We will now show that Φ is itself an inductive subset of Δ. (The relevance of this is made clear below.) First of all, we clearly have $\Phi \subseteq \Delta$. To show that Φ is inductive, we must show that it obeys Properties 1 and 2 above. To show it obeys Property 1, we must show that $\varnothing \in \Phi$. Now for each $A \in \Omega$, we have $\varnothing \in A$ because A is inductive. Therefore, by the definition of the intersection of a family of sets, we have $\varnothing \in \cap_{A \in \Omega} A$; that is $\varnothing \in \Phi$. To show Property 2 we must show that $B \in \Phi \Rightarrow \{B\} \in \Phi$. So, suppose $B \in \Phi$. Then by the definition of the intersection of a family of sets, we have $B \in A$ for each $A \in \Omega$. Now, because each member of Ω is inductive, we have $\{B\} \in A$ for each $A \in \Omega$. Hence, by the definition of the intersection of a family of sets, we have $\{B\} \in \Phi$. We have shown that Φ is inductive.

The next step is to show that we can define a successor function on Φ. Define $s\colon \Phi \to \Phi$ by $s(B) = \{B\} \;\forall B \in \Phi$. Note the fact that Φ is inductive ensures that this function makes sense. In other words, we need the assurance that if B is a member of Φ, then so is $\{B\}$. Now we need an initial element. Let us try $\varnothing$ as our initial element. We already know that $\varnothing \in \Phi$ because Φ is inductive.

We begin by verifying that Peano's postulate Property 1 holds. To show that s is one-to-one, suppose we have $s(B) = s(C)$ for some $B, C \in \Phi$. Then $\{B\} = \{C\}$. Because both $\{B\}$ and $\{C\}$ each have only one element, we must have $B = C$. This shows that s is one-to-one, so Property 1 holds.

To show Property 2, it will be shown that $\forall B \in \Phi$, $s(B) \neq \varnothing$. We have $s(B) = \{B\}$, so the set $s(B)$ cannot be the empty set. Therefore, we cannot have $s(B) = \varnothing$ (because $\varnothing$ *is* the empty set).

Finally, we must show that Property 3 holds. Suppose Γ is a subset of Φ such that

1) $\varnothing \in \Gamma$, and

2) $\forall B \in \Phi$, $B \in \Gamma \Rightarrow s(B) \in \Gamma$ (which means $\forall B \in \Phi$, $B \in \Gamma \Rightarrow \{B\} \in \Gamma$).

We must show that $\Gamma = \Phi$. We already know that $\Gamma \subseteq \Phi$, so we are done if we can show that $\Phi \subseteq \Gamma$. Now, numbers 1 and 2 shown above immediately tell us that Γ is an inductive subset of Δ. Recall that Φ is defined as the intersection of the family of *all* inductive subsets of Δ. Therefore, $x \in \Phi$ implies that x is a member of each set in this family. Because Γ is in this family, we must have $x \in \Gamma$ for each $x \in \Phi$. We have shown that $x \in \Phi \Rightarrow x \in \Gamma$, which means $\Phi \subseteq \Gamma$. Hence, $(\Phi, s, \varnothing)$ is a Peano space. ∎

There are some instructive concepts to note about the proof of Theorem 3.1. First, recall how we defined the set Φ by $\Phi = \cap_{A \in \Omega} A$. But before defining Φ, we made sure that the family of sets Ω was nonempty. We will not be interested in taking the intersection of an empty family of sets.

The reader may wonder why we needed to define Φ in the first place. Why not just use $(\Delta, s, \varnothing)$ as the candidate Peano space? The answer is that we cannot prove that $(\Delta, s, \varnothing)$ is a Peano space. We need to "remove" members of Δ that are not of the form $\varnothing$, $\{\varnothing\}$, $\{\{\varnothing\}\}$, $\{\{\{\varnothing\}\}\}$, and so on. It is entirely possible that Δ contains other members besides the ones that we want. This is why we took the intersection of all inductive subsets of Δ. This intersection gives us the sets of the above-mentioned form that we want. The collection of sets of the form $\varnothing$, $\{\varnothing\}$, $\{\{\varnothing\}\}$, $\{\{\{\varnothing\}\}\}$, and so on, is the *smallest* inductive subset of Δ in the sense that any inductive subset of Δ must contain this collection.

The point is worth generalizing because it occurs so often in mathematics. Suppose we have a family F of sets such that each member of this family has a certain property that we care about; let us refer to this property as property X. Define I by $I = \bigcap_{A \in F} A$. Suppose it turns out that I also has property X (so that $I \in F$). Then I is the *smallest* member of the family in the sense that we have $I \subseteq A$ for each member A of the family.

Now, we discuss a more subtle point about Peano spaces. We showed that the set Φ defined in the proof of Theorem 3.1 qualifies. Why not go ahead and define the set $\mathbf{N}$ of natural numbers to be Φ? Logically, we could. In fact, we show in Section 3.4 that all Peano spaces are essentially the same. Any one has all the properties that we associate with our conception of what positive integers are. If all we cared about were the positive integers, then there would be no problem defining $\mathbf{N}$ to be Φ. But we care about more than this. We also want to construct the set of integers $\mathbf{Z}$, the set of rational numbers $\mathbf{Q}$, and the set of all real numbers $\mathbf{R}$. When we have completed these constructions, we want to make sure that $\mathbf{N} \subset \mathbf{Z} \subset \mathbf{Q} \subset \mathbf{R}$. Requiring that these sets be subsets of each other will dictate the choice of which Peano Space to use to define $\mathbf{N}$. This concept is made more clear in Sections 3.6 and 3.7, where we show how to construct a set that has all the properties that we associate with our conception of what $\mathbf{Z}$ is supposed to be.

We now prove some theorems about Peano spaces that begin to make it apparent that the set P of a Peano space (P, s, e) behaves the way we think positive integers ought to behave. We start with the above-promised uniqueness proof. From now on (after the proof), we shall normally refer to 1 as the initial element for a given Peano space.

Theorem 3.2 Suppose (P, s, e) is a Peano space. Suppose $f \in P$ with $f \neq e$. Then f is not an initial element with respect to s.

Proof: We directly show that any element of P other than e is not an initial element with respect to s. First of all, note that if i is *any* initial element of P with respect to s, then Property 2 of Definition 3.1 mandates that $\forall n \in P$, $s(n) \neq i$. That is, every initial element with respect to s has the property of not being in the image of the function s. We now employ a "sneaky" mathematical induction proof (i.e., we employ Property 3 in a "sneaky" way). Define the set A by $A = \{n \in P | n \neq e \Rightarrow n$ is not an initial element with respect to $s\}$. We have $e \in A$ by default! This is an immediate consequence of the definition of implication. We now take the inductive step and assume that $n \in A$ for some $n \in P$, and then try to show $s(n) \in A$. To show $s(n) \in A$, we have to show $s(n) \neq e \Rightarrow s(n)$ is not an initial element with respect to s. This is trivial because $s(n)$ is clearly in the image of s, and we just said that every initial element with respect to s has the property of not being in the image of the function s. We have shown that $n \in A \Rightarrow s(n) \in A$. Hence, we have $A = P$ by Property 3. This immediately shows that e is the only initial element with respect to s.   ∎

Intuitively, Theorem 3.3 below says that $n + 1 \neq n$ for every positive integer n.

Theorem 3.3

Suppose $(P, s, 1)$ is a Peano space.
    Then $s(n) \neq n \ \forall n \in P$.

Proof: We will establish the result by mathematical induction (i.e., by Property 3.) Let $A = \{n \in P | s(n) \neq n\}$. Now Property 2 forces $1 \in A$. Suppose $n \in A$ for some $n \in P$. We want to show that $s(n) \in A$; that is, we want to show that $s(s(n)) \neq s(n)$. Because $n \in A$, we have $s(n) \neq n$ by the definition of A. Suppose (preparing to use proof by contradiction) that our desired conclusion is not true, so that $s(s(n)) = s(n)$. Then using the fact that s is one-to-one (mandated by Property 1), this gives us $s(n) = n$. This is a contradiction because we have already concluded that $s(n) \neq n$. Hence, $s(s(n)) \neq s(n)$. We have shown that $n \in A \Rightarrow s(n) \in A$. Hence, we have $A = P$ by Property 3, which gives us our desired result that $s(n) \neq n$ for each $n \in P$.   ∎

Intuitively, the next theorem says that every positive integer n with $n \neq 1$ can be written as $n = m + 1$ for some positive integer m. (Of course, we know intuitively that this integer m is $n - 1$.)

Theorem 3.4

Suppose $(P, s, 1)$ is a Peano space. Then $\forall n \in P$, $n \neq 1 \Rightarrow n = s(m)$ for some $m \in P$.

Proof: Before we begin the proof, it is important that the logic of the theorem statement be read correctly. We are claiming that for each $n \in P$ other than 1, there exists an $m \in P$ with a certain property. This m may depend on n. In other words, we are not claiming that the same m value holds for all n.

   We now prove the theorem by a mathematical induction proof. Define the set A by $A = \{n \in P | n \neq 1 \Rightarrow n = s(m)$ for some $m \in P\}$. We have $1 \in A$

by default. We now take the inductive step and assume that $n \in A$ for some $n \in P$, and then try to show $s(n) \in A$. There are two cases to consider:

Case 1: $n = 1$. Does $s(1) = s(m)$ for some $m \in P$? Of course, setting m to 1 works, so $s(1) \in A$.

Case 2: $n \neq 1$. By the inductive step, we have $n = s(m)$ for some $m \in P$. Hence, $s(n) = s(s(m))$.

In either case, we have $s(n) \in A$. Thus, $A = P$ by Property 3, and this gives us our desired result. ∎

The student should see by now that being faithful to the "mandate" of only being allowed to use prior concepts in proofs can create a great deal of work. Is all this work really worth it? After all, you have intuitively "known" what positive integers are since you learned to count. There are several answers to this. The first one is that many mathematicians are not likely to be satisfied with a "proof" unless it can be demonstrated to be based on prior concepts. The second answer is that by going through the arguments of this section, you are being exposed to the type of mathematical arguments that you will use time and time again.

Finally, it should be pointed out that although we have chosen to build up the integers "starting" with 1, many books employ the same procedure "starting" with 0. This is simply a matter of taste.

EXERCISES

1. Suppose $(P, s, 1)$ is a Peano space. Define the function $s': P - \{1\} \to P - \{1\}$ by $s'(n) = s(n)$ $\forall n \in P - \{1\}$. Show that $(P - \{1\}, s', s(1))$ is a Peano space.

2. Suppose $(P, s, 1)$ is a Peano space. Suppose there is an object $a \notin P$. Define the function $s': P \cup \{a\} \to \{P\} \cup \{a\}$ by $s'(n) = s(n)$ $\forall n \in P$ and $s'(a) = 1$. Show that $(P \cup \{a\}, s', a)$ is a Peano space. Intuitively, what does this have to do with "starting" from 0?

3. Let M denote the set of the first two letters of the alphabet, so that $M = \{a, b\}$. Define $s: M \to M$ by $s(a) = b$ and $s(b) = b$. In attempting to define (M, s, a) as a Peano space, which of Peano's postulates hold? Which do not?

4. Suppose $(P, s, 1)$ is a Peano space. Show that $s(s(n)) \neq n$ $\forall n \in P$.

Section 3.2   The Recursion Theorem, Addition, and Multiplication

In this section we define the binary operations of addition and multiplication on the set P of a Peano space $(P, s, 1)$, and we prove many of properties of these operations that you have taken for granted throughout your mathematics education. The student is advised to review Section 2.5 if necessary. The student may also want to review the discussion of recursion in Section 2.6. Throughout this section, we work with a given Peano space $(P, s, 1)$.

To define addition and multiplication on P, we need to first state the **Recursion Theorem**. To motivate this, we review the basic concepts behind it that were discussed in Section 2.6. To invoke your intuition, pretend (for the moment) that we already have developed the properties of the set $\mathbf{N}$ of natural numbers and the set $\mathbf{R}$ of real numbers.

Let S be a nonempty set. The basic idea of defining a function $f: \mathbf{N} \to S$ recursively is as follows: First, we specify the value of the function at 1; that is, we specify $f(1)$. Next, we show how to determine the value of the function at $n + 1$ from the value of the function at n. That is, assuming we know $f(n)$, we specify how to obtain $f(n + 1)$ from $f(n)$. As we discussed in Section 2.6, the intuitive justification for this method is mathematical induction, but the argument in Section 2.6 is not sufficiently rigorous (which is why we present Theorem 3.5 that follows). We use Example 3.1 below as our motivating example of recursion. For this example, the nonempty set S is the set $\mathbf{R}$.

Example 3.1

Let $a \in \mathbf{R}$. We "want" to define the exponential function $f: \mathbf{N} \to \mathbf{R}$ that we all "know" to be given by $f(n) = a^n$ for each $n \in \mathbf{N}$. To do this via recursion, we first define $f(1)$ by $f(1) = a$. Next, to define $f(n + 1)$ in terms of $f(n)$, we define $f(n + 1)$ by $f(n + 1) = af(n)$.

∎

As a first step in making things more precise, we need to concretize the concept of "obtaining" $f(n + 1)$ from $f(n)$. One way to do this is to find a "suitable" function $g: S \to S$ such that $f(n + 1) = g(f(n))$. In the case of Example 3.1, what would this be? For this example, let us try the function $g: \mathbf{R} \to \mathbf{R}$ defined by $g(x) = ax \; \forall x \in \mathbf{R}$. Then $g(f(n)) = af(n) \forall n \in \mathbf{N}$, so this "works."

We now are armed with sufficient intuition to understand the statement of Theorem 3.5, and we are ready to resume the formal development of our subject. In the theorem statement, we do not yet have addition, so we express $n + 1$ in terms of the successor function; that is, $s(n)$. The proof of Theorem 3.5 is rarely shown in mainstream mathematics books. This is a shame because this proof uses many of the proof techniques that we discussed in Section 1.4 and Section 2.6. This includes proof by mathematical induction, proof by contradiction, and uniqueness proofs. The proof uses all of Properties 1 through 3 of Definition 3.1.

Admittedly, the proof is a little lengthy. In fact, because of its length, it is postponed until the end of this section, and it has been designated as optional. (Of course, you are encouraged to read it.) We want to get right down to business using the theorem to define addition and multiplication.

The student who does not want to wait may proceed directly to the proof now. The student may want to review Definition 2.21 before reading the proof.

<table><tr><td>Theorem 3.5</td></tr></table>

(The Recursion Theorem)

Suppose $(P, s, 1)$ is a Peano space. Suppose S is a nonempty set with $a \in S$. Let $g: S \to S$. Then there is a unique function $f: P \to S$ such that $f(1) = a$ and $f(s(n)) = g(f(n)) \forall n \in P$. ∎

Throughout your mathematical career, you will be seeing functions defined recursively many times. It is rare that you will ever see the function g of Theorem 3.5 explicitly listed. However, if you ever have any doubts concerning the veracity of a proof that you are reading or creating that involves the recursive definition of a function, it may be useful for you to explicitly produce this function g to convince yourself that recursion is being invoked properly.

In the definition of addition below, we use the Recursion Theorem, in which the function g of Theorem 3.5 is the successor function s in this case, and the set S is P in this case. The reader should intuitively interpret $m + n$ to be obtained by applying the successor function n times to m.

Definition 3.2

Suppose $(P, s, 1)$ is a Peano space. For each $m \in P$, we define the function $f_m: P \to P$ as follows (by using Theorem 3.5): $f_m(1) = s(m)$, and $f_m(s(n)) = s(f_m(n)) \forall n \in P$. We define a binary operation on P, which we call **addition** and denote by '+'. For each $(m, n) \in P \times P$, we define $m + n$ to be $f_m(n)$. We also call $m + n$ the **sum** of m and n.

Note that in Definition 3.2, we invoke the Recursion Theorem (Theorem 3.5) to ensure the existence and uniqueness of the function $f_m: P \to P$ for each $m \in P$. Now, armed with the definition of '+', the statement $f_m(1) = s(m)$ says $m + 1 = s(m)$, and the statement $f_m((s(n)) = s(f_m(n))$ says $m + s(n) = s(m + n)$. That is, the latter statement says that $m + (n + 1) = (m + n) + 1$. We are about to make much use of this statement for the next few results, so we write this again using the letters a and b instead of m and n (because we are going to be making a lot of use of the letters m and n in the proofs). For reference purposes, we call this Equation 1.

$$a + (b + 1) = (a + b) + 1 \ \forall a, b \in P. \tag{1}$$

The next step is to move toward proving that the above-defined binary operation of addition on P is commutative and associative. To prove that addition is commutative, we first prove two results that we will need. A theorem that is mainly used to prove a more important theorem is often called a **lemma**. From now on, we will normally write $n + 1$ instead of $s(n)$ in induction proofs.

Lemma 3.1 $n + 1 = 1 + n \ \forall n \in P.$

Proof: Let $A = \{n \in P | n + 1 = 1 + n\}$. We use induction to show that $A = P$. Clearly, $1 + 1 = 1 + 1$, so $1 \in A$. Now, suppose $n \in A$ for some $n \in P$, so that $n + 1 = 1 + n$. We must show that $n + 1 \in A$. That is, we must show that $(n + 1) + 1 = 1 + (n + 1)$. Now,

$$(n + 1) + 1 = (1 + n) + 1 \quad \text{(by the inductive step)}$$
$$= 1 + (n + 1) \quad \text{(by Equation 1 with } a = 1 \text{ and } b = n).$$

Thus, $n + 1 \in A$, and we are done.

Lemma 3.2 $(m + n) + 1 = (m + 1) + n \ \forall m, n \in P.$

Proof: Let $A = \{n \in P | (m + n) + 1 = (m + 1) + n \ \forall m \in P\}$. We will use induction to show that $A = P$. Clearly $1 \in A$ because $(m + 1) + 1 = (m + 1) + 1 \ \forall m \in P$. Now, suppose $n \in A$ for some $n \in P$, so that $(m + n) + 1 = (m + 1) + n \ \forall m \in P$. We must show that $n + 1 \in A$. That is, we must show that $[m + (n + 1)] + 1 = (m + 1) + (n + 1) \ \forall m \in P$. Now,

$$[m + (n + 1)] + 1 = [(m + n) + 1] + 1$$
$$\text{(by Equation 1 with } a = m \text{ and } b = n)$$
$$= [(m + 1) + n] + 1 \quad \text{(by the inductive step)}$$
$$= (m + 1) + (n + 1)$$
$$\text{(by Equation 1 with } a = m + 1 \text{ and } b = n),$$

so we are done.

Note: From now on, we frequently will be a bit more informal in our induction proofs in order to prepare the student for more advanced texts. For example, instead of formally defining the set A as we did in the proof of Lemma 3.2, we frequently will say "the proof is by induction on n," or something similar to that. The context will always be clear, and the student should always keep the set A in mind and be able to produce it if necessary.

Theorem 3.6 **(Commutative Law of Addition)**

$m + n = n + m \; \forall m, n \in P.$

Proof: The proof is by induction on m. The result holds for 1 by Lemma 3.1. Assume the inductive step, so that $m + n = n + m \; \forall n \in P$ for some $m \in P$. We must show that this holds true for $m + 1$. That is, we must show that $(m + 1) + n = n + (m + 1) \; \forall n \in P$. Now,

$$n + (m + 1) = (n + m) + 1 \quad \text{(by Equation 1 with } a = n \text{ and } b = m)$$
$$= (m + n) + 1 \quad \text{(by the inductive step)}$$
$$= (m + 1) + n \quad \text{(by Lemma 3.2)},$$

so we are done.

Theorem 3.7 **(Associative Law of Addition)**

$k + (m + n) = (k + m) + n \; \forall k, m, n \in P.$

Proof: The proof is by induction on n. To show that the result holds for 1, we must show that $k + (m + 1) = (k + m) + 1$. This holds by Equation 1 with $a = k$ and $b = m$. We now take the inductive step and assume that the result holds for some $n \in P$, so that $k + (m + n) = (k + m) + n \; \forall k, m \in P$. We must show that the result holds for $n + 1$. That is, we must show that $k + [m + (n + 1)] = (k + m) + (n + 1) \; \forall k, m \in P$. Now,

$$k + [m + (n + 1)] = k + [(m + n) + 1]$$
$$\text{(by Equation 1 with } a = m \text{ and } b = n)$$
$$= [k + (m + n)] + 1$$
$$\text{(by Equation 1 with } a = k \text{ and } b = m + n)$$
$$= [(k + m) + n] + 1 \quad \text{(by the inductive step)}$$
$$= (k + m) + (n + 1)$$
$$\text{(by Equation 1 with } a = k + m \text{ and } b = n),$$

so we are done.

We can define 2 to be $1 + 1$, 3 to be $2 + 1$, 4 to be $3 + 1$, 5 to be $4 + 1$, 6 to be $5 + 1$, and so on. Now, the reader is asked to redo his or her addition tables from grade school.

Example 3.2   Show that $3 + 2 = 5$.

Solution: We have

$$3 + 2 = 3 + (1 + 1) \quad \text{(by the definition of 2)}$$
$$= (3 + 1) + 1 \quad \text{(by Theorem 3.7)}$$
$$= 4 + 1 \quad \text{(by the definition of 4)}$$
$$= 5 \quad \text{(by the definition of 5)}.$$

Next on the agenda is the definition of multiplication, which uses the definition of addition.

Definition 3.3

Suppose $(P, s, 1)$ is a Peano space. For each $m \in P$, we define the function $g_m: P \to P$ by $g_m(x) = x + m$ $\forall x \in P$. Now, for each $m \in P$, we define the function $k_m: P \to P$ as follows (by using Theorem 3.5): $k_m(1) = m$, and $k_m(n + 1) = g_m(k_m(n))$ $\forall n \in P$. We define a binary operation on P, which we call **multiplication** and denote by placing members of P (to be multiplied) next to each other. For each $(m, n) \in P \times P$, we define mn to be $k_m(n)$. We also call mn the **product** of m and n.

We sometimes write mn as $m \cdot n$ to improve readability. It should also be noted that we adhere to what is often called the **rules of precedence**. These are the rules specifying which operations get performed first in the absence of parentheses. For example, when we write $ab + c$, it is assumed that the student knows that what is meant is $(ab) + c$ rather than $a(b + c)$. As students of computer science know, rules of precedence are usually specified in the description of computer programming languages.

Note that the statements $k_m(1) = m$ and $k_m(n + 1) = g_m(k_m(n))$ can now be written as $m \cdot 1 = m$ and $m(n + 1) = mn + m$, respectively. We are about to make much use of these statements for the next few results, so we write them again using the letters a and b instead of m and n (because we are going to be making a lot of use of the letters m and n in the proofs).

$$a \cdot 1 = a \ \forall a \in P. \tag{2}$$

$$a(b + 1) = ab + a \ \forall a, b \in P. \tag{3}$$

To prove that multiplication is commutative, we first prove two lemmas that we will use.

Lemma 3.3 $1 \cdot n = n \ \forall n \in P.$

Proof: The proof is by induction on n. The result holds for 1 by Equation 2. We now take the inductive step and assume that the result holds for some $n \in P$. We must show that the result holds for $n + 1$. That is, we must show that $1 \cdot (n + 1) = n + 1$. Now

$$1 \cdot (n + 1) = 1 \cdot n + 1 \quad \text{(by Equation 3 with } a = 1 \text{ and } b = n)$$
$$= n + 1 \quad \text{(by the inductive step),}$$

and we are done.

Lemma 3.4 $(m + 1)n = mn + n \ \forall m, n \in P.$

Proof: The proof is by induction on n. To show that the result holds for 1, we must show that $(m + 1) \cdot 1 = m \cdot 1 + 1 \ \forall m \in P$. It is clear that this holds by Equation 2. We now take the inductive step and assume that the result holds for some $n \in P$. We must show that the result holds for $n + 1$. That is, we must show that $(m + 1)(n + 1) = m(n + 1) + (n + 1)$. Now,

$$(m + 1)(n + 1) = (m + 1)n + (m + 1)$$
$$\text{by Equation 3 with } a = m + 1 \text{ and } b = n)$$
$$= (mn + n) + (m + 1) \quad \text{(by the inductive step)}$$
$$= mn + [n + (m + 1)] \quad \text{(by Theorem 3.7)}$$
$$= mn + [(m + 1) + n] \quad \text{(by Theorem 3.6)}$$
$$= mn + [m + (1 + n)] \quad \text{(by Theorem 3.7)}$$
$$= mn + [m + (n + 1)] \quad \text{(by Theorem 3.6)}$$
$$= (mn + m) + (n + 1) \quad \text{(by Theorem 3.7)}$$
$$= m(n + 1) + (n + 1)$$
$$\text{(by Equation 3 with } a = m \text{ and } b = n),$$

so we are done.

Theorem 3.8 **(Commutative Law of Multiplication)**

$nm = mn \ \forall m, n \in P.$

Proof: The proof is by induction on m. To show that the result holds for 1, we must show that $n \cdot 1 = 1 \cdot n \ \forall n \in P$, but this clearly holds by Equation 2 and Lemma 3.3. We now take the inductive step and assume that the result holds for some $m \in P$. We must show that the result holds for $m + 1$. That is, we must show that $n(m + 1) = (m + 1)n \; \forall n \in P$. Now,

$$n(m + 1) = nm + n \quad \text{(by Equation 3 with } a = n \text{ and } b = m)$$
$$= mn + n \quad \text{(by the inductive step)}$$
$$= (m + 1)n \quad \text{(by Lemma 3.4),}$$

and we are done.

Next, we show the **distributive law**, which ties addition and multiplication together. We then use the Distributive Law to prove the **associative law of multiplication**. We begin the practice of sometimes lumping several steps together when it should be fairly obvious to the student (with a little work) how to obtain the result.

Theorem 3.9 **(Distributive Law)**

$m(n + p) = mn + mp \; \forall m, n, p \in P.$

Proof: The proof is by induction on m. The result holds for 1 by Lemma 3.3. We now take the inductive step and assume that the result holds for some $m \in P$. We must show that the result holds for $m + 1$. That is, we must show that $(m + 1)(n + p) = (m + 1)n + (m + 1)p \; \forall n, p \in P$.

$$(m + 1)(n + p) = m(n + p) + (n + p) \quad \text{(by Lemma 3.4)}$$
$$= (mn + mp) + (n + p) \quad \text{(by the inductive step)}$$
$$= (mn + n) + (mp + p) \quad \text{(by Theorems 3.6 and 3.7)}$$
$$= (m + 1)n + (m + 1)p \quad \text{(by Lemma 3.4),}$$

and we are done.

By the way, there is a nice way to visualize the distributive law geometrically. The following argument is completely unrigorous, but highly intuitive. Consider the "big" rectangle in Figure 3.1, which is divided into

Figure 3.1
An "area" argument for the distributive law

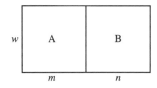

rectangles A and B. Suppose rectangles A and B have width w. Suppose rectangle A has length m, and rectangle B has length n. Then the length of the big rectangle is $m + n$. Hence, the area of the "big" rectangle is $w(m + n)$. Now, the area of rectangle A is wm, and the area of rectangle B is wn. Because the area of the big rectangle is the sum of the areas of rectangles A and B, we have $w(m + n) = wm + wn$.

Theorem 3.10   **(Associative Law of Multiplication)**

$m(np) = (mn)p \; \forall m, n, p \in P.$

Proof: The proof is by induction on m. The result clearly holds for 1. We now take the inductive step and assume that the result holds for some $m \in P$. We must show that the result holds for $m + 1$. That is, we must show that $(m + 1)(np) = [(m + 1)n]p \; \forall n, p \in P$. Now,

$$
\begin{aligned}
(m + 1)(np) &= m(np) + np && \text{(by Theorems 3.8 and 3.9)} \\
&= (mn)p + np && \text{(by the inductive step)} \\
&= (mn + n)p && \text{(by Theorems 3.8 and 3.9)} \\
&= [m(n + 1)]p && \text{(by Theorem 3.9),}
\end{aligned}
$$

and we are done.

Notational Note: Having used the notation f_m, k_m, and g_m in Definition 3.2 and Definition 3.3 to define addition and multiplication, we now release these symbols to be used again.

We have one more theorem to present in this section. This is the **cancellation law of addition** (for positive integers). The proof is left to the reader.

Theorem 3.11   **(Cancellation Law of Addition)**

$n + p = n + q \Rightarrow p = q \; \forall n, p, q \in P.$

Intuitively, if we have the equation $n + p = n + q$, we obtain the result $p = q$ by subtracting n from both sides of the equation.

Proof of Theorem 3.5 [Optional]

We first show that there exists such a function and then show that this function uniquely satisfies the given conditions.

Let H be the collection of all subsets h of $P \times S$ with the following properties:

1) $(1, a) \in h$, and

2) $(u, v) \in h \Rightarrow (s(u), g(v)) \in h$, for each $(u, v) \in P \times S$.

Note that H is nonempty because we clearly have $P \times S \in H$. Now, we define the subset f of $P \times S$ by $f = \cap_{h \in H} h$. We will show that f is a function that satisfies the required conditions with domain P and range S. Note that by the definition of intersection, we have $f \subseteq h$ for each $h \in H$. First, we will show that $f \in H$ by showing that f satisfies conditions 1 and 2 above.

Because $(1, a) \in h$ for each $h \in H$, we have $(1, a) \in f$ (by the definition of intersection), so condition 1 is satisfied. Now, suppose $(u, v) \in f$ for a given $(u, v) \in P \times S$. Then, $(u, v) \in h$ for each $h \in H$. Hence, by the definition of H, we have $(s(u), g(v)) \in h$ for each $h \in H$. Hence, we have $(s(u), g(v)) \in f$. We have shown that $(u, v) \in f \Rightarrow (s(u), g(v)) \in f$, for each $(u, v) \in P \times S$. This demonstrates that f satisfies condition 2. We conclude that $f \in H$. We will make use of this information shortly.

Because this proof is rather long, let us list again the two properties about f that we have shown and will use shortly: These are

1) $f \subseteq h$ for each $h \in H$, and

2) $f \in H$.

To demonstrate that f is a function with domain P and range S, we must show that for each $u \in P$, there is a unique $v \in S$ such that $(u, v) \in f$. We do this by mathematical induction (i.e., Property 3 of Definition 3.1). Define A by $A = \{u \in P | \exists \text{ unique } v \in S \text{ such that } (u, v) \in f\}$. We will show that $A = P$.

We first show that $1 \in A$. To do this, we must show that there is a unique $x \in S$ such that $(1, x) \in f$. We already know that $(1, a) \in f$ because $f \in H$. We now assume that there is a $w \in S$ with $w \neq a$ such that $(1, w) \in f$, and we will strive to arrive at a contradiction. It will now be shown that this assumption forces the conclusion that $f - \{(1, w)\} \in H$. Recall from the definition of H that we must show Properties 1 and 2 above. Now by assumption, $w \neq a$, so that $(1, a) \neq (1, w)$. Hence, $(1, a) \in f - \{(1, w)\}$, so we have Property 1. To show Property 2, we must show that $(u, v) \in f - \{(1, w)\} \Rightarrow (s(u), g(v)) \in f - \{(1, w)\}$ for each $(u, v) \in P \times S$. Hence, consider a $(u, v) \in P \times S$ with $(u, v) \in f - \{(1, w)\}$. Because $f \in H$, we must have $(s(u), g(v)) \in f$. Now, by Property 2 of Definition 3.1, $s(u) \neq 1$, so $(s(u), g(v)) \neq (1, w)$. Thus, $(s(u), g(v)) \in f - \{(1, w)\}$, so we have established Property 2. Hence, we have shown that $f - \{(1, w)\} \in H$. However, recall that $f \subseteq h$ for each $h \in H$, so in particular, we must have $f \subseteq f - \{(1, w)\}$, and here we have our contradiction! We conclude that $1 \in A$.

We now take the inductive step and consider an $n \in P$ such that $n \in A$, and we must show that this forces the conclusion that $s(n) \in A$. Now $n \in A$ implies that there is a unique element $z \in S$ such that $(n, z) \in f$. We must show that there is a unique $y \in S$ such that $(s(n), y) \in f$. We know that we have $(s(n), g(z)) \in f$ (because f is a member of H). Hence, to arrive at a contradiction, we assume nonuniqueness. That is, we assume that there is a $t \in S$ with $t \neq g(z)$ such that $(s(n), t) \in f$. It will now be shown that this forces the conclusion that $f - \{(s(n), t)\} \in H$. Recall from the definition of H that we must show Properties 1 and 2 above. By Property 2 of Definition 3.1, $s(n) \neq 1$, so $(1, a) \neq (s(n), t)$. Hence, $(1, a) \in f - \{(s(n), t)\}$, so we have established Property 1. To show Property 2, we must show that $(u, v) \in f - \{(s(n), t)\} \Rightarrow (s(u), g(v)) \in f - \{(s(n), t)\}$ for each $(u, v) \in P \times S$. Hence, consider a $(u, v) \in P \times S$ with $(u, v) \in f - \{(s(n), t)\}$. Because $f \in H$, we must have $(s(u), g(v)) \in f$. We want to show that $(s(u), g(v)) \neq (s(n), t)$. To do this, we assume that $(s(u), g(v)) = (s(n), t)$ and arrive at a contradiction. If this equality holds, then we must have $s(u) = s(n)$ and $g(v) = t$. By Property 1 of Definition 3.1, we must have $u = n$. Hence, we have $(n, v) \in f$. However, using the uniqueness of the element z (from the inductive step), we must have $v = z$, so $t = g(z)$. But this is a contradiction because we already said that $t \neq g(z)$. Hence, we have $(s(u), g(v)) \neq (s(n), t)$, so $(s(u), g(v)) \in f - \{(s(n), t)\}$, and we have established Property 2. This shows that $f - \{(s(n), t)\} \in H$. However, recall that $f \subseteq h$ for each $h \in H$, so in particular, we must have $f \subseteq f - \{(s(n), t)\}$, and here we have our contradiction to the assumption that there is a $t \in S$ with $t \neq g(z)$ such that $(s(n), t) \in f$! We conclude that $s(n) \in A$. This completes our lengthy induction.

We have shown that we have a function $f: P \rightarrow S$. By recalling the definition of H, the student should be able to see that $f(1) = a$ and $f(s(n)) = g(f(n)) \forall n \in P$. Our final task is to show the uniqueness of f. To do this, suppose we have $k: P \rightarrow S$ such that $k(1) = a$ and $k(s(n)) = g(k(n)) \forall n \in P$. We will show that $k = f$. To do this, define T by $T = \{n \in P \mid k(n) = f(n)\}$. We will show that $T = P$ by mathematical induction. We have $f(1) = a = k(1)$, so $1 \in T$. Consider an $n \in P$ with $n \in T$. Then $k(n) = f(n)$. We want to show that $s(n) \in T$; that is, we want to show that $k(s(n)) = f(s(n))$. However, $k(s(n)) = g(k(n)) = g(f(n)) = f(s(n))$, and we are done.

EXERCISES

1. Prove Theorem 3.11. Make sure that you correctly state your induction argument.

2. Show that $n + m \neq n \ \forall n, m \in P$.

3. Prove (by justifying each step) that $3 \cdot 2 = 6$.

4. Try to show the **cancellation law of multiplication** (for positive integers), which is stated as follows: $np = nq \Rightarrow p = q \ \forall n, p, q \in P$. Do not feel bad if you cannot do it. It is not that easy yet.

Section 3.3   An Order on the Positive Integers and the Second Principle of Mathematical Induction

This section builds on the concept of a linear order, which was introduced in Section 2.4. We discuss some properties of linear orders that are of particular importance to Peano spaces. Following this, we define a relation on the set P of a Peano space $(P, s, 1)$ and demonstrate that it is a linear order. Properties of this linear order will then be derived. This linear order is nothing other than the "less than or equal to" linear order that we have all used since at least middle school. Finally, we use this order to prove a variant of the principle of mathematical induction.

Least and Greatest Elements

For this subsection, suppose we have a given linear order $\leq$ on a nonempty set S.

Definition 3.4

Let $A \subseteq S$. An element $b \in A$ is said to be a **least element** of A with respect to $\leq$ if $b \leq a \ \forall a \in A$. An element $c \in A$ is said to be a **greatest element** of A with respect to $\leq$ if $a \leq c \ \forall a \in A$. A least element of A is also called a **minimum** of A, and a greatest element of A is also called a **maximum** of A.

As illustrated in Example 3.3 below, there are linear orders in which certain subsets fail to have least elements and/or greatest elements. It should be noted that the expression "with respect to $\leq$" is often omitted when it is obvious which linear order is being referred to.

Theorem 3.12

Let $A \subseteq S$. A least element of A with respect to $\leq$ is unique if it exists. A greatest element of A with respect to $\leq$ is unique if it exists.

Proof: We show that the least element of A is unique if it exists. The reader can use the same type of argument to show that the greatest element of A is unique if it exists. Suppose that u and v are least elements of A. We will show that $u = v$. Because u is a least element of A, we have $u \leq a \ \forall a \in A$. In particular, we have $u \leq v$ because $v \in A$. Because v is a least element of A, we have $v \leq a \ \forall a \in A$. In particular, we have $v \leq u$ because $u \in A$. Hence, because we have $u \leq v$ and $v \leq u$, we must have $u = v$ because linear orders obey the antisymmetric property. ∎

Example 3.3

Consider the "normal" "less than or equal to" linear order $\leq$ on **R** with which we are all familiar. Define sets A_1 and A_2 by $A_1 = \{x \in \mathbf{R} | 0 < x \leq 1\}$ and $A_2 = \{x \in \mathbf{R} | -3 \leq x < 100\}$. Clearly -3 is a least element of A_2, and 1 is a greatest element of A_1. It will now be shown that A_1 does not have a least element by proof by contradiction. Suppose that A_1 has a least element. Let us call it ε. Then we must have $0 < \varepsilon \leq 1$ and $\varepsilon \leq x \ \forall x \in A_1$. But we know (from high school algebra) that $0 < \frac{\varepsilon}{2} < \varepsilon \leq 1$. Hence, $\frac{\varepsilon}{2}$ is also a least element of A_1. Hence, by Theorem 3.12, we must have $\frac{\varepsilon}{2} = \varepsilon$, and this is clearly a contradiction. A similar type of argument can be employed to show that A_2 does not have a greatest element.

Note: The reader should verify that Definition 3.4 makes sense and that Theorem 3.12 holds even if we make the weaker assumption that $\leq$ is a **partial order**. (For the definition of a partial order, see Exercise 1 of Section 2.4.)

An Important Linear Order

For the remainder of this section, we work with a fixed Peano space $(P, s, 1)$. How can we define the "usual" order $\leq$ on P? We start by defining the relation $<$. Intuitively, what does the expression $m < n$ mean? Intuitively, this means that n is further to the right than m on a number line. In other words, we can obtain n from m by adding a positive integer to m because adding a positive integer means "moving further to the right." This is the motivation for Definition 3.5. From now on, we use the results of Theorems 3.6 through 3.10 without explicitly referencing these theorems.

Definition 3.5

1) We define a relation $<$ on P as follows: We specify that $m < n$, provided there exists an element $k \in P$ such that $n = m + k$. We can also express $m < n$ by writing $n > m$.
2) We define a relation $\leq$ on P as follows: We specify that $m \leq n$, provided $m < n$ or $m = n$. We can also express $m \leq n$ by writing $n \geq m$.

Of course, the statement $m < n$ is read "m is less than n," and the statement $n > m$ (which has the same meaning as $m < n$) is read "n is greater than m." The statement $m \leq n$ is read "m is less than or equal to n," and the statement $n \geq m$ (which has the same meaning as $m \leq n$) is read "n is greater than or equal to m." Note that by Definition 3.5, we have $m < m + k \ \forall m, k \in P$.

There is some work to be done. For example, at this point, we have not yet shown that it is impossible to have both $m < n$ and $m = n$. In the following theorems, we shall usually use the symbols $<$ and $\leq$. It should be clear that any statement involving $<$ and $\leq$ could be written in terms of $>$ and $\geq$, respectively, and the student is asked to make this connection without the need for us to write it out formally.

Theorem 3.13   Let $m, n \in P$. Then, at most, one of the following statements is true:

1) $m = n$.
2) $m < n$.
3) $n < m$.

Proof: Suppose we have $m = n$. It will be shown that the other two statements are false. Suppose (to reach a contradiction) that we have $m < n$. Then, by definition, we have $n = m + k$ for some $k \in P$. Now, because $m = n$, we have $n = n + k$. However, this contradicts Exercise 2 of Section 3.2. Hence, we have shown that $m = n \Rightarrow \neg(m < n)$. The same type of argument shows that $m = n \Rightarrow \neg(n < m)$. The contrapositive of these two implications shows that $m < n \Rightarrow m \neq n$ and $n < m \Rightarrow m \neq n$, respectively.

To complete the proof, it must be shown that $m < n \Rightarrow \neg(n < m)$ and $n < m \Rightarrow \neg(m < n)$. To show that $m < n \Rightarrow \neg(n < m)$, suppose that $m < n$. Then, we have $n = m + k$ for some $k \in P$. Suppose (to reach a contradiction) that we also have $n < m$. Then, by definition, we have $m = n + p$ for some $p \in P$. Hence, we have $n = (n + p) + k = n + (p + k)$. However, this contradicts Exercise 2 of Section 3.2. Hence, we have shown that $m < n \Rightarrow \neg(n < m)$. The contrapositive of this statement gives us $n < m \Rightarrow \neg(m < n)$.

It will shortly be shown that exactly one of the statements listed in Theorem 3.13 is true for any given $m, n \in P$.

Theorem 3.14   Let $m, n, p \in P$.

1) Suppose $m < n$ and $n < p$. Then $m < p$.
2) Suppose $m \leq n$ and $n \leq p$. Then $m \leq p$.

Proof: We prove Part 1 and leave Part 2 to the reader. Suppose that $m < n$ and $n < p$. Then there exists $k_1, k_2 \in P$ such that $n = m + k_1$ and $p = n + k_2$. Thus, $p = (m + k_1) + k_2 = m + (k_1 + k_2)$. Hence, $m < p$.

Note that Theorem 3.14 indicates that both of the relations $<$ and $\leq$ are transitive. Of course, $\leq$ is also reflexive, but $<$ fails to be reflexive. (Verify!)

Theorem 3.15   Let m, n, $k \in P$.

1) $m < n \Leftrightarrow m + k < n + k$.

2) $m \leq n \Leftrightarrow m + k \leq n + k$.

Proof: We show Part 1 and leave Part 2 to the reader. To show Part 1, we first show that $m < n \Rightarrow m + k < n + k$, and then show that $m + k < n + k \Rightarrow m < n$. To show the former implication, suppose $m < n$. Then there is a $p \in P$ such that $n = m + p$. Hence, $n + k = (m + p) + k = (m + k) + p$, so $m + k < n + k$.

To show that $m + k < n + k \Rightarrow m < n$, suppose $m + k < n + k$. Then there is a $q \in P$ so that $n + k = (m + k) + q = (m + q) + k$. From the equation $n + k = (m + q) + k$, we invoke the cancellation law of addition (Theorem 3.11) to get $n = m + q$. Hence, $m < n$.

A note on the proof of Theorem 3.15 is in order. The proofs of each of the two parts of Theorem 3.15 involve proving the equivalence of statements. Let p and q denote given statements. Our general method for proving $p \Leftrightarrow q$ has been to show $p \Rightarrow q$ and then show $q \Rightarrow p$. In the interest of brevity and conformance with advanced mathematics texts, we often word the setup for the proof of $q \Rightarrow p$ by simply writing something similar to "conversely, suppose q," and then we will proceed to prove p. The justification for this is that the statement $q \Rightarrow p$ is the converse of the statement $p \Rightarrow q$. (Review Section 1.2 if necessary.)

The student should have no trouble proving the next theorem.

Theorem 3.16   The relation $\leq$ on P is antisymmetric.

Because we now know that $\leq$ is reflexive, antisymmetric, and transitive, showing that any two elements of P are comparable will complete the proof that $\leq$ is a linear order on P. It should be clear that the following theorem demonstrates that any two elements of P are comparable.

Theorem 3.17   Let m, $n \in P$. Then exactly one of the following statements is true:

1. $m = n$.

2. $m < n$.

3. $n < m$.

Proof: It has already been shown (Theorem 3.13) that, at most, one of the listed statements is true. Therefore, it suffices to show that at least one of these statements is true. This will be proved by mathematical induction. It is important to state this induction carefully. For each $n \in P$, let $S(n)$ be the statement "For each $m \in P$, we have $m = n$ or $m < n$ or $n < m$." We will establish the result by showing $S(1)$ and by showing $S(n) \Rightarrow S(n + 1)$.

To show $S(1)$, let $m \in P$. We must show that $m = 1$ or $m < 1$ or $1 < m$. Now, if $m = 1$, then we are done showing $S(1)$, so consider the case in which $m \neq 1$. We must show that $m < 1$ or $1 < m$. We show that $1 < m$. By Theorem 3.4, there is a $k \in P$ such that $m = k + 1$, so that $1 < m$. We have established $S(1)$. (Note that because of this and Theorem 3.13, the statement $m < 1$ is false for all $m \in P$.)

We now take the inductive step and assume $S(n)$. We will show $S(n + 1)$. Let $m \in P$. We must show that $m = n + 1$ or $m < n + 1$ or $n + 1 < m$. Now, by the inductive step, we have $m = n$ or $m < n$ or $n < m$. We analyze each of these cases.

Case 1: $m = n$. Then clearly $m < n + 1$ because $n < n + 1$.

Case 2: $m < n$. Then $m < n + 1$ by Theorem 3.14.

Case 3: $n < m$. Then, by definition, there is a $w \in P$ such that $m = n + w$.
     Subcase 3a: $w = 1$. Then $m = n + 1$.
     Subcase 3b: $w \neq 1$. Then by Theorem 3.4, there is a $u \in P$ such that $w = u + 1$. Hence, $m = n + (u + 1) = (n + 1) + u$, so $n + 1 < m$.

Note that in all cases, we have $S(n + 1)$.   ∎

The student should have no difficulty in seeing that instead of invoking the "informal" induction proof using the statement $S(n)$, we could have been more formal by defining the set T specified by $T = \{n \in P | \forall m \in P, m = n$ or $m < n$ or $n < m\}$. We would then have shown that $1 \in T$ and then have shown that $n \in T \Rightarrow n + 1 \in T \ \forall n \in P$. Sometimes it is more readable and more easily understood to use statements such as $S(n)$ in induction proofs.

Theorem 3.18 below simply says that if we have a given inequality between positive integers, and if we multiply both sides of this inequality by a positive integer, then the direction of the resultant inequality is the same as the direction of the original inequality. So, for example, we know that $2 < 3$. Hence, we can state that $2 \cdot 5 < 3 \cdot 5$.

Theorem 3.18   Let $m, n, p \in P$.

1) $m < n \Leftrightarrow mp < np$.
2) $m \leq n \Leftrightarrow mp \leq np$.

Proof: We show Part 1 and leave Part 2 to the reader. Suppose $m < n$. Then, by definition, we have $n = m + k$ for some $k \in P$. Hence, $np = (m + k)p = mp + kp$, so $mp < np$. Conversely, suppose $mp < np$. Now, by Theorem 3.17, exactly one of the following statements is true:

1) $m = n$;
2) $m < n$;
3) $n < m$.

Now, the statement $m = n$ cannot be true because this statement implies that $mp = np$, which we know is false (by Theorem 3.17) because we are given $mp < np$. Now, suppose (to reach a contradiction) we have $n < m$. Then using the implication $m < n \Rightarrow mp < np$ (that we have just proved) and switching m and n immediately gives us $np < mp$, which contradicts our assumption $mp < np$ (using Theorem 3.17). Hence, we must have $m < n$.

The proof of the next theorem is left to the reader. This theorem is the Cancellation Law of Multiplication (for positive integers). The student who had difficulty trying to show this for Exercise 4 of Section 3.2 should find it easier now.

Theorem 3.19   **(Cancellation Law of Multiplication)**

Let $n, p, q \in P$.
$np = nq \Rightarrow p = q$.

The student is invited to determine what Theorem 3.20 below "says" intuitively.

Theorem 3.20   Let $m, n \in P$. Then
$m < n \Rightarrow m + 1 \leq n$.

Proof: Suppose $m < n$. Then, by definition, we have $n = m + k$ for some $k \in P$. We have already remarked in the proof of Theorem 3.17 that no member of P is less than 1. Hence, $1 \leq k$. By Theorem 3.15, $m + 1 \leq m + k$. That is, $m + 1 \leq n$.

The next theorem says that for each $n \in P$, there is no member of P that is strictly between n and $n + 1$. The (easy) proof is left to the reader.

Theorem 3.21   For each $n \in P$, there does not exist a $k \in P$ such that $n < k < n + 1$. ∎

Notational Note: Of course, the expression $n < k < n + 1$ means $n < k$ and $k < n + 1$. Similarly, an expression such as $a < b \leq c$ means $a < b$ and $b \leq c$. The student should interpret similar statements appropriately.

The Well-Ordering Property and the Second Principle of Mathematical Induction

We continue to work with a Peano space $(P, s, 1)$. Next on our agenda, we will derive a result called the **Well-Ordering Property** and use it to prove a variant form of mathematical induction that is often useful. The Well-Ordering Property is very important, and you will see it used often in advanced mathematics courses. We will also make use of it in later chapters. The Well-Ordering Property simply states that every nonempty subset of P has a least element (with respect to the linear order $\leq$ that we defined above). The proof employs a "tricky" use of mathematical induction.

Theorem 3.22   **(Well-Ordering Property)**

Every nonempty subset of P has a least element.

Proof: For each $n \in P$, let $S(n)$ be the statement:
For every nonempty subset A of P, $n \in A \Rightarrow A$ has a least element. Our proof method is to use mathematical induction to show the truth of $S(n)$ for each $n \in P$. It should be clear that this suffices to prove the theorem. (Why?)

To verify $S(1)$, we must show that any subset of P having 1 as a member must have a least element. However, because 1 is the least element of the whole set P (i.e., $1 \leq n \ \forall n \in P$), it is clear that 1 is the least element of any subset of P having 1 as a member.

We now take the inductive step and assume $S(n)$. We will show $S(n + 1)$. Let A be any subset of P such that $n + 1 \in A$. We must verify that A has a least element. We examine the possible cases:

Case 1: $n \in A$. Then A has a least element by the inductive step.

Case 2: $n \notin A$. Let $B = A \cup \{n\}$. Because $n \in B$, B has a least element by the inductive step. Let us call this least element k. Hence, $k \leq a$ $\forall a \in A$, and $k \leq n$.

Subcase 2a: $k \in A$. Then k is the least element of A, and we are done.

Subcase 2b: $k \notin A$. Then we must have $k = n$. Hence, we must have $n < a \ \forall a \in A$. By Theorem 3.20, $n + 1 \le a \ \forall a \in A$. Because we also have $n + 1 \in A$, this shows that $n + 1$ is the least element of A.

Thus, having shown that A has a least element, this verifies $S(n + 1)$.

———

We now state and prove a variant of mathematical induction that is frequently referred to as the **Second Principle of Mathematical Induction**. Suppose that we have a statement $S(n)$ for each $n \in P$, and we wish to show that $S(n)$ is true for each $n \in P$. As with "ordinary" mathematical induction, we begin by showing the truth of $S(1)$. Now, as you know, in "ordinary" mathematical induction, we next show that $S(n) \Rightarrow S(n + 1)$ for each $n \in P$. Herein lies the difference. For this variant form of mathematical induction, we show that $(S(1) \wedge S(2) \wedge \Lambda \wedge S(n)) \Rightarrow S(n + 1)$ for each $n \in P$. (Recall from Chapter 1 that the symbol "$\wedge$" means *and*.) In other words, our induction hypothesis is the truth of the statement $\forall m \in \{k \in P | k \le n\} \ (S(m))$.

For increased clarity, let us write this in terms of sets. First of all, note that the expression $\{1, 2, \ldots, n\}$ is just a shorthand way to write the set $\{k \in P | k \le n\}$. Suppose T is a subset of P with the following two properties:

1) $1 \in T$;
2) For each $n \in P$, $\{1, 2, \ldots, n\} \subseteq T \Rightarrow n + 1 \in T$.

Then, the upcoming theorem states that $T = P$.

The reason for the usefulness of this variant is that sometimes it is easier to show that $\{1, 2, \ldots, n\} \subseteq T \Rightarrow n + 1 \in T$ instead of showing $n \in T \Rightarrow n + 1 \in T$ (as is done in "ordinary" mathematical induction proofs). We will have occasion to use this in the next chapter when we show that every positive integer greater than 1 can be expressed as a product of prime numbers.

Theorem 3.23 **(Second Principle of Mathematical Induction)**

Suppose T is a subset of P with the following two properties:

1) $1 \in T$;
2) For each $n \in P$, $\{1, 2, \ldots, n\} \subseteq T \Rightarrow n + 1 \in T$.

Then $T = P$.

Proof: We do a proof by contradiction. Suppose the result is false, so that $T \ne P$. Let $A = \{k \in P | k \notin T\}$. Then A is nonempty. By the Well-Ordering Property (Theorem 3.22), the set A must have a least element. Let us call this least element n. Hence, by definition, n is the least member of P that fails to be a member of T.

The first thing to note is that $n \neq 1$ because we are given that $1 \in T$ by Property 1. Hence, by Theorem 3.4 there is an $m \in P$ such that $n = m + 1$. Thus, $m < n$. Therefore, we must have $\{1, 2, \ldots, m\} \subseteq T$ because n is the least member of P that fails to be a member of T. Consequently, Property 2 mandates that $m + 1 \in T$. But $m + 1 = n$, and we already concluded that $n \notin T$. Hence, we have our contradiction.

EXERCISES

1. Prove that the set A_2 of Example 3.3 does not have a greatest element.

2. Consider the power set $\mathbf{P}(S)$ of Example 2.15. With regard to the partial order defined in that example, does the set $\mathbf{P}(S)$ have a least element? If so, what is it? How about a greatest element?

3. Prove Part 2 of Theorem 3.14.

4. Prove Part 2 of Theorem 3.15.

5. Prove Theorem 3.16.

6. Prove Part 2 of Theorem 3.18.

7. Prove Theorem 3.19.

8. Prove Theorem 3.21.

9. Let $(P, s, 1)$ be a Peano space. Let $m, n, p \in P$. Suppose $m < n \leq p$. Show that $m < p$.

10. Let $(P, s, 1)$ be a Peano space. Let $m, n, p \in P$. Suppose $m \leq n < p$. Show that $m < p$.

11. Let $(P, s, 1)$ be a Peano space. Let $p \in P$. Suppose $S \subseteq P$ such that
    a. $p \in S$; and
    b. $\forall n \in P$, we have $n \in S \Rightarrow n + 1 \in S$.
    Show that $\{m \in P \mid m \geq p\} \subseteq S$.

12. Suppose $(P, s, 1)$ is a Peano space with $m, n \in P$. Suppose $m < n$. Show that there is a *unique* $x \in P$ such that $n = m + x$. [This provides us with our definition of subtraction: We denote this x by $n - m$. Hence, we have $n = m + (n - m)$.]

Section 3.4 | # The Uniqueness of Peano Spaces and the Beginnings of Counting

By now you should be really convinced that the set P of a Peano space $(P, s, 1)$ has the properties that you would "expect" of the set of positive integers. However, there are a few more issues to be settled. What is the most basic and ancient function of positive integers? Positive integers were invented (or perhaps discovered) initially for the purpose of assigning quantitative measures to discrete aspects of reality in order to *count* things. We will lay down the foundations of using Peano spaces for counting in this section.

Before we do this, we settle an issue that is of important theoretical concern. In doing so, you will get a strong introduction to the concept of an **isomorphism**. Basically, an isomorphism is a bijection between two sets that demonstrates that the two sets have the same behavior with respect to properties of concern. The concept of an isomorphism is vital to advanced mathematics. We will show that if we have Peano spaces $(P, s, 1)$ and $(P', s', 1')$, then there is an isomorphism $f \colon P \to P'$ that demonstrates that P and P' have the "same structure." We often speak of this identical structure by saying that Peano spaces are unique. In other words, the unique properties that all Peano spaces share make it immaterial as to which Peano space is being used; one is as good as the other.

Peano Space Isomorphisms

We motivate the discussion with a little story. Consider Farmer A talking to Farmer B. Farmer A tells Farmer B with pride that she has 6 sheep. Farmer B asks Farmer A which Peano space is being used to do the counting! Farmer A is a little taken aback by this question. After a little thought, Farmer A tells Farmer B that it does not matter.

Indeed, consider any Peano space $(P, s, 1)$ that utilizes the notation of Example 3.2 and Exercise 3 of Section 3.2. Now consider a different Peano space (P', s', I). Suppose for (P', s', I), the binary operation of addition defined in Definition 3.2 is written as $\oplus$, and the binary operation of multiplication defined in Definition 3.3 is written as $\otimes$. Furthermore, suppose we define the elements II, III, IV, V, and VI of P' by II $= s'(I)$, III $= s'(II)$, IV $= s'(III)$, V $= s'(IV)$, and VI $= s'(V)$.

Now consider the subsets X and Y of P and P', respectively, defined by $X = \{1, 2, 3, 4, 5, 6\}$ and $Y = \{I, II, III, IV, V, VI\}$. Now, do we have $X = Y$? Not necessarily. Depending on how the sets P and P' are defined, the elements of X could be very different from the elements of Y. However, the set X has been constructed from its Peano space $(P, s, 1)$ in the same way that the set Y has been constructed from its Peano space $(P', s', 1)$. In fact, the reader should see that the same arguments compelling us to accept the equations $3 + 2 = 5$ and $3 \cdot 2 = 6$ also compel us to accept the equations III $\oplus$ II $=$ V and III $\otimes$ II $=$ VI.

We will now begin to make precise our intuitive feeling that the sets X and Y are, in essence, the "same." We define the function $f \colon X \to Y$ by $f(1) = $ I, $f(2) = $ II, $f(3) = $ III, $f(4) = $ IV, $f(5) = $ V, and $f(6) = $ VI. It is immediately clear that this mapping is a bijection. Now note that the equation

Figure 3.2
The mapping from
X to Y

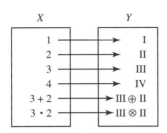

III $\oplus$ II = V can also be written $f(3) \oplus f(2) = f(5)$. Also, the equation III $\otimes$ II = VI can be written $f(3) \otimes f(2) = f(6)$. We can also write these equations as $f(3) \oplus f(2) = f(3 + 2)$ and $f(3) \otimes f(2) = f(3 \cdot 2)$. Continuing with this argument, we can see that $f(1)$ "behaves" like 1, $f(2)$ "behaves" like 2, and so on. Indeed, we can "view" the set Y as a carbon copy of X with the exception that the names of the elements may have been changed. This is why Farmer A told Farmer B that the choice of Peano space is irrelevant. This mapping is illustrated in Figure 3.2. Hopefully, this has provided sufficient motivation for the following definition.

Definition 3.6

Let $(P, s, 1)$ and $(P', s', 1')$ be Peano spaces. Suppose that for each $m, n \in P$, the operations of addition and multiplication (corresponding to Definitions 3.2 and 3.3, respectively) are denoted by $m + n$ and $m \cdot n$ respectively. Suppose that for each $x, y \in P'$, the operations of addition and multiplication (corresponding to Definitions 3.2 and 3.3, respectively) are denoted by $x \oplus y$ and $x \otimes y$, respectively. Then a bijection $f: P \rightarrow P'$ is called **a Peano space isomorphism** between P and P', provided $f(m + n) = f(m) \oplus f(n)$ and $f(m \cdot n) = f(m) \otimes f(n)$ for each $m, n \in P$.

The reader is asked to mull over this definition in light of the above motivation until he or she reaches the point of concluding that the definition "captures" the intuitive notion that the existence of such an isomorphism shows that the two Peano spaces "behave" the same (with respect to addition and multiplication). The following theorem should help.

Theorem 3.24   Let $(P, s, 1)$ and $(P', s', 1')$ be Peano spaces. Suppose $f: P \rightarrow P'$ is a Peano space isomorphism between P and P'. Then $f(1) = 1'$.

Proof: We have $f(1) = f(1 \cdot 1) = f(1) \otimes f(1)$. We also have $f(1) \otimes 1' = f(1)$. Hence, we have $f(1) \otimes f(1) = f(1) \otimes 1'$. The desired result follows from the cancellation law of multiplication (Theorem 3.19) applied to the Peano space $(P', s', 1')$.

———————

The next theorem is the point of this subsection. It states that any two Peano spaces "behave" the same (with respect to addition and multiplication). This theorem makes use of the Recursion Theorem (Theorem 3.5). The proof is a little lengthy, but very instructive.

Theorem 3.25

Let $(P, s, 1)$ and $(P', s', 1')$ be any Peano spaces. Then there exists a Peano space isomorphism between P and P'.

Proof: Let us use the notation of Definition 3.6 to represent the binary operations of addition and multiplication for the Peano spaces $(P, s, 1)$ and $(P', s', 1')$. We will produce a bijection $f: P \to P'$ with the desired properties.

By the Recursion Theorem, there is a unique function $f: P \to P'$ such that $f(1) = 1'$ and $f(s(n)) = s'(f(n))$ $\forall n \in P$. That is, $f(n + 1) = f(n) \oplus 1'$ $\forall n \in P$. This equation will be used frequently, so we call it Equation 1 for the remainder of this proof.

$$f(k + 1) = f(k) \oplus 1' \; \forall k \in P. \tag{1}$$

We first show (by induction on n) that $f(m + n) = f(m) \oplus f(n)$ $\forall m, n \in P$. To show that the result holds for 1, we must show that $f(m + 1) = f(m) \oplus f(1)$ $\forall m \in P$. Because we have $f(1) = 1'$, this is given by Equation 1.

We now take the inductive step and assume that the result holds for some $n \in P$. We must show that this causes the result to hold for $n + 1$. That is, we must show that $f(m + (n + 1)) = f(m) \oplus f(n + 1)$ $\forall m \in P$. Now, we have

$$\begin{aligned}
f(m + (n + 1)) &= f((m + n) + 1) \\
&= f(m + n) \oplus 1' && \text{(by Equation 1)} \\
&= (f(m) \oplus f(n)) \oplus 1' && \text{(by the inductive step)} \\
&= f(m) \oplus (f(n) \oplus 1') && \text{(Why?)} \\
&= f(m) \oplus f(n + 1) && \text{(by Equation 1).}
\end{aligned}$$

Next, we show (by induction on m) that $f(m \cdot n) = f(m) \otimes f(n)$ $\forall m, n \in P$. To show that the result holds for 1, we must show that $f(1 \cdot n) = f(1) \otimes f(n)$ $\forall n \in P$. Now, $f(1 \cdot n) = f(n) = 1' \otimes f(n) = f(1) \otimes f(n)$.

We now take the inductive step and assume that the result holds for some $m \in P$. We must show that this causes the result to hold for $m + 1$. That is, we must show that $f((m + 1) \cdot n) = f(m + 1) \otimes f(n)$ $\forall n \in P$. Now, we have

$$\begin{aligned}
f((m + 1) \cdot n) &= f(m \cdot n + n) \\
&= f(m \cdot n) \oplus f(n) && \text{(Why?)} \\
&= (f(m) \otimes f(n)) \oplus f(n) && \text{(by the inductive step)} \\
&= (f(m) \oplus 1') \otimes f(n) && \text{(Why?)} \\
&= f(m + 1) \otimes f(n) && \text{(by Equation 1).}
\end{aligned}$$

Is the proof complete? No, because we have not yet proved that the function f is a bijection. To demonstrate that f is onto, we show that for each $k \in P'$, there is an $m \in P$ such that $k = f(m)$. We do this by performing an induction on k. [Note that this induction is with respect to the Peano space $(P', s', 1')$]. It is immediate that the result holds for $1'$ because we have $f(1) = 1'$.

We now take the inductive step and assume that the result holds for some $k \in P'$. We must show that the result holds for $k \oplus 1'$. That is, we show the existence of some $w \in P$ such that $k \oplus 1' = f(w)$. By the inductive step, there is an $n \in P$ such that $k = f(n)$. Now, $k \oplus 1' = f(n) \oplus 1' + f(n + 1)$. Hence, $n + 1$ "does the trick." This shows that f is onto.

Finally, we have to prove that f is one-to-one. We will do this directly by showing that $\forall m, n \in P, f(m) = f(n) \Rightarrow m = n$. Let $<$ denote the relation on P that is defined in Part 1 of Definition 3.5. Now, suppose $f(m) = f(n)$. We know that $m < n$ or $n < m$ or $m = n$. We will complete the proof by showing that it is not possible to have $m < n$ or $n < m$. Suppose (to reach a contradiction) that $m < n$. Then we have $n = m + k$ for some $k \in P$. Hence, we have $f(m) = f(n) = f(m + k) = f(m) \oplus f(k)$. However, this contradicts Exercise 2 of Section 3.2 as applied to the Peano space $(P', s', 1')$. A similar argument shows the impossibility of $n < m$. Hence, $m = n$. This concludes the proof.

The Beginnings of Counting

Theorem 3.25 demonstrates that from a theoretical viewpoint, one Peano space is as good as another. However, there are still a few issues to be solved once and for all before we are "free" to apply Peano spaces to counting. For the remainder of the section, we work with a given Peano space $(P, s, 1)$ and employ the "usual" notation (i.e., $m + n$, mn or $m \cdot n$, $m < n$, and $m \leq n$, etc.).

It will be beneficial to introduce some useful notation and to make a few observations about one-to-one and onto functions before continuing. The reader may want to review portions of Section 2.5 dealing with these topics before proceeding.

Definition 3.7

Let X and Y be nonempty sets. Suppose there exists a one-to-one correspondence $f \colon X \to Y$. Then we say that X is **set equivalent** to Y. We can also say that X and Y are set equivalent. We can also indicate this by writing $X \approx Y$.

A good way to intuitively view the statement $X \approx Y$ is that the elements of X and Y can be "paired up" such that each element of Y is "assigned" to one, and only one, element of X in the "pairing." The reader should note the following properties for nonempty sets X, Y, and Z:

1) $X \approx Y \Leftrightarrow Y \approx X$.

2) $(X \approx Y$ and $Y \approx Z) \Rightarrow X \approx Z$.

3) Suppose $X \approx Y$, and suppose there exists a one-to-one function with domain Y and range Z. Then there exists a one-to-one function with domain X and range Z.

4) Suppose there exists a one-to-one function with domain X and range Y, and suppose $Y \approx Z$. Then there exists a one-to-one function with domain X and range Z.

5) Suppose $X \approx Y$, and suppose there exists an onto function with domain Y and range Z. Then there exists an onto function with domain X and range Z.

6) Suppose there exists an onto function with domain X and range Y, and suppose $Y \approx Z$. Then there exists an onto function with domain X and range Z.

Property 1 can be immediately concluded from Exercise 7 of Section 2.5. The remaining properties follow immediately from Theorem 2.11.

It will also be helpful to formally define substraction of appropriate elements of P. Suppose $m, n \in P$ with $m < n$. Then, by definition, there exists an $x \in P$ such that $n = m + x$. Furthermore, this x is unique. Why? Because suppose there is a $y \in P$ such that $n = m + y$. Then we have $m + x = m + y$. Then the cancellation law of addition forces us to conclude that $x = y$. This provides the motivation for the following definition.

Definition 3.8

Let $m, n \in P$ with $m < n$. Then $n - m$ is defined to be that unique $x \in P$ such that $n = m + x$. (Of course, $n - m$ is often called n **minus** m. It is also called the **subtraction** of m from n.)

Note that if $m < n$, then we have $m + (n - m) = n$. Furthermore, suppose we have $a, b, c \in P$ such that $a + b = c$. Then $a < c$, so we have $b = c - a$. Also, $b < c$, so $a = c - b$. The following definition will also prove useful.

Definition 3.9

For each $n \in P$, $\mathbf{I}_n$ is defined by $\mathbf{I}_n = \{m \in P | m \leq n\}$.

Another way to write $\mathbf{I}_n$ is by $\mathbf{I}_n = \{1, 2, \ldots, n\}$. Of course, $\mathbf{I}_n$ will be defined with respect to the Peano space that is being used, but the above work shows that this consideration is not of practical importance.

Another definition will prove useful. Let S and T be any nonempty sets, and consider a function $f\colon S \to T$. Let $x \in P$, and let $y = f(x)$. Recall that y is often referred to as the image of x with respect to f. This motivates Definition 3.10.

Definition 3.10

Let S and T be any nonempty sets, and consider a function $f\colon S \to T$. The **image** of f, written $\mathbf{im}(f)$, is defined to be the set $\{f(x) \,|\, x \in S\}$.

Note that $T = \mathbf{im}(f)$ iff f is onto. Also note that $T = \mathbf{im}(f)$ iff $T = f(S)$. (Recall Definition 2.22.)

Now, we are ready to start counting.

Definition 3.11

Let S be a nonempty set. Suppose that there exists an $n \in P$ such that $\mathbf{I}_n \approx S$. Then we say that S is of **size** n. This also can be expressed by saying S has n **distinct elements**.

Let us now turn back to Farmer A and Farmer B for a moment. When Farmer A said that she had 6 sheep, she meant that there exists a one-to-one correspondence between the set $\mathbf{I}_6 = \{1, 2, 3, 4, 5, 6\}$ and the set S consisting of her sheep. This bijection could be expressed by Farmer A labeling her sheep. She could designate one sheep as sheep number 1, another sheep as sheep number 2, and so on.

Now, Farmer B (who just could not leave well enough alone) asked her how she knew that there did not also exist a one-to-one correspondence between $\mathbf{I}_n$ and S, for some $n \neq 6$. Indeed, if there was such an n, we would have $\mathbf{I}_n \approx S$ and $\mathbf{I}_6 \approx S$. Hence, by the remarks at the beginning of this subsection on bijections, this would imply that $\mathbf{I}_n \approx \mathbf{I}_6$. Such a situation would cast serious doubt on the usefulness of Peano spaces for quantification (i.e., indicating *how many* of a quantity).

Farmer A was very annoyed with this question because it was completely obvious to her that there could not exist such an $n = 6$. However, she even became more annoyed when Farmer B asked her to prove this, and she was not initially able to provide this proof. We will go to work on providing this proof. We will actually prove some stronger results. The following lemma says that if we have a set of size $n + 1$, and we remove one element from this set, the resultant set is of size n.

Lemma 3.5

Let $n \in P$, and suppose A is a set of size $n + 1$. Then for any $a \in A$, $A - \{a\}$ is of size n.

Proof: Because $\mathbf{I}_{n+1} \approx A$, there exists a bijection $f: \mathbf{I}_{n+1} \to A$. Hence, there is an $i \in \mathbf{I}_{n+1}$ such that $a = f(i)$. Now, it is easy to see that $\mathbf{I}_{n+1} - \{i\} \approx A - \{f(i)\}$. Hence, it suffices to show that $\mathbf{I}_{n+1} - \{j\} \approx \mathbf{I}_n$ for each $j \in \mathbf{I}_{n+1}$. (Why?) We will prove this statement is true by an induction on n.

To show the statement is true for 1, we must show that $\mathbf{I}_2 - \{j\} \approx \mathbf{I}_1$ for each $j \in \mathbf{I}_2$. Corresponding to $j = 1$, the statement says that $\{2\} \approx \{1\}$, which is clearly true. Corresponding to $j = 2$, the statement says that $\{1\} \approx \{1\}$, which is also clearly true.

We now take the inductive step and assume that the result holds for some $n \in P$. We must show that the result holds for $n + 1$. That is, we must show that $\mathbf{I}_{n+2} - \{j\} \approx \mathbf{I}_{n+1}$ for each $j \in \mathbf{I}_{n+2}$.

Case 1: $j = n + 2$. Then $\mathbf{I}_{n+2} - \{j\} = \mathbf{I}_{n+2} - \{n + 2\} = \mathbf{I}_{n+1}$, so the result is obvious.

Case 2: $j < n + 2$. Hence, $j \leq n + 1$. (Why?) Now, $\mathbf{I}_{n+2} - \{j, n + 2\} = \mathbf{I}_{n+1} - \{j\}$. By the induction hypotheses, we have $\mathbf{I}_{n+1} - \{j\} \approx \mathbf{I}_n$. Hence, there must exist a bijection $f: \mathbf{I}_{n+1} - \{j\} \to \mathbf{I}_n$. Define $g: \mathbf{I}_{n+2} - \{j\} \to \mathbf{I}_{n+1}$ to be the extension of f to $\mathbf{I}_{n+1} - \{j\}$ with $g(n + 2) = n + 1$. Clearly, g is bijective, so we have $\mathbf{I}_{n+2} - \{j\} \approx \mathbf{I}_{n+1}$. ∎

Theorem 3.26 below immediately implies that if a set is of size n for some $n \in P$, then it is not of any other size. Of course, the reader already knows this intuitively.

Theorem 3.26

Suppose $m, n \in P$ with $m < n$.

1) Every function with domain $\mathbf{I}_m$ and range $\mathbf{I}_n$ fails to be onto.

2) Suppose A and B are sets with $\mathbf{I}_m \approx A$ and $\mathbf{I}_n \approx B$. Then every function with domain A and range B fails to be onto.

Proof: We first show that Part 1 $\Rightarrow$ Part 2. Then we will show that Part 1 is true. To show that Part 1 $\Rightarrow$ Part 2, we show that the contrapositive of this statement is true. Suppose Part 2 fails to be true. Then there must exist an onto function $f: A \to B$. Because $\mathbf{I}_m \approx A$, there must exist an onto function $g: \mathbf{I}_m \to B$. (Here we have used the remarks at the beginning of this subsection dealing with set equivalent sets.) We also have $B \approx \mathbf{I}_n$ (because $\mathbf{I}_n \approx B$). Hence (using the remarks at the beginning of this subsection dealing with set equivalent sets), we have the existence of an onto function with domain $\mathbf{I}_m$ and range $\mathbf{I}_n$, and this says that Part 1 is false. Thus, we have Part 1 $\Rightarrow$ Part 2.

The remainder of the proof is dedicated to showing the truth of Part 1 via induction on m. To show that the result holds for 1, consider any function $f: \mathbf{I}_1 \to \mathbf{I}_n$ in which $n > 1$. Hence, $n \geq 2$, so $\{1, 2\} \subseteq \mathbf{I}_n$. Note that $\mathbf{I}_1 = \{1\}$.

Case 1: $f(1) = 1$. Then f is not onto because $2 \notin \mathbf{im}(f)$.

Case 2: $f(1) \neq 1$. Then f is not onto because $1 \notin \mathbf{im}(f)$.

We now take the inductive step and assume that the result holds for some $m \in P$. We must show that the result holds for $m + 1$. That is, we must show that if $f: \mathbf{I}_{m+1} \to \mathbf{I}_n$ with $m + 1 < n$, then f is not onto. Consider such a function f. Suppose (to reach a contradiction) that f is onto. Let $w = f(m + 1)$. Now consider the restriction $f|\mathbf{I}_m$ of f to $\mathbf{I}_m$ with $f|\mathbf{I}_m: \mathbf{I}_m \to \mathbf{I}_n$. Then, by the induction hypothesis, f is not onto. Hence, w is the only element of $\mathbf{I}_n$ that is not a member of $\mathbf{im}(f|\mathbf{I}_m)$. (Why?) Now, let g be the same function as $f|\mathbf{I}_m$, except having range $\mathbf{I}_n - \{w\}$, so that we have $g: \mathbf{I}_m \to \mathbf{I}_n - \{w\}$. Then g is onto. However, we have $\mathbf{I}_n - \{w\} \approx \mathbf{I}_{n-1}$ by Lemma 3.5. Hence, there must exist an onto function with domain $\mathbf{I}_m$ and range $\mathbf{I}_{n-1}$. However, this contradicts the induction hypothesis because we have $m < n - 1$ (as we were given $m + 1 < n$). This concludes the proof.

Note that due to Theorem 3.26, we have already achieved our goal of showing that if a set A is on size n then it is not of size m if $m \neq n$. (Why?) We record this as a theorem because it is very important.

Theorem 3.27   Suppose that A is a set of size n for some $n \in P$. Then if $m \in P$ with $m \neq n$, A is not of size m.

Having met our above-mentioned goal, we could stop here. However, we can also prove another result that turns out to be very useful in higher mathematics.

Theorem 3.28   Suppose $m, n \in P$ with $m > n$.

1) Every function with domain $\mathbf{I}_m$ and range $\mathbf{I}_n$ fails to be one-to-one.

2) Suppose A and B are sets with $\mathbf{I}_m \approx A$ and $\mathbf{I}_n \approx B$. Then every function with domain A and range B fails to be one-to-one.

Proof: The reader is asked to show that Part 1 $\Rightarrow$ Part 2. We now show that Part 1 is true by an induction on n. To show this holds for 1, let $f: \mathbf{I}_m \to \mathbf{I}_1$ with $m > 1$. Hence, $m \geq 2$, so $\{1, 2\} \subseteq \mathbf{I}_m$. Therefore, we must have $f(1) = 1 = f(2)$, so f is not one-to-one. We now take the inductive step and assume that the result holds for some $n \in P$. We must show that the result holds for $n + 1$. That is, we must show that if $f: \mathbf{I}_m \to \mathbf{I}_{n+1}$ with $m > n + 1$, then f is not one-to-one. Now assume that f is one-to-one (to reach a contradiction). Let $y = f(m)$, and consider the restriction of f to $\mathbf{I}_{m-1}$ with range $\mathbf{I}_{n+1} - \{y\}$. That is, we have the function $f|\mathbf{I}_{m-1}: \mathbf{I}_{m-1} \to \mathbf{I}_{n+1} - \{y\}$. (How have we used the assumption that f is one-to-one?) Note that $f|\mathbf{I}_{m-1}$ must also be one-to-one. Also, by Lemma 3.5, we have $\mathbf{I}_{n+1} - \{y\} \approx \mathbf{I}_n$. Hence, there exists a one-to-one function with domain $\mathbf{I}_{m-1}$ and range $\mathbf{I}_n$. Noting that $m - 1 > n$, this contradicts the inductive step. Hence, the desired result.

In mathematical circles, Theorem 3.28 is often called the **Pigeon Hole Principle**. Imagine a collection of $n + 1$ pigeons and a collection of n cages to hold them all. Then the Pigeon Hole Principle says that if every pigeon is held, then at least one cage must contain more than 1 pigeon (i.e., at least 2 pigeons). The Pigeon Hole Principle turns out to have a lot of applications in mathematics.

Recall that Theorem 3.27 tells us that if a nonempty set has a size, then this size is unique. Hence, we can talk about *the* size of the set. Suppose the set S is of size n. We can also express this by saying that the **cardinality** of S is n. We can also write this by the equation **card**$(S) = n$. Now we can use the term "**card**(S)" in formulae. For any nonempty set A, we say that *the cardinality of A is finite* provided that there is an $n \in P$ such that **card**$(S) = n$. In other words, the cardinality of A is finite iff there is an $n \in P$ such that $\mathbf{I}_n \approx A$.

Before concluding this section, we present a few more results. The proof of the next theorem provides a good opportunity to make doubly sure that the reader is understanding and creating mathematical induction proofs correctly. Hence, we will go over the induction carefully (one last time). We will have much more to say about counting in the next chapter.

Theorem 3.29

Suppose A is a nonempty set such that the cardinality of A is finite. Suppose $B \subset A$ with B nonempty. Then the cardinality of B is finite, and we have **card**$(B) <$ **card**(A).

Proof: To prove this by induction, define the statement $S(n)$ as follows for each $n \in P$: The statement $S(n)$ is defined to mean that "for *any* set X with **card**$(X) = n$, if Y is *any* nonempty proper subset of X, then the cardinality of Y is finite, and **card**$(Y) < n$."

As usual, we begin by establishing $S(1)$. Consider any set X with **card**$(S) = 1$. What do we have to show? We have to show that Y is *any* nonempty proper subset of X $\Rightarrow$ (cardinality of Y is finite and **card**$(Y) < 1$). However, we have this statement by default. This is because it is not possible for a set of size 1 to have any proper subsets! This establishes $S(1)$.

We now take the inductive step and assume that $S(n)$ holds for some $n \in P$. We must show that $S(n + 1)$ holds. To do that, consider any set W such that **card**$(W) = n + 1$. Now, let Z be any nonempty proper subset of W. Then we must show that the cardinality of Z is finite, and **card**$(Z) < n + 1$. We now break into cases.

Case 1: Z contains exactly 1 element. Clearly we have **card**$(Z) = 1 < n + 1$.

Case 2: Z does not contain exactly 1 element. Then, there must exist z_1, $z_2 \in Z$ with $z_1 \neq z_2$. Hence, $Z - \{z_1\}$ is nonempty. Because $Z \subset W$, we must have $Z - \{z_1\} \subset W - \{z_1\}$. By Lemma 3.5, we have **card**$(W - \{z_1\}) = n$. Note that we can now apply the inductive step. Doing so, there must exist $m \in P$ with $m < n$ such that **card**$(Z - \{z_1\}) = m$. The reader should be able to show that this implies that **card**$(Z) = m + 1$. Because $m + 1 < n + 1$, we have shown $S(n + 1)$.

Note that in both cases, we have shown $S(n + 1)$.

⬛

Finally, we conclude the section with a result that follows immediately from Theorem 3.29. The student should have no problem supplying the proof. A result that follows easily and directly from a theorem is often called a **corollary** to the theorem in higher mathematics. Of course, a corollary is also a theorem in its own right.

Corollary 3.1

Suppose A is a nonempty set such that the cardinality of A is finite. Suppose $B \subseteq A$ with **Card**$(B) = $ **Card**(A). Then $B = A$.

EXERCISES

1. Let $(P, s, 1)$ and $(P', s', 1')$ be any Peano spaces. Show that there exists only one Peano space isomorphism between P and P'.

2. Let $(P, s, 1)$ and $(P', s', 1')$ be any Peano spaces with Peano space isomorphism $f : P \rightarrow P'$. Suppose relations $<$ and $\leq$ on P are defined as in Definition 3.5, and let $\hat{<}$ and $\hat{\leq}$ denote the corresponding relations on P'. Show that for all $m, n \in P$, we have $m < n \Leftrightarrow f(m) \hat{<} f(n)$, and $m \leq n \Leftrightarrow f(m) \hat{\leq} f(n)$.

3. Complete the proof of Theorem 3.28 by showing that that Part 1 $\Rightarrow$ Part 2.

4. Suppose Dr. Rosen's discrete math class contains 367 students. Show that at least two students have birthdays falling on the same day.

5. Suppose sets A and B are disjoint, and suppose that both sets are of finite cardinality. Show that $\mathbf{card}(A \cup B) = \mathbf{card}(A) + \mathbf{card}(B)$.

6. Suppose set A is of finite cardinality, and suppose that the proper subset B of A is nonempty. Show $\mathbf{card}(A - B) = \mathbf{card}(A) - \mathbf{card}(B)$.

7. Let $(P, s, 1)$ be a Peano space. Suppose that $n \in P$ and $f: P \to \mathbf{I}_n$ Prove that f is not one-to-one.

8. Let $(P, s, 1)$ be a Peano space. Suppose that $n \in P$. Suppose that S is a nonempty set with $f: \mathbf{I}_n \to S$. Show that $f(\mathbf{I}_n)$ is of finite cardinality, with $\mathbf{card}(f(\mathbf{I}_n)) \leq n$.

9. Let $(P, s, 1)$ be a Peano space. Suppose that $n \in P$ and $f: \mathbf{I}_n \to P$. Prove that f is not onto.

Section 3.5 Introduction to Rings

At this point, we could go ahead and construct the set of integers (having already constructed the natural numbers). However, before doing this, our work will be made easier if we first develop a little bit of the theory of **rings**. As you will hopefully discover in later courses, the theory of rings is an important part of abstract algebra. You will shortly see that the sets **Z**, **Q**, and **R** constitute rings (with the appropriately defined binary operations of addition and multiplication). Therefore, any theorems that hold true for all rings will automatically hold for these special rings. Hence, there are results that we will only have to prove once, instead of proving that the results hold for each of these three special rings.

Suppose that we have elements a, b, c, d, e of any set. We define a **5-tuple**, denoted by (a, b, c, d, e), to be the set $((a, b, c), (d, e))$. That is, a 5-tuple is an ordered pair with the first component a triplet and the second component an ordered pair. We are now ready to define a ring.

Definition 3.12

Let S be a set with at least two distinct elements that we shall denote by 0 and 1. Suppose that '+' and '·' are distinct binary operations on S. Then the 5-tuple $(S, +, \cdot, 0, 1)$ is called a **ring**, provided all of the following properties hold:

1) The operation '+' is commutative and associative.

2) The operation '·' is associative.

3) $a + 0 = a \ \forall a \in S$.

4) $a \cdot 1 = 1 \cdot a = a \ \forall a \in S$.

5) For each $a \in S$, there exists an element of S, which we denote by $-a$, such that we have $a + -a = 0 \ \forall a \in S$.

6) $\forall a, b, c \in S$, $a \cdot (b + c) = (a \cdot b) + (a \cdot c)$, and $(b + c) \cdot a = (b \cdot a) + (c \cdot a)$.

Some comments/conventions are in order.

1) The binary operation '+' is called addition, and the binary operation '·' is called multiplication, even though these operations may have nothing to do with the addition and multiplication of real numbers. We often write ab instead of $a \cdot b$. Also, the elements 0 and 1 may have nothing to do with the real numbers 0 and 1, respectively. Although these elements could be called anything desired, they have been called 0 and 1 because their interactions with other ring elements are analogous to the interactions of the real numbers 0 and 1 with other real numbers.

2) We adhere to the same rules of precedence that apply to Peano spaces. (See the remarks following Definition 3.3.) Hence, for example, Property 6 of Definition 3.12 could be written $\forall a, b, c \in S$, $a(b + c) = ab + ac$, and $(b + c)a = ba + ca$.

3) Because addition is commutative, we have $0 + a = a \, \forall a \in S$, and $-a + a = 0 \, \forall a \in S$.

4) The element $-a$ is called an **additive inverse** of a.

5) You may have noticed that the definition of a ring does not specify that multiplication is commutative. There is, in fact, a whole branch of abstract algebra devoted to rings in which this operation fails to be commutative. Note that we specify that *both* $a \cdot 1$ and $1 \cdot a$ are equal to a for each $a \in S$.

6) Most advanced mathematics books do not define a ring to be a 5-tuple. Instead, they informally define a ring to be a set along with the specified properties. Therefore, our definition is more rigorous because we make clear that two rings are identical iff the corresponding sets are equal, the corresponding binary operations are equal, and the corresponding 0 and 1 elements are equal. Given a ring $(S, +, \cdot, 0, 1)$, advanced texts usually refer to the set S as the ring. We will sometimes be "guilty" of this as well.

7) The reader should take note of the fact that if $(S, +, \cdot, 0, 1)$ is a ring, then $(S, +, 0)$ is a group. (See Definition 2.31.)

We will shortly provide some examples of rings. Before this, we present some simple but important consequences of the definition.

Theorem 3.30   Suppose $(S, +, \cdot, 0, 1)$ is a ring. Suppose the element $0' \in S$ obeys Property 3 of Definition 3.12, so that $a + 0' = a \, \forall a \in S$. Then $0' = 0$.

Proof: Applying Property 3 with $a = 0'$, we have $0' + 0 = 0'$. Because $0'$ also satisfies Property 3, applying $0'$ to Property 3 with $a = 0$ gives $0 + 0' =$

0. Applying Property 1 to this latter equation results in $0' + 0 = 0$. Because we just showed that $0' + 0 = 0'$, we must have $0' = 0$.

| Theorem 3.31 |

Suppose $(S, +, \cdot, 0, 1)$ is a ring. Suppose the element $1' \in S$ obeys the property that $a \cdot 1' = a \ \forall a \in S$. Then $1' = 1$.

Proof: Applying Property 4 gives us $1' \cdot 1 = 1 \cdot 1' = 1'$. Applying the property listed in the theorem with $a = 1$ gives us $1 \cdot 1' = 1$. The result follows immediately.

| Theorem 3.32 |

Suppose $(S, +, \cdot, 0, 1)$ is a ring. Suppose $a \in S$. Suppose $b \in S$ with $a + b = 0$. Then $b = -a$.

Proof: Add $-a$ to both sides of the equation $a + b = 0$. This gives us $-a + (a + b) = -a + 0$, which we call Equation 1. By Property 3, the right-hand side of Equation 1 is $-a$. By the associativity of addition, the left-hand side of Equation 1 is $(-a + a) + b$, which is $0 + b$ by Property 5, which is b by Property 3 and the commutativity of addition. The result follows.

Theorem 3.30 shows that the element 0 is unique. It is frequently called the **zero element** of the ring. Theorem 3.31 shows that the element 1 is unique. It is frequently called the **unit element** of the ring. Theorem 3.32 shows that each $a \in S$ has a unique additive inverse. Therefore, $-a$ can be called *the* additive inverse of a rather than *an* additive inverse of a.

| Theorem 3.33 |

Suppose $(S, +, \cdot, 0, 1)$ is a ring. Then $a \cdot 0 = 0 \cdot a = 0 \ \forall a \in S$.

Proof: We show that $a \cdot 0 = 0$. The proof that $0 \cdot a = 0$ is left to the reader. We have $0 = 0 + 0$ by Property 3. Hence, $a \cdot 0 = a(0 + 0)$, which is $a \cdot 0 + a \cdot 0$ by Property 6. Hence, so far we have the equation $a \cdot 0 = a \cdot 0 + a \cdot 0$. Thus, $-(a \cdot 0) + a \cdot 0 = -(a \cdot 0) + (a \cdot 0 + a \cdot 0)$. The left-hand side of this equation is 0 by Property 5. The right-hand side is $(-(a \cdot 0) + a \cdot 0) + a \cdot 0$ (by Property 1), which is $0 + a \cdot 0$ (by Property 5), which is $a \cdot 0$ (by Property 3). We have shown that $0 = a \cdot 0$.

The proof of the next theorem is left to the reader.

Theorem 3.34 Suppose $(S, +, \cdot, 0, 1)$ is a ring. Then $-(-a) = a \ \forall a \in S$.

Theorem 3.35 Suppose $(S, +, \cdot, 0, 1)$ is a ring. Then $(-1) \cdot a = a \cdot (-1) = -a \ \forall a \in S$.

Proof: We show that $(-1) \cdot a = -a$. The proof that $a \cdot (-1) = -a$ is left to the reader.

$$\begin{aligned}
(-1) \cdot a + a &= (-1) \cdot a + 1 \cdot a &&\text{(by Property 4)} \\
&= (-1 + 1)a &&\text{(by Property 6)} \\
&= 0 \cdot a &&\text{(by Property 5)} \\
&= 0 &&\text{(by Theorem 3.33)}.
\end{aligned}$$

We have shown that $(-1) \cdot a + a = 0$. Now adding $-a$ to both sides of this equation gives the equation $((-1) \cdot a + a) + -a = 0 + -a$. Now the right-hand side of this latter equation is $-a$ by Properties 1 and 3. The left-hand side is $(-1) \cdot a + (a + -a)$ (by Property 1), which is $(-1) \cdot a + 0$ (by Property 5), which is $(-1) \cdot a$ (by Property 3).

Theorem 3.36 Suppose $(S, +, \cdot, 0, 1)$ is a ring. Then $(-a)b = a(-b) = -(ab) \ \forall a, b \in S$.

Proof: We show that $(-a)b = -(ab)$. The proof that $a(-b) = -(ab)$ is left to the reader. We provide the steps to show $(-a)b = -(ab)$, but we ask the reader to fill in the justifications. $0 = 0 \cdot b = (a + -a) \cdot b = ab + (-a)b$. The result follows by adding $-(ab)$ to both sides of the equation.

The proof of the following theorem is left to the reader.

Theorem 3.37 Suppose $(S, +, \cdot, 0, 1)$ is a ring. Then $(-a)(-b) = ab \ \forall a, b \in S$.

Now, just as it is useful to have the concept of subtraction of real numbers, it is also useful to have this concept for rings in general.

Definition 3.13

Suppose $(S, +, \cdot, 0, 1)$ is a ring. Then we define $a - b$ to be $a + -b$ $\forall a, b \in S$.

The proof of the following theorem is left to the reader.

Theorem 3.38 Suppose $(S, +, \cdot, 0, 1)$ is a ring. Then $a - (b + c) = (a - b) - c$ $\forall a, b, c \in S$.

Before presenting some examples of rings, we want to make an important point concerning equations. Often in mathematics you will be presented with an equation that contains an unknown, and you will attempt to solve for this unknown. For example, suppose $(S, +, \cdot, 0, 1)$ is a ring, with $a, b \in S$. Consider the equation with unknown value x: $a + x = b$. In this context, the unknown x is frequently referred to as a **variable**. (Of course, we could use other symbols besides x to denote the variable.) To **solve** the equation means to provide the set of values that can be substituted for the variable that makes the equation hold true. Any such value is called a **solution**. The set of such values is called the **solution set**.

Let A denote the solution set for the equation $a + x = b$. Then $A = \{x \in S | a + x = b\}$. Now, we might attempt to solve the equation as follows: $-a + (a + x) = -a + b$, so $(-a + a) + x = -a + b$, so $0 + x = -a + b$, so $x = -a + b$. Does this prove that $-a + b$ is the one and only solution to the equation? No, it does not. It proves *that if there is a solution*, then the solution is unique with value $-a + b$. In other words, it proves the following implication: x a solution $\Rightarrow x = -a + b$. We have yet to show the converse. In other words, we have not yet shown that $x = -a + b \Rightarrow x$ is a solution. To do this, we must "plug" this value of x into the equation and show this solves the equation. That is, we must show that $a + (-a + b) = b$. The reader can easily show that this equation holds. Only then can we conclude that the solution set is $A = \{-a + b\}$. Of course, our example equation was very simple, but the same principle applies for more complicated equations.

Now we provide some examples of rings.

Example 3.4 Let $S = \{a, b\}$ (i.e., S is the set consisting of the first two letters of the alphabet). We define binary operations on S as follows: $a + a = a$, $a + b = b$, $b + a = b$, $b + b = a$, $a \cdot a = a$, $a \cdot b = a$, $b \cdot a = a$, and $b \cdot b = b$. The reader is now asked to work through all the details to show that this forms a ring with zero element a and unit element b.

Example 3.5

The reader should immediately see that **Z**, **Q**, and **R** constitute rings with respect to the "usual" addition and multiplication.

Example 3.6

(For students who have had linear algebra; otherwise, disregard and do not worry.)

Let $n \in \mathbf{N}$, and let M_n denote the set of real $n \times n$ matrices. Then the student should see that M_n forms a ring with respect to matrix addition and multiplication. The ring's zero element is the matrix

$$\begin{bmatrix} 0 & 0 & 0 & \cdots & 0 \\ 0 & 0 & 0 & \cdots & 0 \\ 0 & 0 & 0 & \cdots & 0 \\ \vdots & \vdots & \vdots & \ddots & 0 \\ 0 & 0 & 0 & \cdots & 0 \end{bmatrix}.$$

The ring's unit element is the matrix

$$\begin{bmatrix} 1 & 0 & 0 & \cdots & 0 \\ 0 & 1 & 0 & \cdots & 0 \\ 0 & 0 & 1 & \cdots & 0 \\ \vdots & \vdots & \vdots & \ddots & 0 \\ 0 & 0 & 0 & \cdots & 1 \end{bmatrix}.$$

The student should also recall that multiplication is noncommutative for $n > 2$ because there exists $A, B \in M_n$ such that $AB \neq BA$. For example,

$$\begin{bmatrix} 1 & 2 \\ 3 & 4 \end{bmatrix}\begin{bmatrix} 5 & 6 \\ 7 & 8 \end{bmatrix} = \begin{bmatrix} 19 & 22 \\ 43 & 50 \end{bmatrix}, \text{ but } \begin{bmatrix} 5 & 6 \\ 7 & 8 \end{bmatrix}\begin{bmatrix} 1 & 2 \\ 3 & 4 \end{bmatrix} = \begin{bmatrix} 23 & 34 \\ 31 & 46 \end{bmatrix}.$$

Example 3.7

Here we provide a whole class of examples. Let $(S, +, \cdot, 0, 1)$ be a ring, and let X be any nonempty set. Let F be the set of all functions with domain X and range S. That is, $F = \{f \,|\, f: X \to S\}$. We define binary operations on F that make it into a ring. For each $f, g \in F$, we define the function $f \oplus g: X \to S$ as follows: $(f \oplus g)(x) = f(x) + g(x) \; \forall x \in X$. For each $f, g \in F$, we define the function $f \otimes g: X \to S$ as follows: $(f \otimes g)(x) = f(x) \cdot g(x) \; \forall x \in X$.

What shall we use for the zero element and unit element? For the zero element, we propose the function $0_F: X \to S$, which we define by $0_F(x) = 0 \; \forall x \in X$. For the unit element, we propose the function $1_F: X \to S$, which we define by $1_F(x) = 1 \; \forall x \in X$. What shall we use for the additive inverse of a given $f \in F$? We propose the function $-f: X \to S$, which we define by $(-f)(x) = -(f(x)) \; \forall x \in X$.

The reader is asked to complete the proof that $(F, \oplus, \otimes, 0_F, 1_F)$ is a ring. We get the reader started by showing that the ring's addition, $\oplus$, is commutative. We must show that $f \oplus g = g \oplus f \; \forall f, g \in F$. We note that the functions $f \oplus g$ and $g \oplus f$ both have domain X and range F. Hence, to complete the proof that $f \oplus g = g \oplus f$, it suffices to show that $(f \oplus g)(x) = (g \oplus f)(x) \; \forall x \in X$. Now,

$$
\begin{aligned}
(f \oplus g)(x) &= f(x) + g(x) && \text{(by definition of } f \oplus g) \\
&= g(x) + f(x) && \text{(because `+' is commutative)} \\
&= (g \oplus f)(x) && \text{(by definition of } g \oplus f).
\end{aligned}
$$

Note how $(F, \oplus, \otimes, 0_F, 1_F)$ "inherits" its ring properties from the ring properties of $(S, +, \cdot, 0, 1)$.

$\blacksquare$

An important notation should be made concerning Example 3.7. Note how we denoted addtion of elements of F by '$\oplus$' to distinguish this binary operation from the binary operation '$+$' of addition of elements of S. Similarly, we denoted multiplication of elements of F by '$\otimes$'. It should be noted that in most advanced mathematics books, $f \oplus g$ would be written as $f + g$. In other words, the symbol '$+$' would be used to denote both binary operations. The authors of these books assume that the reader can determine from the context which binary operation is meant. Similarly, $f \otimes g$ would be written as fg or $f \cdot g$. Although such "abuse of notation" could possibly lead to confusion, in general the context is strong enough to prevent confusion, and readability is enhanced. Therefore, the student is strongly encouraged to get used to these types of "abuses."

Next, we introduce the concept of a **ring isomorphism**.

Definition 3.14

Suppose we have any two rings $(S, +, \cdot, 0, 1)$ and $(S', \oplus, \otimes, 0', 1')$. Suppose that there is a bijective function $f: S \to S'$ such that $f(a + b) = f(a) \oplus f(b)$ and $f(ab) = f(a) \otimes f(b) \; \forall a, b \in S$. Then f is said to be a **ring isomorphism**, and the rings $(S, +, \cdot, 0, 1)$ and $(S', \oplus, \otimes, 0', 1')$ are said to be **isomorphic**.

Intuitively, saying that two rings are isomorphic means that they have the same structure with respect to their additions and multiplications. They are, in essence, the same ring, in the sense that they "behave" the same except that the elements of S and S' may be different. The reader should immediately see the analogy with Peano space isomorphisms (see Definition 3.6). Given this analogy, the following theorem should come as no surprise. In the Exercises, the reader is asked to show that if $f: S \to S'$ is a ring isomorphism, then so is $f^{-1}: S' \to S$. We start using ring isomorphisms in the next section.

Theorem 3.39 Suppose the rings $(S, +, \cdot, 0, 1)$ and $(S', \oplus, \otimes, 0', 1')$ are isomorphic with ring isomorphism $f: S \to S'$. Then

1) $f(0) = 0'$.
2) $f(1) = 1'$.
3) $f(-a) = -f(a) \; \forall a \in S$.

Proof: We show that $f(1) = 1'$, and the rest is left to the reader. By Theorem 3.31, it suffices to show that $b \otimes f(1) = f(1) \otimes b = b \; \forall b \in S'$. Let $b \in S'$. Then, because f is onto, there exists an $a \in S$ such that $b = f(a)$.

Now, $b \otimes f(1) = f(a) \otimes f(1) = f(a \cdot 1) = f(a) = b$. Also, we have $f(1) \otimes b = f(1) \otimes f(a) = f(1 \cdot a) = f(a) = b$. ∎

It should be noted that most advanced mathematics books also "abuse" notation with regard to isomorphisms by writing $f(a + b) = f(a) + f(b)$ and $f(ab) = f(a) \cdot f(b)$ instead of $f(a + b) = f(a) \oplus f(b)$ and $f(ab) = f(a) \otimes f(b)$, respectively. The authors of these books assume that the reader can determine from the context which binary operation is meant.

Next, we consider the concept of a **subring**. To motivate the discussion, let S be any nonempty set, and consider any binary operation on S. Let us denote one by '*'. Let A be any nonempty subset of S. Recall that * is (by definition) a mapping with domain $S \times S$ and range S. Now consider the restriction of * to $A \times A$ (which we can write as $*|A \times A$). Is this restriction a binary operation on A? Not necessarily. Why? Because it is possible that there exists $a, b \in A$ such that $a * b \notin A$.

However, if we have the situation in which we have $a * b \in A$ for all $a, b \in A$, then $*|A \times A$ is a binary operation on A. We express this by saying that the binary operation * is **closed** on A. We list this as a formal definition.

Definition 3.15

Suppose * is a binary operation on a nonempty set S, and suppose that A is a nonempty subset of S. Then we say that * is **closed** on A provided $a * b \in A$ for all $a, b \in A$. We can also express this by saying that A is closed with respect to *.

Example 3.8 Consider the "normal" addition and multiplication defined on $\mathbf{Z}$. Let $A = \{2k \mid k \in \mathbf{Z}\}$. Then we see that A is closed with respect to this addition because the sum of two even integers is also even. We also see that A is closed with respect to multiplication. Let $B = \mathbf{Z} - \{0\}$. Then B is not closed with respect to addition because $2, -2 \in B$, but $2 + -2 = 0 \notin B$. However, B is closed with respect to multiplication. Let $C = \{0, 1\}$. Note that C is closed with respect to multiplication, but not with respect to addition.

Theorem 3.40   Suppose $(S, +, \cdot, 0, 1)$ is a ring, and suppose that T is a nonempty subset of S satisfying the following conditions:

1) T is closed with respect to addition.
2) T is closed with respect to multiplication.
3) $0 \in T$.
4) $1 \in T$.
5) $-t \in T$ for each $t \in T$.

Let $+ \,|\, T \times T$ and $\cdot \,|\, T \times T$ denote the restrictions of addition and multiplication, respectively, to $T \times T$. Then $(T, + \,|\, T \times T, \cdot \,|\, T \times T, 0, 1)$ is a ring. It is called a **subring** of $(S, +, \cdot, 0, 1)$.

Proof: The reader should easily see that each of Properties 1 through 6 of Definition 3.12 is satisfied by $(T, + \,|\, T \times T, \cdot \,|\, T \times T, 0, 1)$.

It should be noted that in almost all advanced mathematics texts, notation is "abused" by saying that T is a subring of S, rather than writing the unwieldly expression that $(T, + \,|\, T \times T, \cdot \,|\, T \times T, 0, 1)$ is a subring of $(S, +, \cdot, 0, 1)$. This "unwieldliness" not withstanding, the main points are that a subring is closed with respect to addition and multiplication, the subring contains 0 and 1, and it contains the additive inverse of each of its members.

Example 3.9   It is clear (once we define these sets) that the rings **Z** and **Q** are subrings of **R** and that **Z** is a subring of **Q**. The set A of Example 3.8 fails to be a ring only because $1 \notin A$.

We conclude with the definitions of two very special and important classes of rings.

Definition 3.16

The ring $(S, +, \cdot, 0, 1)$ is said to be a **commutative ring** if it satisfies Property 7: Multiplication is commutative.

Definition 3.17

The commutative ring $(D, +, \cdot, 0, 1)$ is said to be an **integral domain** if it satisfies Property 8: $\forall a, b \in D$, $ab = 0 \Rightarrow (a = 0$ or $b = 0)$.

Hence, a commutative ring is an integral domain provided the only way that a product of elements can be zero is if at least one of the elements in the product is zero. Note that all rings obey Properties 1 through 6, commutative rings obey Properties 1 through 7, and integral domains obey Properties 1 through 8. (Recall that Properties 1–6 are given in Definition 3.12, Property 7 is given in Definition 3.16, and Property 8 is given in Definition 3.17). It is clear that any subring of a commutative ring is a commutative ring. It is also clear that any subring of an integral domain is also an integral domain. Most rings of concern to us in the remainder of this book will be integral domains.

Example 3.10

The rings $\mathbf{Z}$, $\mathbf{Q}$, and $\mathbf{R}$ are integral domains.

Example 3.11

(For students who have had linear algebra; otherwise, disregard and do not worry.)

The ring M_n of Example 3.6 is not an integral domain for $n > 1$. Not only does it fail to satisfy Property 7, but it also fails to satisfy Property 8. For example, we have

$$\begin{bmatrix} 0 & 1 \\ 0 & -1 \end{bmatrix}\begin{bmatrix} 1 & 0 \\ 0 & 0 \end{bmatrix} = \begin{bmatrix} 0 & 0 \\ 0 & 0 \end{bmatrix},$$

even though neither of

$$\begin{bmatrix} 0 & 1 \\ 0 & -1 \end{bmatrix} \text{ and } \begin{bmatrix} 1 & 0 \\ 0 & 0 \end{bmatrix}$$

is the zero of the ring.

The proof of the following theorem is left to the reader. Note that the theorem applies to integral domains and not all rings.

Theorem 3.41   (Cancellation Law of Multiplication)

Suppose $(D, +, \cdot, 0, 1)$ is an integral domain. Suppose $a, b, c \in D$ such that $ac = bc$ and $c \neq 0$. Then $a = b$. ∎

EXERCISES

1. Prove Theorem 3.34.

2. Complete the proof of Theorem 3.35.

3. Complete the proof of Theorem 3.36.

4. Prove Theorem 3.37.

5. Prove Theorem 3.38.

6. Complete the proof that $(F, \oplus, \otimes, 0_F, 1_F)$ of Example 3.7 is a ring.

7. Suppose the rings $(S, +, \cdot, 0, 1)$ and $(S', \oplus, \otimes, 0', 1')$ are isomorphic and $(S, +, \cdot, 0, 1)$ is an integral domain. Show that $(S', \oplus, \otimes, 0', 1')$ is an integral domain.

8. Suppose the rings $(S, +, \cdot, 0, 1)$ and $(S', \oplus, \otimes, 0', 1')$ are isomorphic with ring isomorphism $f: S \to S'$. Show that the inverse mapping $f^{-1}: S' \to S$ is a ring isomorphism.

9. Complete the proof of Theorem 3.39.

10. Prove Theorem 3.41.

11. Suppose $(S, +, \cdot, 0, 1)$ is an integral domain, and suppose $a, b \in S$ with $a \neq 0$. Suppose the equation $ax = b$ with variable x has a solution. Show that this solution is unique.

12. Consider the ring **R**. Define the set S by $S = \{m + n\sqrt{2} | m, n \in \mathbf{Z}\}$. Show (using your prior knowledge of real numbers) that S is a subring of **R**.

Section 3.6   Postulates for the Integers and Initial Properties

In this section, we begin by introducing some preliminary concepts concerning orders and integral domains. Using some of these concepts, we derive many of the important properties of inequalities that the student is used to. For example, from your work with real numbers, you are "used to" the result that for any real numbers a, b, and c, if $a \leq b$, then $a + c \leq b + c$.

Next, we list the postulates obeyed by integers. These are "starting points" that suffice for the purpose of deriving all of the theorems about integers that we need. Next, we derive some initial properties of integers using these postulates. We show that any set obeying these postulates must contain a subset that obeys the Peano Postulates. (This subset is the positive integers, of course.) We also prove an important isomorphism theorem.

In Section 3.7, using our previous work with Peano spaces as a starting point, we demonstrate that there actually exists a set that satisfies these postulates for the integers. Doing this keeps our logical integrity. That is, we do not need to "axiomitize" into existence a set that obeys these postulates. We already have the ammunition to demonstrate that such a set exists. This proof of the existence of such a set will not be needed anywhere else in the book, so Section 3.7 has been designated "optional." However, the student is encouraged to read the proof as it makes elegant usage of the concept of equivalence classes.

Lower and Upper Bounds with Respect to an Order

Before proceeding, it may be helpful to the student to review the first part of Section 3.3 (i.e., the portion concerned with linear orders). We introduce additional concepts about linear orders here, some of which are of immediate use to us in the construction of the integers. Also, all of these concepts prove vital later.

Definition 3.18

Suppose $\leq$ is a linear order on a nonempty set S, and let $A \subseteq S$ with A nonempty. An element $l \in S$ is called a **lower bound** of A, provided $l \leq a \ \forall a \in A$. An element $u \in S$ is called an **upper bound** of A, provided $a \leq u \ \forall a \in A$. If A has a lower bound, it is said to be **bounded below**. Similarly, if A has an upper bound, it is said to be **bounded above**. If A is both bounded below and bounded above, it is said to be **bounded**.

Of course, by writing "a lower bound of A" and "an upper bound of A," we mean "with respect to $\leq$." If l is a lower bound of A, we often say that A is bounded below by l. If u is an upper bound of A, we often say that A is bounded above by u. The first thing to notice about Definition 3.18 is that we do not require that lower and upper bounds of A be members of A (although they must be elements of S, of course).

Example 3.12

Consider the normal "less than or equal to" linear order $\leq$ on $\mathbf{R}$ that we are all familiar with. Define sets A_1, A_2, and A_3 by $A_1 = \{x \in \mathbf{R} | 0 < x \leq 1\}$, $A_2 = \{x \in \mathbf{R} | x \leq -3\}$, and $A_3 = \{x \in \mathbf{R} | x > 100\}$. Note that 0 is a lower bound of A_1, although $0 \notin A_1$. Of course, a set could potentially have many lower bounds. For example, -75 is also a lower bound of A_1. Note that 1 is an upper bound of A_1, and we also have $1 \in A_1$. Of course, a set could potentially have many upper bounds. For example, 900 is also an upper bound of A_1. Note that A_1 is bounded, as it is both bounded below and bounded above.

Note that A_2 is not bounded below, but it is bounded above. Indeed, -3 is an upper bound. Note that A_3 is bounded below. Indeed, 100 is a lower bound. Note that A_3 is not bounded above.

In Example 3.12, note that 1 is the maximum (i.e., greatest element) of A_1. This is a simple consequence of the following theorem. The proof is trivial and is left to the reader.

Theorem 3.42 Suppose $\leq$ is a linear order on a nonempty set S, and let $A \subseteq S$ with A non-empty. Suppose $l \in S$ is a lower bound of A and $l \in A$. Then l is the minimum of A. Suppose $u \in S$ is an upper bound of A and $u \in A$. Then u is the maximum of A. ∎

Notational Notes:

1) From now on, whenever we are working with a linear order on some set S, if we use the symbol $\leq$ to denote the linear order, then by the expression $u < v$ (for $u, v \in S$), we mean $u \leq v$ and $u \neq v$. We can also write $u \leq v$ as $v \geq u$, and we can write $u < v$ as $v > u$. Note that the definition of $<$ and $>$, along with the definition of a linear order, allows us to assert the following: $\forall u, v \in S$, exactly one of the following three conditions hold: 1) $u < v$, 2) $u > v$, or 3) $u = v$. (Why?) We will use this result over and over without formal reference.

2) From now on, whenever we are working with a linear order $\leq$ on some set S, then if $u, v, w \in S$, by the expression $u \leq v \leq w$, we mean $u \leq v$ and $v \leq w$. This makes sense because if $u \leq v$ and $v \leq w$, then we also have $u \leq w$ by the transitive property of linear orders. Similarly, by the expression $u \leq v < w$, we mean $u \leq v$ and $v < w$. Note that if $u \leq v < w$, then $u < w$. (Why?) By the expression $u < v \leq w$, we mean $u < v$ and $v \leq w$. The reader should prove that $u < v \leq w \Rightarrow u < w$. Finally, by the expression $u < v < w$, we mean $u < v$ and $v < w$. The reader should prove that $u < v < w \Rightarrow u < w$. We will use these results over and over without formal reference.

Ordered Integral Domains

Before listing the postulates for integers, it will be useful to introduce one more concept.

Definition 3.19

Suppose $(D, +, \cdot, 0, 1)$ is an integral domain with the additional property that there is a linear order $\leq$ on D such that

$\forall a, b, c \in D, a \leq b \Rightarrow a + c \leq b + c.$

$\forall a, b, c \in D, (a \leq b \text{ and } c > 0) \Rightarrow ac \leq bc.$

Then $(D, +, \cdot, 0, 1)$ is called an **ordered integral domain** (with respect to $\leq$, of course).

It will be clear (once we define them) that the rings **Z**, **Q**, and **R** are ordered integral domains. Therefore, any theorem that we prove about arbitrary ordered integral domains will hold for these three special ordered integral domains. The proofs of Theorems 3.44, 3.48, and 3.50 are left to the reader.

Theorem 3.43   Suppose $(D, +, \cdot, 0, 1)$ is an ordered integral domain with respect to $\leq$, and suppose $a, b, c \in D$. Then $a < b$ iff $a + c < b + c$.

Proof: Suppose $a < b$. Then $a \leq b$, so $a + c \leq b + c$. Suppose (to reach a contradiction) that $a + c = b + c$. Then, we can add $-c$ to both sides of this equation and simplify to get $a = b$, which contradicts the statement $a < b$. Hence, $a + c < b + c$.

   Conversely (i.e., to prove the converse), suppose $a + c < b + c$. Then, by the first part of this proof, we have $(a + c) + (-c) < (b + c) + (-c)$. The result follows by simplifying this expression.

Theorem 3.44   Suppose $(D, +, \cdot, 0, 1)$ is an ordered integral domain with respect to $\leq$, and suppose $a, b, c \in D$. Then $a \leq b$ iff $a + c \leq b + c$.

Theorem 3.45   Suppose $(D, +, \cdot, 0, 1)$ is an ordered integral domain with respect to $\leq$, and suppose $a, b, c \in D$. If $a \leq b$ and $c > 0$, then $a < b + c$.

Proof: Because $0 < c$, we have $b + 0 < b + c$, by Theorem 3.43. So, $b < b + c$. Hence, we have $a \leq b < b + c$, so the result follows.

The proof of the next theorem provides us with an opportunity to introduce another proof technique. This probably could have been introduced earlier, but this seems to be a good place to do it. Suppose we have three statements P_1, P_2, P_3, and we want to show that they are all logically equivalent to one another. Then we have to show that $P_1 \Rightarrow P_2$, $P_2 \Rightarrow P_1$, $P_1 \Rightarrow P_3$, $P_3 \Rightarrow P_1$, $P_2 \Rightarrow P_3$, and $P_3 \Rightarrow P_2$. However, we might be able to avoid some of this work! It suffices to show that $P_1 \Rightarrow P_2$, $P_2 \Rightarrow P_3$, and $P_3 \Rightarrow P_1$.

   The reason why this works is because, for any statements A, B, C, we have $(A \Rightarrow B$ and $B \Rightarrow C) \Rightarrow (A \Rightarrow C)$. So, for example, if we have shown that $P_1 \Rightarrow P_2$, $P_2 \Rightarrow P_3$, and $P_3 \Rightarrow P_1$, why can we conclude that $P_3 \Rightarrow P_2$? By using the argument $(P_3 \Rightarrow P_1$ and $P_1 \Rightarrow P_2) \Rightarrow (P_3 \Rightarrow P_2)$. Using this reasoning, the reader should work through the argument to be convinced that the remaining statements needed are also true. The beauty of this is that it works for any number of statements. For example, if we want to show the logical equivalence of statements P_1, P_2, P_3, P_4, it suffices to show that $P_1 \Rightarrow P_2$, $P_2 \Rightarrow P_3$, and $P_3 \Rightarrow P_4$, and $P_4 \Rightarrow P_1$.

Theorem 3.46

Suppose $(D, +, \cdot, 0, 1)$ is an ordered integral domain with respect to $\leq$, and suppose $a, b \in D$. Then the following three statements are equivalent:

1) $a < b$
2) $0 < b - a$.
3) $-b < -a$.

Proof: To show that Part 1 $\Rightarrow$ Part 2, suppose $a < b$. Recall the definition of subtraction (Definition 3.13). By Theorem 3.43, we have $a - a < b - a$, so $0 < b - a$. To show that Part 2 $\Rightarrow$ Part 3, suppose $0 < b - a$. Then, by Theorem 3.43, we have $-b + 0 < -b + (b - a)$, which simplifies to $-b < -a$. Finally, we show that Part 3 $\Rightarrow$ Part 1. Suppose $-b < -a$. First add b to both sides (using Theorem 3.43) to get $0 < b - a$. Now, invoke Theorem 3.43 again, this time adding a to both sides to get $a < b$.

Theorem 3.47

Suppose $(D, +, \cdot, 0, 1)$ is an ordered integral domain with respect to $\leq$, and suppose $a, b, c \in D$ with $c > 0$. Then $a < b$ iff $ac < bc$.

Proof: Suppose $a < b$. Then, of course we have $a \leq b$. By Definition 3.19, $ac \leq bc$. Hence, $ac < bc$ or $ac = bc$. Suppose (to reach a contradiction) we have $ac = bc$. Then $a = b$ by Theorem 3.41. However, this contradicts the earlier statement that $a < b$. Thus, $ac < bc$.

Conversely, suppose that $ac < bc$. Now there are only three possibilities: $a > b$, $a = b$, $a < b$. If $a > b$, then by the first part of the proof, we have $ac > bc$. (Why?) Because this contradicts the statement $ac < bc$, the possibility that $a > b$ can be eliminated. If $a = b$, then $ac = bc$; also a contradiction. Hence, we must have $a < b$.

Theorem 3.48

Suppose $(D, +, \cdot, 0, 1)$ is an ordered integral domain with respect to $\leq$, and suppose $a, b, c \in D$ with $c > 0$. Then $a \leq b$ iff $ac \leq bc$.

| **Theorem 3.49** | Suppose $(D, +, \cdot, 0, 1)$ is an ordered integral domain with respect to $\leq$, and suppose $a, b \in D$ with $a > 0$ and $b > 0$. Then $ab > 0$. |

Proof: Note that we have $0 < a$ and $b > 0$. Hence, by Theorem 3.47, we have $0 \cdot b < ab$. Because $0 \cdot b = 0$ (by Theorem 3.33), the desired result follows.

| **Theorem 3.50** | Suppose $(D, +, \cdot, 0, 1)$ is an ordered integral domain with respect to $\leq$, and suppose $a, b, c, d \in D$ with $0 < a < b$ and $0 < c < d$. Then $ac < bd$. |

| **Theorem 3.51** | Suppose $(D, +, \cdot, 0, 1)$ is an ordered integral domain with respect to $\leq$, and suppose $a, b \in D$ with $a < 0$ and $b < 0$. Then $ab > 0$. |

Proof: By Theorem 3.46, we have $-a > 0$ and $-b > 0$. Hence, $(-a)(-b) > 0$ by Theorem 3.49. The result follows by Theorem 3.37.

| **Theorem 3.52** | Suppose $(D, +, \cdot, 0, 1)$ is an ordered integral domain with respect to $\leq$, and suppose $a \in D$ with $a \neq 0$. Then $a \cdot a > 0$. |

Proof: Because $a \neq 0$, either $a > 0$ or $a < 0$. If $a > 0$, the result follows by Theorem 3.49. If $a < 0$, the result follows by Theorem 3.51.

It should be noted that after exponents are defined (in the next chapter), we normally write $a \cdot a$ by a^2. It should be noted that Theorem 3.52 guarantees that $1 > 0$. (Why?) Finally, suppose $(D, +, \cdot, 0, 1)$ is an ordered integral domain with respect to $\leq$, and suppose $a \in D$. It should come as no surprise to the reader that we refer to a as **positive** if $a > 0$ and as **negative** if $a < 0$.

Integer Spaces

We are now ready to specify the postulates that identify the integers. In fact, we have already listed most of them. We need only identify the integers as a special ordered integral domain by adding one more property. This additional property will guarantee that for each integer n, there are no integers between n and $n + 1$. This property is to be contrasted with the property of the set real numbers in which there is a real number between any two given real numbers.

Definition 3.20

An ordered integral domain $(I, +, \cdot, 0, 1)$ (with respect to $\leq$) is called an **integer space**, provided that it satisfies the additional property that any nonempty subset of I that is bounded below has a minimum (i.e., a least element).

The reader should revisit Example 3.12 to see that this property fails to hold for the set of real numbers. For example, the subset A_1 is bounded below, but A_1 does not have a minimum.

From now on, we usually use the properties of ordered integral domains derived above without specific reference. We also stop writing "with respect to $\leq$," except when there is the possibility of ambiguity.

It turns out that any nonempty subset of I that is bounded above has a maximum (greatest element). This should not be surprising to the student who has hopefully worked with integers for many years. For example, the set $\{\ldots, -1, 0, 1, 2\}$, which is clearly bounded above, has the number 2 as its maximum.

Theorem 3.53

Suppose $(I, +, \cdot, 0, 1)$ is an integer space. Then every nonempty subset of I that is bounded above has a maximum.

Proof: Suppose A is a nonempty subset of I that is bounded above. Then by definition, there exists a $u \in I$ such that $a \leq u$ for all $a \in A$. Now define $B \subseteq I$ by $B = \{-a | a \in A\}$. Note that B is nonempty because A is. Now suppose $b \in B$. Then, by definition, $b = -a$ for some $a \in A$. Now, because $a \leq u$, we have $-u \leq -a$; that is, $-u \leq b$. This shows that $-u$ is a lower bound of B. Because B is a nonempty subset of I that is bounded below, B has a minimum (by the definition of an integer space). Let us denote this minimum by m.

Hence, by the definition of minimum (Definition 3.4), $m \in B$, and $m \leq b \;\forall b \in B$. Note that this is equivalent to saying that $-m \in A$ and $a \leq -m \;\forall a \in A$. Hence, $-m$ is the maximum of A.

To prove that for each integer n, there are no integers between n and $n + 1$, we first establish the following lemma.

Lemma 3.6

Suppose $(I, +, \cdot, 0 \; 1)$ is an integer space. Then for each $n \in I$, $n > 0 \Rightarrow n \geq 1$.

Proof: To do a proof by contradiction, assume that there exists an $n \in I$ such that $n > 0$, but $n < 1$; that is, $0 < n < 1$. Define the set A by

$A = \{n \in I | 0 < n < 1\}$. Then A is nonempty. Because A is bounded below (by 0), it must have a minimum, which we shall denote by m. Because $m \in A$, we must have $0 < m < 1$. Then $0 < m \cdot m < m < 1$. That is, $m \cdot m \in A$ with $m \cdot m < m$. However, this contradicts the fact that m is the minimum of A. Thus, A must be the empty set, and the result follows.

Theorem 3.54 Suppose $(I, +, \cdot, 0, 1)$ is an integer space. Then for each $n \in I$, there does not exist an $m \in I$ such that $n < m < n + 1$.

Proof: Let $n \in I$. Suppose (to reach a contradiction) there does exist an $m \in I$ such that $n < m < n + 1$. Then we have $0 < m - n < 1$. This contradicts Lemma 3.6.

The proof of the following corollary is immediate and is left to the reader.

Corollary 3.2

Suppose $(I, +, \cdot, 0, 1)$ is an integer space. Then for each $m, n \in I$, $n < m \Rightarrow n + 1 \leq m$.

The next theorem extends the principle of mathematical induction. To illuminate it, suppose we are trying to prove that some result holds for all integers ≥ -6. Suppose that we can show the following two properties:

1) The result holds for -6.
2) For each $n \in I$, if the result holds for n, then it also holds for $n + 1$.

Theorem 3.55 tells us that indeed the result holds for all integers ≥ -6.

Theorem 3.55 Suppose $(I, +, \cdot, 0, 1)$ is an integer space, and let $k \in I$. Suppose $A \subseteq I$ such that

1) $k \in A$, and
2) For each $n \in I$, $n \in A \Rightarrow n + 1 \in A$.

Then $n \in A$ for all $n \geq k$.

Proof: We assume that the result is false, and we obtain a contradiction. Define B by $B = \{n \in I | n \geq k \text{ and } n \notin A\}$. Then by our assumption, B is nonempty. Hence, B is bounded below, so B must have a minimum. Let us denote this minimum by b. Because b is a member of B, we have both $b \geq k$ and $b \notin A$ (by the definition of B). However, $b \neq k$ because we have both $k \in A$ and $b \notin A$. Hence, $b > k$. But this implies that $b - 1 \geq k$ (by an immediate application of Corollary 3.2). Now, because b is the minimum of B, this tell us that $b - 1 \notin B$, so $b - 1 \in A$ (by the definition of B). By Property 2 of the statement of the theorem, we must have $(b - 1) + 1 \in A$. But this says that $b \in A$. This contradicts our earlier conclusion that $b \notin A$, so we are done.

We now begin to "wrap up" all the pieces of this chapter into one neat package. The student will recall that we began this chapter with Peano spaces. What does an integer space have to do with a Peano space?

Given an integer space $(I, +, \cdot, 0, 1)$, define P by $P = \{n \in I | n \geq 1\}$. Is there a simple way in which to define a successor function $s: P \to P$ to make $(P, s, 1)$ into a Peano space? Absolutely! Theorem 3.56 below claims that the function $s: P \to P$ defined by $s(n) = n + 1 \; \forall n \in P$ does the trick.

Theorem 3.56 Suppose $(I, +, \cdot, 0, 1)$ is an integer space. Define P by $P = \{n \in I | n \geq 1\}$. Define $s: P \to P$ by $s(n) = n + 1 \; \forall n \in P$. Then $(P, s, 1)$ is a Peano space.

Proof: We must show that Properties 1 through 3 of Definition 3.1 hold.

Property 1: To show that s is one-to-one, suppose that $s(n) = s(m)$ with m, $n \in P$. Then $n + 1 = m + 1$. Subtracting 1 from both sides of this equation gives $n = m$.

Property 2: Suppose (to reach a contradiction) we have $s(n) = 1$ for some $n \in P$. Then $n + 1 = 1$, so $n = 0$. But $0 \notin P$ because all elements of P are strictly larger than zero.

Property 3: Suppose $A \subseteq P$ such that 1) $1 \in A$, and 2) we have $n \in A \Rightarrow n + 1 \in A \; \forall n \in P$. Our goal is to show that $A = P$. To invoke Theorem 3.55, we must show that we have $n \in A \Rightarrow n + 1 \in A \; \forall n \in I$. To do this, let $n \in I$. To show that $n \in A \Rightarrow n + 1 \in A$, suppose n is also a member of A. Now, because $A \subseteq P$, we must have $n \in P$, so now we simply use the above statement that $n \in A \Rightarrow n + 1 \in A \; \forall n \in P$. Hence, by Theorem 3.55, we have $n \in A \; \forall n \geq 1$; that is, $P \subseteq A$. Because we have both $A \subseteq P$ and $P \subseteq A$, we have established that $A = P$.

There are several important points to be made about Theorem 3.56. Note that the successor function s was defined in terms of '+', which is the addition defined on the integer space I. However, you will recall that once we have established a Peano space, we can define an addition and multiplication on it via Definitions 3.2 and 3.3, respectively. To distinguish these operations from the operations of $(I, +, \cdot, 0, 1)$, let us denote the operations corresponding to Definitions 3.2 and 3.3 by $\oplus$ and $\otimes$, respectively (for the moment).

Let P be as in Theorem 3.56, and suppose $m \in P$. It will now be shown (via the mathematical induction of Theorem 3.55) that $m \oplus n = m + n$ for all $n \in P$. To show the result holds for 1, we note that $s(m) = m + 1$ (by definition). We note from Definition 3.2 that $m \oplus 1$ is defined to be $s(m)$. Hence, $m \oplus 1 = m + 1$. Now, we take the inductive step and assume that $m \oplus n = m + n$ for some $n \in P$. We will be done if we can show that this assumption forces the conclusion that $m \oplus (n + 1) = m + (n + 1)$. However, noting that we have already shown that $n + 1 = n \oplus 1$, we see that $m \oplus (n + 1) = m \oplus (n \oplus 1)$. Now, by Definition 3.2, we have $m \oplus (n \oplus 1) = s(m \oplus n)$. In turn, we invoke our induction hypothesis to state that $s(m \oplus n) = s(m + n)$. Now, by the definition of s, we have $s(m + n) = (m + n) + 1$, which is the same as $m + (n + 1)$ (by the associativity of '+', of course), so we have the desired result.

Another way to state the above result is that the binary operation $\oplus$ is the restriction of the binary operation '+' to the Cartesian product $P \times P$. Similarly, the binary operation $\otimes$ is the restriction of the binary operation '$\cdot$' to $P \times P$, and the reader is asked to show this.

To make our last point on this topic, let $\hat{<}$ and $\hat{\leq}$ denote the relations on the Peano space $(P, s, 1)$ of Theorem 3.56 that are defined in Definition 3.5, Parts 1 and 2, respectively. Let $m, n \in P$. The reader is asked to show that $m < n \Leftrightarrow m \hat{<} n$ and $m \leq n \Leftrightarrow m \hat{\leq} n$. We will use all of these results without formal reference.

At this point the astute reader may be wondering why we bothered to examine Peano spaces before examining integer spaces; that is, why not start with Definition 3.20 and then simply derive the properties of the positive elements? It appears that this could have reduced our workload. The problem with such a method is that it would have been difficult to actually prove that an integer space exists had we taken this approach. In the next section (which is optional), we demonstrate the existence of an integer space by using the fact that we have already proven the existence of a Peano space. In this proof, we will need many of the properties of Peano spaces that we derived in Sections 3.1 through 3.3.

Integer Space Isomorphisms [Optional]

It would be a strange world indeed if it mattered which integer space we used; that is, it would be strange if there did not exist a ring isomorphism between any two integer spaces. In this subsection, we prove that there does exist such an isomorphism.

Throughout this subsection, we work with two fixed integer spaces. We denote the first one by $(I, +, \cdot, 0, 1)$, and we denote the second one by $(I', +, \cdot, 0', 1')$. At this point the reader may object because we have used the same symbols to represent addition and multiplication, respectively, for both integer spaces. However, as we said before, this "abuse of notation" is often the case in advanced mathematics books. Indeed, most of them go a step further and refer to both 0 and $0'$ as 0, and both 1 and $1'$ as 1.

Also, we use the symbols $\leq$ and $<$ (along with their usual meanings) for both integer spaces. Also, we use the symbol '$-$' for both additive inverses and subtraction for both integer spaces. By the way, the astute reader may have noticed that we already "reused" the symbol '$-$' to denote the additive inverses for different rings simultaneously in the statement of Part 3 of Theorem 3.39.

The reason, of course, for this "reusing" of symbols in advanced mathematics is for ease in both writing and reading mathematics. **In all cases, the context should make clear which meaning is assigned to the symbols.** That is, there should never be any actual ambiguity introduced in the situation. If you are ever reading an advanced mathematics text and you feel that the "reuse" of notation is causing potential ambiguity, then you should rename symbols for yourself to your satisfaction.

In this subsection, we demonstrate the existence of a function $g: I \to I'$ with the following properties:

1) g is one-to-one and onto;
2) $g(m + n) = g(m) + g(n)$ $\forall m, n \in I$;
3) $g(mn) = g(m) \cdot g(n)$ $\forall m, n \in I$;
4) $m \leq n \Leftrightarrow g(m) \leq g(n)$ $\forall m, n \in I$, and $m < n \Leftrightarrow g(m) < g(n)$ $\forall m, n \in I$.

Note that if there is such a function g, then we have $g(0) = 0'$, $g(1) = 1'$, and $g(-m) = -g(m)$ $\forall m \in I$, by Theorem 3.39.

We prove the result in stages, and we leave some of the pieces to the reader. By the way, all of the work that we do here is worth it because we will get to use the mapping g again when we show the uniqueness of the real number system.

To begin, we let P and P' denote the positive elements of I and I', respectively. We know from Theorem 3.56 that P and P' are Peano spaces. (Again, here is an "abuse of notation." For example, P is not technically a Peano space; the Peano space referred to is actually $(P, s, 1)$ with the function s as defined in Theorem 3.56. It is important that the student get used to these "abuses.")

We know from Theorem 3.25 that there is a Peano space isomorphism $f: P \to P'$. Hence, recall that for all $m, n \in P$, we have $f(m + n) = f(m) + f(n)$ and $f(mn) = f(m) \cdot f(n)$. Also, recall that f is one-to-one and onto. Finally, recall Exercise 2 of Section 3.4. We now define the function $g: I \to I'$ to be used throughout this subsection. We define g by:

$$g(m) = \begin{bmatrix} f(m) & \text{if} & m \geq 1. \\ 0' & \text{if} & m = 0. \\ -f(-m) & \text{if} & m \leq -1. \end{bmatrix}$$

Note that by this definition, we have $g(m)$ is positive (i.e., a member of P') iff m is positive (i.e., a member of P).

We first show that g is onto. The reader is asked to show that g is one-to-one. Let $w \in I'$. We must show that there exists an $x \in I$ such that $w = g(x)$. We break the situation into cases.

Case 1: $w \geq 1'$. Then there is an $m \in P$ such that $w = f(m)$ because f is onto. It is clear that $g(m) = f(m)$, so $w = g(m)$. Hence, setting x to be m works.

Case 2: $w = 0'$. Then $w = g(0)$. Hence, setting x to be 0 works.

Case 3: $w \leq -1'$. Hence, $-w \geq 1'$. Because f is onto, there is an $n \in P$ so that $-w = f(n)$. Now $-n \leq -1$, so $g(-n) = -f(-(-n)) = -f(n) = w$. Hence, setting x to be $-n$ works.

Next, we show that $g(mn) = g(m) \cdot g(n)$ $\forall m, n \in I$. Let $m, n \in I$.

Case 1: $m, n \in P$. Then $mn \in P$, so $g(mn) = f(mn) = f(m) \cdot f(n) = g(m) \cdot g(n)$.

Case 2: $m = 0$ or $n = 0$. Then $g(m) = 0'$ or $g(n) = 0'$, so $g(m) \cdot g(n) = 0'$. Also $mn = 0$, so $g(mn) = 0' = g(m) \cdot g(n)$.

Case 3: $m \leq -1$ and $n \geq 1$. Note that $g(m) = -f(-m)$, and $g(n) = f(n)$, so $g(m) \cdot g(n) = (-f(-m)) \cdot f(n) = -[f(-m) \cdot f(n)]$. Now, $mn \leq -1$, so $g(mn) = -f(-(mn)) = -f((-m) \cdot n) = -[f(-m) \cdot f(n)] = g(m) \cdot g(n)$.

Case 4: $m \geq 1$ and $n \leq -1$. Note that this case is the same as the previous case, with m and n switched. This switching is justified by the commutative law of multiplication.

Case 5: $m \leq -1$ and $n \leq -1$. Left for the reader.

Next, it must be shown that $g(m + n) = g(m) + g(n)$ $\forall m, n \in I$. We only do this for the difficult cases, and we leave the other cases for the reader. For the reader's convenience, we number some equations. Let $m, n \in I$.

Case 1: $m \leq -1$ and $n \geq 1$ and $m + n > 0$. We have:

$$g(m + n) = f(m + n) \tag{1}$$

Now,

$$g(m) + g(n) = -f(-m) + f(n) = f(n) - f(-m). \tag{2}$$

Because $m + n > 0$, we have $n > -m$, so $f(n) > f(-m)$, so $f(n) - f(-m) > 0'$.

Now, noting that $f(n) - f(-m) \in P'$, and using the fact that f is onto, there is a $w \in P$ so that

$$f(w) = f(n) - f(-m). \tag{3}$$

Hence, $f(n) = f(w) + f(-m) = f(w - m)$. Because f is also one-to-one, we have $n = w - m$, so $w = m + n$. Plugging this value of w back into Equation 3 gives

$$f(m + n) = f(n) - f(-m). \tag{4}$$

Hence, by Equations 1, 2, and 4, we have

$$g(m + n) = f(m + n) = f(n) - f(-m) = g(m) + g(n).$$

Case 2: $m \leq -1$ and $n \geq 1$ and $m + n < 0$. We have:

$$g(m + n) = -f(-m - n). \tag{5}$$

Now,

$$g(m) + g(n) = -f(-m) + f(n) = f(n) - f(-m). \tag{6}$$

Since $m + n < 0$, we have $-m > n$, so $f(-m) > f(n)$, so $f(-m) - f(n) > 0'$. Now, noting that $f(-m) - f(n) \in P'$, and using the fact that f is onto, there is a $x \in P$ so that

$$f(x) = f(-m) - f(n). \tag{7}$$

Hence, $f(-m) = f(x) + f(n) = f(x + n)$. Because f is also one-to-one, we have $-m = x + n$, so $x = -m - n$. Plugging this value of x into Equation 7 gives

$$f(-m - n) = f(-m) - f(n). \tag{8}$$

Hence, by Equations 5, 6, and 8 we have

$$g(m + n) = -f(-m - n) = f(n) - f(-m) = g(m) + g(n).$$

Case 3: $m \leq -1$ and $n \geq 1$ and $m + n = 0$. We have $m = -n$. Now, $g(m + n) = g(0) = 0'$. Also, $g(m) + g(n) = g(-n) + g(n) = -f(n) + f(n) = 0'$, so we have $g(m + n) = g(m) + g(n)$.

Finally, suppose $m, n \in I$. We will show that $m \leq n \Leftrightarrow g(m) \leq g(n)$. The reader is asked to show that $m < n \Leftrightarrow g(m) < g(n)$. First, we note that by Theorem 3.39, Part 3, we have $g(-a) = -g(a)$ for any $a \in I$ because we have already shown that g is a ring isomorphism (once the reader fills in the missing cases above). Hence, for any $a, b \in I$, we have $g(b - a) = g(b + -a) = g(b) + g(-a) = g(b) + -g(a) = g(b) - g(a)$.

To show that $m \leq n \Rightarrow g(m) \leq g(n)$ for any $m, n \in I$, suppose $m, n \in I$ with $m \leq n$.

Case 1: $m = n$. Then the result is obvious.

Case 2: $m < n$. Then $n - m > 0$, so $g(n - m) = f(n - m) > 0'$. Also, $g(n - m) = g(n) - g(m)$. Thus, $g(n) - g(m) > 0'$, so $g(m) < g(n)$.

To prove the converse, suppose $m, n \in I$ with $g(m) \leq g(n)$. Assume (to reach a contradiction) that $n < m$. Then (by what we have just shown above), we have $g(n) < g(m)$, which contradicts the statement $g(m) \leq g(n)$. We are forced to conclude that $m \leq n$.

EXERCISES

1. Suppose $\leq$ is a linear order on a nonempty set S. Write out the formal proof of the following: For each $u, v \in S$, exactly one of the following holds:
   a. $u < v$;
   b. $u > v$;
   c. $u = v$.

2. Suppose $\leq$ is a linear order on a nonempty set S. Write out the formal proof of the following: For each $u, v, w \in S$,
   a. $u \leq v < w \Rightarrow u < w$.
   b. $u < v \leq w \Rightarrow u < w$.
   c. $u < v < w \Rightarrow u < w$.

3. Prove Theorem 3.44.

4. Suppose $(D, +, \cdot, 0, 1)$ is an ordered integral domain with respect to $\leq$, and suppose $w, x, y, z \in D$ with $x \leq y$ and $z \leq w$. Show $x + z \leq y + w$.

5. Suppose $(D, +, \cdot, 0, 1)$ is an ordered integral domain with respect to $\leq$, and suppose $w, x, y, z \in D$ with $x \leq y$ and $z < w$. Show $x + z < y + w$.

6. Prove Theorem 3.48.

7. Prove Theorem 3.50.

8. Suppose $(D, +, \cdot, 0, 1)$ is an ordered integral domain with respect to $\leq$, and suppose $a, b, c, d \in D$ with $0 \leq a \leq b$ and $0 < c < d$. Show $ac < bd$.

9. Suppose $(D, +, \cdot, 0, 1)$ is an ordered integral domain with respect to $\leq$, and suppose $a, b, c, d \in D$ with $0 \leq a \leq b$ and $0 < c \leq d$. Show $ac \leq bd$.

10. Suppose $(D, +, \cdot, 0, 1)$ is an ordered integral domain with respect to $\leq$. Prove $1 > 0$.

11. Suppose $(D, +, \cdot, 0, 1)$ is an ordered integral domain with respect to $\leq$, and suppose that there exists an $a \in D$ such that $0 < a < 1$. Show that $a^2 < a$ (where a^2 is defined to be $a \cdot a$).

12. Prove Corollary 3.2.

13. Consider the binary operations of '$\cdot$' and $\otimes$ discussed in the remarks after the proof of Theorem 3.56. Show that $\otimes$ is the restriction of the binary operation '$\cdot$' to $P \times P$.

14. Consider the relations $\hat{<}$ and $\hat{\leq}$ discussed in the remarks after the proof of Theorem 3.56. Show that for any $m, n \in P$, we have $m < n \Leftrightarrow m \hat{<} n$, and $m \leq n \Leftrightarrow m \hat{\leq} n$.

15. Fill in all of the "missing pieces" of the proof in the subsection "Integer Space Isomorphisms."

16. Consider a Peano space $(P, s, 1)$ with the "usual" order $\leq$. Show that any nonempty subset of P that is bounded above has a maximum. [Hint: For each

$n \in P$, let $S(n)$ denote the statement "Every nonempty subset of P bounded above by n has a maximum." Demonstrate that $S(n)$ is true for each $n \in P$ by using mathematical induction.]

Section 3.7 [Optional] Proof of the Existence of the Integers

In this section, the existence of an integer space is demonstrated by constructing one via a Peano space. Recall (from Theorem 3.1) that we have already demonstrated the existence of a Peano space. Throughout this section, we work with a fixed Peano space $(P, s, 1)$ with the usual notation used for addition, multiplication, and order. We will reuse the symbols for addition, multiplication, and order on this constructed integer space. (See the discussion on the "abuse of notation" following Example 3.7.) Some of the easy but tedious details of the construction will be assigned to the reader. Before studying this section, the reader is encouraged to review the portion of Section 2.4 concerned with equivalence relations.

To motivate the discussion, we begin with a property that we intuitively know the integers must have once they are defined. It should be obvious from your previous exposure with integers since grade school that any integer can be expressed as the difference of two positive integers. That is, given any integer i, there are positive integers m and n such that $i = m - n$. For example, the integer 12 can be expressed as $12 = 15 - 3$. As another example, the integer -23 can be expressed as $-23 = 4 - 27$. As a third example, the integer 0 can be expressed as $0 = 5 - 5$.

Recall from Definition 3.8 that $m - n$ is defined provided that $m, n \in P$ with $n < m$. However, $m - n$ is not defined if $n \geq m$. One possible way to "represent" this subtraction is the ordered pair (m, n). That is, it seems possible to represent all of the integers by the set $P \times P$ of all ordered pairs of elements of P.

However, there is an immediate problem with this. Consider the ordered pair $(3, 4)$ (with 3 and 4 in P, of course). This represents the integer $3 - 4 = -1$. Also note that the ordered pair $(5, 6)$ also "represents" the integer -1. However, the problem is that $(3, 4) \neq (5, 6)$ because the only way two ordered pairs are equal is if both components are equal. To "fix" this, we shall define a relation on $P \times P$ that we shall denote by $\cong$. [Note that this means that $\cong$ will be a subset of $(P \times P) \times (P \times P)$.] Then it will be demonstrated that $\cong$ is an equivalence relation. The set of equivalence classes of this equivalence relation will be shown to possess the properties of the integers that were listed in the last section.

For given (m, n), $(p, q) \in P \times P$, we would intuitively like to have $(m, n) \cong (p, q)$, provided $m - n = p - q$. Of course, it has already been mentioned that these subtractions may not be defined. However, assume for the moment that these subtractions did make sense. Adding n and q to both sides of the equation would give the equation $m + q = n + p$, and this always makes sense because the addition of elements of P is always defined. It will turn out that this will work for us. Hence, we formally define the relation $\cong$ as follows: For each (m, n), $(p, q) \in P \times P$, we define $(m, n) \cong (p, q)$, provided $m + q = n + p$. That is, an ordered pair is declared related to another ordered pair provided the sum of the first component of the former ordered pair with the second component of the latter ordered pair is equal to the sum of the second component of the former ordered pair with the first component of the latter ordered pair.

We now show that the relation $\cong$ is an equivalence relation.

Reflexive Property   Suppose $(m, n) \in P \times P$. We must show that $(m, n) \cong (m, n)$; that is, we must show that $m + n = n + m$. However, we know this is true because addition is a commutative binary operation on P (i.e., Theorem 3.6).

Symmetric Property   Suppose (m, n), $(p, q) \in P \times P$ with $(m, n) \cong (p, q)$. We must show that $(p, q) \cong (m, n)$. That is, our goal is to show that $p + n = q + m$. Now, because we have $(m, n) \cong (p, q)$ we know that $m + q = n + p$. The desired result again follows by the commutativity of addition.

Transitivity Property   Suppose (m, n), (p, q), $(r, s) \in P \times P$ with $(m, n) \cong (p, q)$ and $(p, q) \cong (r, s)$. We must show that $(m, n) \cong (r, s)$. That is, our goal is to show that $m + s = n + r$. Now, $(m, n) \cong (p, q)$, so we have $m + q = n + p$. Also, $(p, q) \cong (r, s)$, so we have $p + s = q + r$. Using these two equations and the simple algebra derived for Peano spaces gives $(m + s) + q = (m + q) + s = (n + p) + s = n + (p + s) = n + (q + r) = (n + r) + q$. Hence, $(m + s) + q = (n + r) + q$. Using the cancellation law of addition (i.e., using Theorem 3.11) along with the commutativity of addition of elements of P delivers the desired result of $m + s = n + r$.

Having demonstrated that $\cong$ is indeed an equivalence relation, let us denote the equivalence class of each $(m, n) \in P \times P$ by $[(m, n)]$. Let us denote the set of equivalence classes by Ω; that is, $\Omega = \{[(m, n)] \,|\, (m, n) \in P \times P\}$. Our next task is to define addition and subtraction on Ω.

Consider equivalence classes $[(m, n)]$ and $[(p, q)]$. Let us define $[(m, n)] + [(p, q)]$ to be the equivalence class $[(m + p, n + q)]$. Recall that, intuitively, $[(m, n)]$ represents $m - n$ and $[(p, q)]$ represents $p - q$. Therefore, we would like to have $[(m + p, n + q)]$ represent $(m - n) + (p - q)$. Indeed, it does because it represents $(m + p) - (n + q)$, which can be simplified to $(m - n) + (p - q)$.

However, the astute student may recognize that there is a potential problem with this definition. The sum of the equivalence classes $[(m, n)]$ and $[(p, q)]$ has been defined in terms of the ordered pairs (m, n) and (p, q). The potential problem is that, for example, $[(m, n)]$ can be written as the equivalence class of many other ordered pairs besides (m, n). For example, $[(m, n)] = [(m + 1, n + 1)]$ because $(m, n) \cong (m + 1, n + 1)$. (See Theorem 2.7.) It must be demonstrated that the definition of the addition of equivalence classes listed above is **well defined** (i.e., is unambiguous).

Specifically, it must be shown that the addition of equivalence classes results in the same equivalence class regardless of which ordered pairs were used to write the equivalence classes that are being added. In other words, the following must be shown: For any (a, b), $(c, d) \in P \times P$ with $(a, b) \cong (m, n)$ and $(c, d) \cong (p, q)$, we must have $[(a + c, b + d)] = [(m + p, n + q)]$. Note that by Theorem 2.7 it suffices to show that $(a + c, b + d) \cong (m + p, n + q)$. Hence, it suffices to show that $(a + c) + (n + q) = (b + d) + (m + p)$. Now, we have $a + n = b + m$ and $c + q = d + p$. It is clear that adding these two latter equations and rearranging terms by using the simple algebra for Peano spaces yields the desired result.

What are we *really* saying by demonstrating that this definition is *well defined*? We are saying that for any $S, T \in \Omega$, there exists a unique $U \in \Omega$ (which we write as $S + T$) such that $U = [(a + c, b + d)]$ for each $(a, b) \in S$ and each $(c, d) \in T$. It should be noted that the issue of showing that a definition is well defined will come up again and again in higher mathematics.

Next, it will be shown that this addition of equivalence classes is an associative binary operation. Suppose $[(a, b)], [(c, d)], [(e, f)] \in \Omega$. The goal is to show that $([(a, b)] + [(c, d)]) + [(e, f)] = [(a, b)] + ([(c, d)] + [(e, f)])$. This is very straightforward. We simply keep using the addition of equivalence classes that we defined above and the associativity of "Peano" addition on P.

$$\begin{aligned}([(a, b)] + [(c, d)]) + [(e, f)] &= [(a + c, b + d)] + [(e, f)] \\ &= [((a + c) + e, (b + d) + f)] \\ &= [(a + (c + e), b + (d + f))] \\ &= [(a, b)] + [(c + e, d + f)] \\ &= [(a, b)] + ([(c, d)] + [(e, f)]).\end{aligned}$$

What should be used for the zero element, which will be denoted by **0**? A very intuitive and correct choice is $\mathbf{0} = [(1, 1)]$. After all, recall that, intuitively, $[(1, 1)]$ represents $1 - 1$. To demonstrate that this "works," let $[(a, b)] \in \Omega$. It will be shown that $[(a, b)] + \mathbf{0} = [(a, b)]$. Now, $[(a, b)] + \mathbf{0} = [(a, b)] + ([1, 1)] = [(a + 1, b + 1)]$. To complete the proof, it suffices to show (by virtue of Theorem 2.7) that $(a + 1, b + 1) \cong (a, b)$. However, the reader should quickly see that this holds because $(a + 1) + b = (b + 1) + a$. The reader is asked to show that this addition of equivalence classes is commutative.

If Ω is to be a ring, every element of Ω must have an additive inverse. What will "work" to be the additive inverse of $[(a, b)] \in \Omega$? We propose $[(b, a)]$. This makes intuitive sense because $[(a, b)]$ represents $a - b$, and $[(b, a)]$ represents $b - a$. Hence, we define $-[(a, b)]$ to be $[(b, a)]$. As before, the astute reader may realize that it is necessary to show that this definition is, in fact, well defined. This is very easy. It is clear that if $(u, v) \cong (a, b)$ for some $(a, b) \in P \times P$, then $[(v, u)] = [(b, a)]$. (Why?) The reader is now asked to show that $-[(a, b)] + [(a, b)] = \mathbf{0}$.

Next on the agenda is the definition of the multiplication of the equivalence classes $[(a, b)]$, $[(c, d)] \in \Omega$. Recalling that $[(a, b)]$ represents $a - b$ and $[(c, d)]$ represents $c - d$, we will want $[(a, b)] \cdot [(c, d)]$ to represent $(a - b)(c - d) = (ac + bd) - (ad + bc)$. Hence, it appears to be "reasonable" to define $[(a, b)] \cdot [(c, d)]$ to be $[(ac + bd, ad + bc)]$ for each $[(a, b)]$, $[(c, d)] \in \Omega$. The proof that this definition is, in fact, well defined is left to the reader. It is a little tricky (but nothing that you cannot handle because a hint is provided). The reader is also asked to verify that this multiplication is commutative.

Next, the associativity of this multiplication is demonstrated. Incidentally, this demonstration is a good example of the forward–backward method of proof that was discussed in Section 1.4. Consider $[(a, b)]$, $[(c, d)]$, $[(e, f)] \in \Omega$. The goal is to show that $([(a, b)] \cdot [(c, d)]) \cdot [(e, f)] = [(a, b)] \cdot ([(c, d)] \cdot [(e, f)])$.

Now, it is easy to see that

$$([(a, b)] \cdot [(c, d)]) \cdot [(e, f)] = [(ac + bd, ad + bc)] \cdot [(e, f)]$$
$$= [(e(ac + bd) + f(ad + bc), f(ac + bd) + e(ad + bc))].$$

It is also easy to see that

$$[(a, b)] \cdot ([(c, d)] \cdot [(e, f)]) = [(a, b)] \cdot [(ce + df, cf + de)]$$
$$= [(a(ce + df) + b(cf + de), a(cf + de) + b(ce + df))].$$

The reader should now have no trouble finishing the proof by showing that $[(e(ac + bd) + f(ad + bc), f(ac + bd) + e(ad + bc))]$ is identical to $[(a(ce + df) + b(cf + de), a(cf + de) + b(ce + df))]$. The reader is also asked to demonstrate that the distributive law holds.

The next step is to define the unit element of the ring that is being constructed. Define $\mathbf{1}$ by $\mathbf{1} = [(1 + 1, 1)]$. For each $[(a, b)] \in \Omega$, we must demonstrate that this "works" by showing $[(a, b)] \cdot \mathbf{1} = [(a, b)]$. That is, we must show that $[(a, b)] \cdot [(1 + 1, 1)] = [(a, b)]$. However, this is equivalent to showing that $(a(1 + 1) + b) + b = (a + b(1 + 1)) + a$, which is clearly true.

Assuming that the reader has worked through Exercises 1 through 5 of this section, it has now been verified that Ω is a commutative ring. Next, it will be shown that Ω is an integral domain. To do this, suppose we have $[(a, b)]$, $[(c, d)] \in \Omega$ with $[(a, b)] \cdot [(c, d)] = \mathbf{0}$. To demonstrate that Ω is an integral domain, we must show that $[(a, b)] = \mathbf{0}$ or $[(c, d)] = \mathbf{0}$. Using the definition of $\mathbf{0}$ as $\mathbf{0} = [(1, 1)]$, the reader should see that it suffices to show that $a = b$ or $c = d$. This is done by expressing the equation $[(a, b)] \cdot [(c, d)] = \mathbf{0}$ as $[(ac + bd, ad + bc)] \cong (1, 1)$. Hence, $ac + bd + 1 = ad + bc + 1$, so $ac + bd = ad + bc$.

Hence, we will have completed the proof of Ω being an integral domain if we can show that $(ac + bd = ad + bc) \Rightarrow (a = b$ or $c = d)$. We will accomplish this by showing that $((ac + bd = ad + bd)$ and $(c \neq d)) \Rightarrow (a = b)$. (The reader should use the basic logic of Chapter 1 to see why this works.) Thus, we assume that $ac + bd = ad + bc$ and $c \neq d$, and these two assumptions will be used to show that $a = b$. Now there are two cases to consider. One of them is left to the reader.

Case 1: $c < d$. Then, there is an $n \in P$ so that $d = c + n$. (Review Definition 3.5 if necessary.) Substituting this into the equation $ac + bd = ad + bc$ and simplifying results in $ac + bc + bn = ac + an + bc$. Canceling terms results in $bn = an$, and the result of $a = b$ follows (by Theorem 3.19, of course.)

Case 2: $d < c$. This is left to the reader.

It has now been established that Ω is an integral domain. The next step in demonstrating that Ω has all the properties of Section 3.6 is to define an order on Ω (which will be denoted by $\leq$, as we acknowledge our "abuse of notation" given that this symbol is already being used to express the "usual" order on P). For $[(a, b)]$, $[(c, d)] \in \Omega$, we specify $[(a, b)] \leq [(c, d)]$ to mean that $a + d \leq b + c$.

Now, of course, we have the immediate problem of showing that this definition is, in fact, well defined. To do this, suppose $(u, v) \cong (a, b)$ and $(w, x) \cong (c, d)$. We must, of course, show that $u + x \leq v + w$. Now, we have $u + b = v + a$ and $w + d = x + c$. How can we use this information, along with the inequality $a + d \leq b + c$, to show that $u + x \leq v + w$? Adding the equations $u + b = v + a$ and $w + d = x + c$ and performing simple Peano space algebra results in the equation $(u + x) + (b + c) = (v + w) + (a + d)$. Now, employing the inequality gives us $(v + w) + (a + d) \leq (v + w) + (b + c)$, so we have shown that $(u + x) + (b + c) \leq (v + w) + (b + c)$. The desired result follows by Theorem 3.15.

It is left to the reader to show that this relation is, in fact, a linear order. It is also left to the reader to show that this order satisfies the conditions of Definition 3.19, thereby showing that Ω is an ordered integral domain. The last item is to show that Definition 3.20 holds for Ω. That is, it must be demonstrated that any nonempty subset of Ω that is bounded below has a minimum (i.e., a least element). The demonstration of this takes some work.

Suppose Γ is a nonempty subset of Ω with lower bound $[(a_1, a_2)] \in \Omega$. Our goal is to demonstrate that Γ has a least element. Define the set Z by $Z = \{(x, y) | [(x, y)] \in \Gamma\}$. Note that Z is simply the set of ordered pairs $(x, y) \in P \times P$ such that the equivalence class that (x, y) belongs to (i.e., $[(x, y)]$) is a member of Γ. Now, because $[(a_1, a_2)]$ is a lower bound of Γ, we must have $[(a_1, a_2)] \leq [(x, y)]$ for each $(x, y) \in Z$. Hence, we have shown that $a_1 + y \leq a_2 + x$ for each $(x, y) \in Z$. Let us refer to this inequality as *the inequality*.

Now define the set X by $X = \{x \in P | (x, y) \in Z$ for some $y \in P\}$. By Theorem 3.22, X has a minimum, which we shall denote by x^*. Define Y^* by $Y^* = \{y \in P | (x^*, y) \in Z\}$. (Why is it clear that Y^* is nonempty?) Note that we have $a_1 + y \leq a_2 + x^*$ for each $y \in Y^*$ by *the inequality* because

Figure 3.3
The construction of
an integer space via
a Peano space

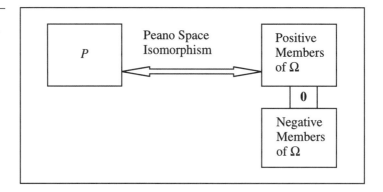

$(x^*, y) \in Z$ for each $y \in Y^*$. Define A_Y by $A_Y = \{a_1 + y | y \in Y^*\}$. Then A_Y is bounded above by $a_2 + x^*$. Hence, by Exercise 16 of Section 3.6, A_Y has a greatest element, which we shall denote by a^*. Clearly, there is a unique $y^* \in Y^*$ such that $a^* = a_1 + y^*$. (Why?) The reader should also verify that y^* is the greatest element of Y^*. Note that $(x^*, y^*) \in Z$.

It will now be shown that $[(x^*, y^*)]$ is the minimum of Γ. To accomplish this, it suffices to show that $[(x^*, y^*)] \leq [(x, y)]$ for all $(x, y) \in Z$. That is, it suffices to show that $x^* + y \leq y^* + x$ for each $(x, y) \in Z$. Suppose (to reach a contradiction) that $x^* + y > y^* + x$ for some $(x, y) \in Z$. Then, by definition, there exists an $n \in P$ such that $x^* + y = (y^* + x) + n$. However, we see that this says that $(x^*, y^* + n) \cong (x, y)$. However, this tells us that $(x^*, y^* + n) \in Z$. (Why?) In turn, this tells us that $y^* + n \in Y^*$. Now, $y^* + n > y^*$. This contradicts the fact that y^* is the greatest element of Y^*. Hence, we conclude that $x^* + y \leq y^* + x$ for each $(x, y) \in Z$, as desired.

Actually, we did not need to take the minimum value x^* of X to work with. Selecting any member of X would have worked. Can you see why?

We have achieved our goal! Let us review on a macro level what we did. We started out with a Peano space, P, [i.e., we mean $(P, s, 1)$, of course]. We then defined a set of equivalence classes Ω that we demonstrated is an integer space. What is the relationship between P and Ω? To see this, let $P' = \{[(m, n)] \in \Omega | m, n \in P \text{ and } m > n\}$. The reader should see that P' is precisely the set of members of Ω that exceed $\mathbf{0}$ (i.e., **the positive members** of Ω). Note that P' is itself a Peano space. Performing this construction, in essence, gave us the number $\mathbf{0}$ and negative integers. Figure 3.3 illustrates this relationship.

EXERCISES

1. Show that $[(a, b)] + [(c, d)] = [(c, d)] + [(a, b)] \;\forall\; [(a, b)], [(c, d)] \in \Omega$.

2. Show that $-[(a, b)] + [(a, b)] = \mathbf{0} \;\forall\; [(a, b)] \in \Omega$.

3. Prove that the equivalence class multiplication defined above is well defined. That is, show that for any $(a, b), (c, d), (u, v), (x, y) \in P \times P$ such that

$(u, v) \cong (a, b)$ and $(x, y) \cong (c, d)$, we must have $(ac + bd, ad + bc) \cong (ux + vy, uy + vx)$. [Hint: First use $(u, v) \cong (a, b)$ to show that $(ac + bd, ad + bc) \cong (uc + vd, ud + vc)$. Then use $(x, y) \cong (c, d)$ to show that $(uc + vd, ud + vc) \cong (ux + vy, uy + vx)$. Finally, invoke the transitivity of $\cong$.]

4. Show that $[(a, b)] \cdot [(c, d)] = [(c, d)] \cdot [(a, b)] \ \forall \ [(a, b)], [(c, d)] \in \Omega$.

5. Show that the distributive law holds. That is, show that $[(a, b)] \cdot ([(c, d)] + [(e, f)]) = [(a, b)] \cdot [(c, d)] + [(a, b)] \cdot [(e, f)] \ \forall \ [(a, b)], [(c, d)], [(e, f)] \in \Omega$.

6. Complete the proof that Ω is an integral domain by filling in "Case 2" listed in the text.

7. Show that the relation $\leq$ on Ω is an order.

8. Complete the proof that Ω is an ordered integral domain with respect to $\leq$.

Chapter 4

Properties and Applications of Integers

Having developed the system of integers, we now unleash them. Properties and applications of integers that the budding mathematician will need to know to proceed to more advanced courses are derived in this chapter. In doing so, some results will be proved in general settings that will automatically apply to the system of real numbers, once the development of this system is completed in Chapter 6. Therefore, such results will not need to be proved over again.

In Section 4.1, we use integers to introduce the concept of a *sequence*. We also generalize the concept of the Cartesian product. In Section 4.2, we demonstrate how integers can be applied to binary applications. This will allow us to accomplish such tasks as showing that if a binary operation is commutative and associative, then the order of the elements in applying this operation a finite number of times does not matter. We also generalize the concept of recursion. In Section 4.3, we develop the beginnings of the subject known as *number theory*. At a very elementary level, number theory is concerned with divisibility. Finally, in Sections 4.4 and 4.5, we increase our knowledge of set theory by studying finite and infinite sets. Infinite sets have some properties that you are likely to regard as surprising.

It has been demonstrated in Chapter 3 that any two integer spaces are isomorphic. Therefore, throughout this chapter it will be assumed that a suitable such integer space has been chosen. (It is shown in Chapter 6 how to choose a suitable one so that the resultant set of integers is a subset of the set of real numbers.) At long last, we now denote our selected set of integers by $\mathbf{Z}$. Then, of course, the set $\mathbf{N}$ of positive integers is defined by $\mathbf{N} = \{n \in \mathbf{Z} | n \geq 1\}$.

Section 4.1   Sequences, Cartesian Products, and the Generalized Recursion Theorem

Of paramount importance to mathematics is the concept of a **sequence**. Fortunately, it is very simple.

Definition 4.1

A **sequence** is a function with domain $\mathbf{N}$.

Hence, if S is any nonempty set and $f: \mathbf{N} \to S$, then f is a sequence by definition. Then, of course, $f(n)$ is the image of n with respect to f; it is often referred to as the nth **term** of the sequence. It is often the case that f_n is written, instead of writing $f(n)$. Frequently, letters such as x, y, a, and b are used to denote sequences.

Suppose x is a sequence, so that the nth term of this sequence is x_n. It is common to denote this sequence by $\{x_n\}_{n=1}^{\infty}$. The use of ∞, the symbol for infinity, reminds us that the domain is the entire set of positive integers, and this set "goes on forever." It is important to realize that ∞ is not an actual number. Also, there is nothing sacred about using the letter n as the subscript. For example, $\{x_i\}_{i=1}^{\infty}$ is the same sequence as is $\{x_n\}_{n=1}^{\infty}$. So is $\{x_k\}_{k=1}^{\infty}$.

Example 4.1

Define $\{x_n\}_{n=1}^{\infty}$ to be the sequence that assigns $-2n$ to each $n \in \mathbf{N}$. Then $x_n = -2n$ for each $n \in \mathbf{N}$. In particular, for example, $x_3 = -6$. ∎

The very astute reader may notice a problem with Example 4.1. Technically, we have not completed the definition of the sequence of the example because we have not specified the range, as called for in Definition 2.21. However, in practice, the precise specification of the range may not matter. Therefore, for example, we may regard the range of the sequence of Example 4.1 as $\mathbf{Z}$. As another example, we could also consider the range to be any set that contains $\mathbf{Z}$ as a subset. If, for some reason, the precise specification of the range matters, then it will be given.

There are some advantages to using the notation $\{x_n\}_{n=1}^{\infty}$ defined above. As one very important example, consider the Recursion Theorem. It can now be stated as follows:

Theorem 4.1 **(Recursion Theorem Restated)**

Suppose S is a nonempty set with $a \in S$. Let $g: S \to S$. Then there exists a unique sequence $\{x_n\}_{n=1}^{\infty}$ with range S such that the following two properties hold:

1) $x_1 = a$.
2) $x_{n+1} = g(x_n)$ for each $n \in \mathbf{N}$.

As mentioned in the paragraph after Theorem 3.5, you will rarely see the function g (of Theorems 3.5 and 4.1) explicitly mentioned. However, you should always think about it lurking behind the scene, and you should be able to produce it if you need to convince yourself that everything is in order.

Frequently, in advanced mathematics texts you will be presented with a "sequence" $\{a_n\}_{n=0}^{\infty}$ that "starts" with 0. Is this really a sequence, as defined by Definition 4.1? Well, it could be regarded as a function with domain $\mathbf{N} \cup \{0\}$. However, another way to look at it is to consider it to be the sequence $\{b_n\}_{n=1}^{\infty}$, in which $b_n = a_{n-1}$ for each positive integer n. Similarly, you will see "sequences" such as $\{a_n\}_{n=p}^{\infty}$ in advanced mathematics texts, in which p is a positive integer. You should regard this as the sequence $\{b_n\}_{n=1}^{\infty}$, in which $b_n = a_{p+n-1}$ for each positive integer n.

It should be noted that a sequence $\{x_n\}_{n=1}^{\infty}$ is not the same as thing as $\{x_1, x_2, \ldots\}$. The former entity is a function with domain $\mathbf{N}$. The latter entity is the set $\bigcup_{n=1}^{\infty}\{x_n\}$. (The reader is asked to review Definition 2.32 and the comments following it.) The following example illustrates this difference.

Example 4.2 Consider the sequence $\{x_n\}_{n=1}^{\infty}$ defined by $x_n = 2$ for each $n \in \mathbf{N}$. Then the set $\{x_1, x_2, \ldots\}$ is the same as the set $\{2\}$. This demonstrates the difference between a sequence and the set of points that constitute the image of the sequence.

For the next concept, the reader is asked to recall Definition 3.9, in which $\mathbf{I}_n$ is defined for each $n \in \mathbf{N}$. The term **finite sequence** is defined in Definition 4.2.

Definition 4.2

A **finite sequence** is a function with domain $\mathbf{I}_n$, for some $n \in \mathbf{N}$.

Unfortunately, this is something of a misnomer because a finite sequence is not a sequence, as defined by Definition 4.1. However, the context will always remove any potential ambiguity.

If x is a finite sequence with domain $\mathbf{I}_n$, it may be denoted by $\{x_i\}_{i=1}^n$. (Of course, there is nothing sacred about using the letter i as the subscript.) Note that $\{x_i\}_{i=1}^n$ is not the same as the set $\{x_1, x_2, \ldots, x_n\}$. The latter entity is the set $\cup_{k \in \mathbf{I}_n}\{x_k\}$.

Now, recall (from Definition 2.32 and the comments after it) that if we have an indexed family of sets $\{A_n\}_{n \in \mathbf{N}}$, then we can write the union of the family as $\cup_{n=1}^\infty A_n$. Similarly, we now employ the notation $\cup_{k=1}^n A_k$ to mean the union of the indexed family of sets $\{A_k\}_{k \in \mathbf{I}_n}$. (Of course, the letter k can be replaced with other letters.) Using this notation, the union $\cup_{k \in \mathbf{I}_n}\{x_k\}$ of the previous paragraph can be written as $\cup_{k=1}^n\{x_k\}$. Analogously, we can use the notation $\cap_{k=1}^n A_k$ to mean $\cap_{k \in \mathbf{I}_n} A_k$.

The reader should note that $\{A_n\}_{n \in \mathbf{N}}$ can also be written as $\{A_n\}_{n=1}^\infty$. Why? Because it is a sequence! It is a sequence of sets, which is a perfectly valid sequence. Similarly, $\{A_k\}_{k \in \mathbf{I}_n}$ can be written as $\{A_k\}_{k=1}^n$ because it is a finite sequence of sets. In addition to having sequences of sets, we can also have sequences of functions. In fact, sequences of functions are used frequently in mathematics.

Example 4.3

Consider the sequence $\{f_n\}_{n=1}^\infty$, in which each $f_n: \mathbf{N} \to \mathbf{N}$ is the function defined by $f_n(x) = n + x \; \forall x \in \mathbf{N}$. Then $\{f_n\}_{n=1}^\infty$ is clearly a sequence of functions.

■

We have provided new notation for expressing sequences and finite sequences. This notation can be used whenever it is convenient. However, the notation does not have to be used all the time. There will be times when we want to express a sequence using the "traditional" functional notation discussed in the second paragraph after Definition 2.21. Readability and ease of expression determine the choice of notation used.

So far in this section, we have introduced some new notation. Next, we begin to move in the direction of generalizing the Recursion Theorem. Before doing so, we will need to generalize the concept of the Cartesian product of a set with itself. The reader is asked to review Definition 2.12 and the first three paragraphs after it.

Recall that the Cartesian product of the set of real numbers with itself, $\mathbf{R} \times \mathbf{R}$, is often written as $\mathbf{R}^2$. This notation can also be used for any set A.

(Recall that $A \times A$ is defined to be the set $\{(a, b) | a, b \in A\}$.) The set $A \times A$ can be written A^2. (If A is the empty set, then $A \times A$ is also the empty set.) The goal now is to generalize this to define A^n for each positive integer n. Intuitively, we "want" A^n to be the set of all ordered collections of n elements of A. Each such collection of n elements is frequently called an **n-tuple**. Readers who have had multivariable calculus will be used to working with the set $\mathbf{R}^n$ for arbitrary positive integers n.

Now, a 1-tuple is just an element of A. A 2-tuple is usually called an ordered pair, as defined in Definition 2.10. A 3-tuple is usually called an ordered triplet, as defined in Definition 2.11. Students who have had three semesters of calculus will be familiar with the set $\mathbf{R}^3 = \{(x, y, z) | x, y, z \in \mathbf{R}\}$. This set represents the set of points in three-dimensional space.

An n-tuple member A^n is expressed as $(a_1, a_2, \ldots, a_n)$, in which $a_i \in A$ for each $i \in \mathbf{I}_n$. (Of course, letters other than a can be used.) There is one exception. When n is 1, parenthesis are not needed.

| **Example 4.4** | Suppose the set A is defined by $A = \{-1, 0, 2, 7, 51\}$. Then the 4-tuple $(2, -1, 0, 2)$ is a valid member of A^4. This is the 4-tuple (b_1, b_2, b_3, b_4), in which $b_1 = 2$, $b_2 = -1$, $b_3 = 0$, and $b_4 = 2$. |

Recall the discussion in Section 2.3 of how to define ordered pairs. We remarked that it did not matter how we defined ordered pairs, so long as two ordered pairs (a, b) and (c, d) are declared to be equal if and only if $a = c$ and $b = d$. Similarly, it does not matter how we define ordered n-tuples, so long as two ordered n-tuples $(a_1, a_2, \ldots, a_n)$ and $(b_1, b_2, \ldots, b_n)$ are declared to be equal if and only if $a_i = b_i$ for each $i \in \mathbf{I}_n$.

Now, the astute reader may reason that a very convenient mechanism of defining A^n for each $n \in \mathbf{N}$ appears to be to use the Recursion Theorem (i.e., Theorem 4.1). In other words, define A^1 to be A, and define A^{n+1} to be $A^n \times A$. The problem with this is that it is not clear as to what the function g of Theorem 4.1 should be. The reader might try experimenting to see if such an appropriate function g can be found. Without invoking the Recursion Theorem, the above method works for defining A^n for any *particular* n (such as $n = 3$), but it does not guarantee that it works for all of $\mathbf{N}$. This is an esoteric point, the likes of which you will not encounter very often in mathematics (unless you specialize in logic and set theory).

Fortunately, there is another way to define A^n for each $n \in \mathbf{N}$ that fulfills our criterion that any two ordered n-tuples $(a_1, a_2, \ldots, a_n)$ and $(b_1, b_2, \ldots, b_n)$ be equal if and only if $a_i = b_i$ for each $i \in \mathbf{I}_n$. Define A^1 to be A, and define A^2 to be $A \times A$. Define A^3 to be $(A \times A) \times A$. (Note that the members of A^3 are the ordered triplets that were defined in Definition 2.11.) For each positive integer n, with $n \geq 4$, simply define A^n to be the set of functions with domain $\mathbf{I}_n$ and range A. This is formally stated in Definition 4.3.

Definition 4.3

Suppose A is a nonempty set. Then A^n is defined as follows for each $n \in \mathbf{N}$:

$$A^n = \begin{cases} A & \text{if } n = 1 \\ A \times A & \text{if } n = 2 \\ (A \times A) \times A & \text{if } n = 3 \\ \{f \mid f : \mathbf{I}_n \to A\} & \text{if } n \geq 4 \end{cases}$$

If A is the empty set, then so is A^n for each $n \in \mathbf{N}$.

Example 4.5

Consider the set A and the 4-tuple $(2, -1, 0, 2)$ of Example 4.4. Then $(2, -1, 0, 2)$ is the function $f : \mathbf{I}_4 \to A$ defined by $f(1) = 2$, $f(2) = -1$, $f(3) = 0$, and $f(4) = 2$.

A few minor points should be made. The reader may wonder why we did not just define A^n to be $\{f \mid f : \mathbf{I}_n \to A\}$ for all $n \in \mathbf{N}$. The reason is that we need the ordered pair and ordered triplet as defined in Definitions 2.10 and 2.11, respectively, to define functions!

The very, very picky reader may observe that Definition 4.3 is inconsistent with the definition of a 5-tuple that was specified in the paragraph before Definition 3.12 in Section 3.5. Technically, this is correct, but it does not matter. If it would make this reader feel better, he or she may change the definition of a 5-tuple as specified in Section 3.5 so as to conform to Definition 4.3. As stated above, it does not really matter how we define ordered n-tuples provided the following theorem is true. The easy proof of Theorem 4.2 is left to the reader.

Theorem 4.2

Suppose A is a set and $n \in \mathbf{N}$. Suppose $(a_1, a_2, \ldots, a_n), (b_1, b_2, \ldots, b_n) \in A^n$. Then $(a_1, a_2, \ldots, a_n) = (b_1, b_2, \ldots, b_n)$ iff $a_i = b_i$ for each $i \in \mathbf{I}_n$.

The final concept of this section is the presentation and proof of the **Generalized Recursion Theorem**. You will recall that the Recursion Theorem provides a means of defining a sequence $\{x_n\}_{n=1}^{\infty}$, in which, 1) the value of x_1 is specified, and 2) for each $n \in \mathbf{N}$, the value of x_{n+1} is specified in terms of the value of x_n.

In mathematics, it will often be necessary to specify the value of x_{n+1}, not necessarily just in terms of x_n but in terms of some or all of the values $x_1, x_2, \ldots, x_n$, and/or the number n. This can be expressed by saying that x_{n+1} is some function of $(x_1, x_2, \ldots, x_n)$. If this function is called g, then we have $x_{n+1} = g(x_1, x_2, \ldots, x_n)$ for each $n \in \mathbf{N}$. Technically, what we mean is that $x_{n+1} = g((x_1, x_2, \ldots, x_n))$. However, it is a frequent "abuse of notation" that $g(x_1, x_2, \ldots, x_n)$ is written instead of $g((x_1, x_2, \ldots, x_n))$. This "abuse" often holds, not just for the upcoming Generalized Recursion Theorem but whenever n-tuples are in the domain of a function. Therefore, the reader should get used to this. Sometimes, the double parentheses are written, and other times not. Let us suppose that we want the range of the sequence $\{x_n\}_{n=1}^{\infty}$ to be in a nonempty set, S. What should the domain of the function g be? Clearly, $(x_1, x_2, \ldots, x_n) \in S^n$ for each $n \in \mathbf{N}$. Hence, it makes sense to insist that S^n is in the domain of g, for each $n \in \mathbf{N}$. To accomplish this, the domain of g can be defined to be $\cup_{n=1}^{\infty} S^n$. Analogous to the function g of the Recursion Theorem, the function g of the Generalized Recursion Theorem is seldom explicitly listed in advanced mathematics texts. Nonetheless, the student should be aware of the necessity of its existence.

The proof of the Generalized Recursion Theorem, which has been designated "optional," is similar to the proof of the Recursion Theorem. In the Exercises, the reader is asked to show how the Recursion Theorem can be derived from the Generalized Recursion Theorem. In other words, the Recursion Theorem is a specialized case of the Generalized Recursion Theorem.

Theorem 4.3 **(Generalized Recursion Theorem)**

Suppose S is a nonempty set with $a \in S$. Suppose $g: \cup_{n=1}^{\infty} S^n \to S$. Then there exists a unique sequence $\{x_n\}_{n=1}^{\infty}$ with range S such that the following two properties hold:

1) $x_1 = a$.
2) $x_{n+1} = g(x_1, x_2, \ldots, x_n)$ for each $n \in \mathbf{N}$.

Example 4.6

One of the simplest examples of using the Generalized Recursion Theorem is the definition of the factorial function. Students who have worked previously with this function will recall that, for each $n \in \mathbf{N}$, $n!$ (read n factorial) is the product $1 \cdot 2 \cdot 3 \cdot \cdots \cdot n$. So, for example, $3! = 1 \cdot 2 \cdot 3 = 6$. (Note that we have not yet *formally* defined arbitrarily long products or sums. This will be done in the next section.)

Formally, we define the sequence $\{x_n\}_{n=1}^{\infty}$ with range $\mathbf{N}$. Define $g: \cup_{n=1}^{\infty} \mathbf{N}^n \to \mathbf{N}$ by $g(a_1, a_2, \ldots, a_n) = (n+1)a_n$ for each $n \in \mathbf{N}$ and each $(a_1, a_2, \ldots, a_n) \in \mathbf{N}^n$. Then, by the Generalized Recursion Theorem, there exists a unique sequence $\{x_n\}_{n=1}^{\infty}$ such that $x_1 = 1$, and $x_{n+1} = g(x_1, x_2, \ldots, x_n)$ for each $n \in \mathbf{N}$. That is, $x_{n+1} = (n+1)x_n$ for each $n \in \mathbf{N}$. Rewriting each x_n as $n!$, this last equation states that $(n+1)! = (n+1)n!$. By plugging in a few examples for n, the student should see that definition "works." Even though the function g is quite simple, the student should see that it would not be a simple matter to employ the Recursion Theorem (Theorem 4.1) to construct this definition. (Try to do so.) Finally, for the record, we extend the domain of this sequence to include 0 by defining $0!$ to be 1. We will explain a reason for this in Section 6.4.

◾

Before proving Theorem 4.3, we make note of a result that is used in the proof but is also important in its own right. Consider any $m, n \in \mathbf{N}$. Note that the number 1 is a lower bound of the set $\{m, n\}$. Hence, by Definition 3.20, the set $\{m, n\}$ has a minimum, which we denote by $\mathbf{min}(\{m, n\})$. Later we generalize this result and write it out formally.

Proof of Theorem 4.3 [Optional]

This proof has been modified slightly from the proof given in Burril (1967).

Uniqueness will be demonstrated first. Suppose $\{x_n\}_{n=1}^{\infty}$ and $\{y_n\}_{n=1}^{\infty}$ are both sequences satisfying the condition of the statement of the theorem. Suppose (to reach a contradiction) that these sequences are not identical. Let $B = \{n \in \mathbf{N} | x_n \neq y_n\}$. Then B is nonempty, so B contains a smallest element (by the Well-Ordering Property, of course), which will be noted by s. Now, by Property 1 of the theorem statement, $x_1 = a = y_1$, so $s \neq 1$. Hence, $s > 1$, so $s - 1 \in \mathbf{N}$. Now, by the definition of s, we have $x_k = y_k$ for each $k \in \{1, 2, \ldots, s - 1\}$. Thus, by Property 2 of the theorem statement, we have $x_s = g(x_1, x_2, \ldots, x_{s-1}) = g(y_1, y_2, \ldots, y_{s-1}) = y_s$. This is our desired contradiction, so uniqueness has been established, provided that such a sequence exists.

Next, we demonstrate that such a sequence exists. Now define the sequence of sets $\{H_n\}_{n=1}^{\infty}$ by

$$H_n = \{(v_1, v_2, \ldots, v_n) \in S^n |$$
$$v_1 = a, \text{ and } v_{m+1} = g(v_1, v_2, \ldots, v_m) \text{ for each } m \in \mathbf{N} \text{ with } m < n\}.$$

Define the set H by $H = \cup_{n=1}^{\infty} H_n$. The first thing to notice about H is that it contains the 1-tuple $a \in S^1$.

Suppose $(v_1, v_2, \ldots, v_m), (w_1, w_2, \ldots, w_n) \in H$, and let $k = \mathbf{min}(\{m, n\})$. We will now show that $(v_1, v_2, \ldots, v_k) = (w_1, w_2, \ldots, w_k)$. To do this, suppose (to reach a contradiction) that $(v_1, v_2, \ldots, v_k) \neq (w_1, w_2, \ldots, w_k)$. Let s denote the smallest member of $\{1, 2, \ldots, k\}$ such that $v_s \neq w_s$. Note that by the definition of H, $v_1 = a = w_1$. Hence, $s > 1$, so $s - 1 \in \mathbf{N}$. This indicates that $(v_1, v_2, \ldots, v_{s-1}) = (w_1, w_2, \ldots, w_{s-1})$. (Why?) Also, note that this $s - 1$-tuple is also a member of H. (Why?) Then, by the definition of H, we have $v_s = g(v_1, v_2, \ldots, v_{s-1}) = g(w_1, w_2, \ldots, w_{s-1}) = w_s$. Thus, we have concluded that $v_s = w_s$, which gives us our desired contradiction, so the desired result is established. For reference purposes, we shall call the result we just established the **minimum principle of H**.

Next, we will show (via induction) that the set H_n contains exactly one element for each $n \in \mathbf{N}$. It is clear that H_1 is precisely the set $\{a\}$. Suppose that the set H_n contains exactly one element for some $n \in \mathbf{N}$. To complete the induction argument, we must show that H_{n+1} contains exactly one element. Let us denote the one member of H_n by $(v_1, v_2, \ldots, v_n)$. Let $x = g(v_1, v_2, \ldots, v_n)$. The reader should be able to see (from the induction hypothesis and the definition of H_{n+1}) that $(v_1, v_2, \ldots, v_n, x) \in H_{n+1}$. To complete the demonstration that H_n contains exactly one element, we must show that if some element $(w_1, w_2, \ldots, w_n, w_{n+1}) \in H_{n+1}$, then $(v_1, v_2, \ldots, v_n, x) = (w_1, w_2, \ldots, w_n, w_{n+1})$. However, this is immediate from the minimum principle of H. This completes this induction argument.

We are ready to define the sequence $\{x_n\}_{n=1}^{\infty}$. For each $n \in \mathbf{N}$, we define x_n to be v_n, in which $(v_1, v_2, \ldots, v_n)$ denotes the unique member of H_n. Now, it is clear that $x_1 = a$. To complete the proof, the reader is asked to show (via induction) that $x_{n+1} = g(x_1, x_2, \ldots, x_n)$ for each $n \in N$.

EXERCISES

1. Prove Theorem 4.2.

2. Show how the Generalized Recursion Theorem can be used to derive the Recursion Theorem.

3. [For readers of the proof of Theorem 4.3]. Complete the proof by showing that $x_{n+1} = g(x_1, x_2, \ldots, x_n)$ for each $n \in \mathbf{N}$.

4. We can define a sequence $\{a_n\}_{n=0}^{\infty}$ of integers via the Generalized Recursion Theorem as follows: $a_0 = 0$, $a_1 = 1$, and $a_{n+1} = a_{n-1} + a_n$ for each $n \in \mathbf{N}$. Find a_{10}. The sequence $\{a_n\}_{n=0}^{\infty}$ is called a **Fibonacci sequence**. It is useful in describing many natural phenomena.

Section 4.2   Applications to Binary Operations

Consider an $n \in \mathbf{N}$ and a finite sequence $\{x_i\}_{i=1}^{n}$ of real numbers. In your previous mathematical training, you have probably been calculating sums such as $x_1 + x_2 + \cdots + x_n$ and products such as $x_1 x_2 \cdots x_n$ without giving much thought to what these sums and products really mean. You also probably know intuitively that it does not matter how you group parentheses or rearrange the order. For example, students have been taught that $(x_1 + x_2) + (x_3 + x_4) = (x_1 + (x_2 + x_3)) + x_4 = (x_3 + x_1) + (x_4 + x_2)$. You probably perform such operations instinctively. However, it takes some work to go from the case of adding/multiplying two or three numbers to the more general case of adding/multiplying n numbers.

In this section, we will consider binary operations on a nonempty set D, in which these operations are commutative and associative. We will then use recursion to demonstrate that such operations can be extended to operate on n items of the set for each $n \in \mathbf{N}$, and it will be established that these "generalized" operations are commutative and associative. These results are called the **generalized commutative law** and the **generalized associative law** respectively. Most mathematics books do not address these issues, and there is very little discussion of them. However, a little focus on them can be very instructive to the student concerning how mathematics works.

Before doing our work in the most general setting, we consider a special case. This special case involves deriving the properties of positive integer exponents for binary operations that are commutative and associative. Because we are working with general commutative and associative binary operations, all the results of this section will carry over to many systems of interest, such as the real number system, so the results will not have to be derived again.

Positive Integer Exponents

Let D be a nonempty set. Let us consider a binary operation on D. For the moment, we will use the "multiplicative" notation. That is, for any $u, v \in D$, we will express the application of the binary operation as $u \cdot v$ or uv. (Shortly afterward we will demonstrate how the notation changes when we use "additive" notation). Let $a \in D$. Definition 4.4 below uses the Recursion Theorem to define what is meant by raising a to the nth power, for each $n \in \mathbf{N}$. That is, a^n is defined for each $n \in \mathbf{N}$.

Definition 4.4

Let D be a nonempty set, and let $a \in D$. Suppose there is a binary operation on D that we denote by "multiplication." We define a^n for each $n \in \mathbf{N}$ as follows via the Recursion Theorem:

1) $a^1 = a$.
2) $a^{n+1} = a \cdot a^n$ for each $n \in \mathbf{N}$.

There are several points that should be made with regard to Definition 4.4. The first one is that we have become a little lazy (as promised). Even though the Recursion Theorem is invoked to make the definition meaningful, we have not invoked it formally; that is, we have not explicitly listed all of the pieces required by Theorem 4.1. The reader is asked to do so in the exercises.

The second point concerning Definition 4.4 is that it was not specified that the binary operation be commutative and associative. The definition "works" in all cases. However, we require that the binary operations under consideration be commutative and associative for the theorems of this section. Although this may not be the most general case, it is the case that we are most interested in for the remainder of this text.

It is now possible to prove some of the laws of exponents in the case where the exponents are positive integers (i.e., as in Definition 4.4).

Theorem 4.4   Let D be a nonempty set, and let $a \in D$. Suppose there is a binary operation on D that is associative and commutative that we denote by "multiplication." Then $a^m \cdot a^n = a^{m+n}$ for each $m, n \in \mathbf{N}$.

Proof: We perform an induction on m. Now, $a^1 \cdot a^n = a \cdot a^n = a^{n+1} = a^{1+n}$ for each $n \in \mathbf{N}$. (Justify each step!) Hence, the result holds for 1. We now take the inductive step and assume that the result holds for some $m \in \mathbf{N}$. That is, we assume that $a^m \cdot a^n = a^{m+n}$ for each $n \in \mathbf{N}$. The goal is to show that the result holds true for $m + 1$. That is, it must be shown that $a^{m+1} \cdot a^n = a^{(m+1)+n}$ for each $n \in \mathbf{N}$. Now,

$$
\begin{aligned}
a^{(m+1)+n} &= a^{(m+n)+1} &&\text{(commutativity and associativity of addition for } \mathbf{N}) \\
&= a \cdot a^{m+n} &&\text{(definition of } a^{(m+n)+1}) \\
&= a \cdot (a^m a^n) &&\text{(inductive step)} \\
&= (a \cdot a^m) a^n &&\text{(associativity of binary operation on } D) \\
&= a^{m+1} a^n &&\text{(definition of } a^{m+1}).
\end{aligned}
$$

Thus, the induction argument is complete.

From now on, proofs of many simple substeps will not have their justifications explicitly stated. The student should increasingly be able to supply these justifications. Of course, the complete justifications for "complicated" leaps will continue to be provided.

Theorem 4.5   Let D be a nonempty set, and let $a, b \in D$. Suppose there is a binary operation on D that is associative and commutative that we denote by "multiplication." Then $(ab)^n = a^n b^n$ for each $n \in \mathbf{N}$.

Proof: We perform an induction on n. It clearly holds for 1. Suppose it holds for some $n \in \mathbf{N}$ (i.e., take the inductive step). We must show that it holds for $n + 1$. That is, we must show that

$$(ab)^{n+1} = a^{n+1} b^{n+1}.$$

Now,

$$
\begin{aligned}
(ab)^{n+1} = (ab)[(ab)^n] = (ab)(a^n b^n) &= [b(aa^n)]b^n \\
&= (ba^{n+1})b^n = a^{n+1}(b \cdot b^n) = a^{n+1}b^{n+1}.
\end{aligned}
$$

The next theorem is left for the reader to prove. Note that in this theorem there are two different "multiplications" sharing the same notation. There is the "multiplication" of elements of D, and the ordinary "multiplication" of positive integers.

Theorem 4.6

Let D be a nonempty set, and let $a \in D$. Suppose there is a binary operation on D that is associative and commutative that we denote by "multiplication." Then $(a^m)^n = a^{mn}$ for each $m, n \in \mathbf{N}$.

Before going on to the next topic, there is an important note that should be made concerning notation. How do we denote "exponentiation" in the case in which the binary operation on D is denoted by something other than multiplication? There is a standard notation frequently used in abstract algebra when the binary operation is denoted by "addition."

Recall that when the "multiplication" notation is used to denote the binary operation, then a^n as defined in Definition 4.4 means the "multiplication" of a by itself n times. When the "addition" notation is used to denote the binary operation, the expression na is used to mean the "addition" of a with itself n times. That is, $na = a + a + \cdots + a$, in which there are n occurrences of a.

How are Theorems 4.4 through 4.6 written using the "additive" notation? Theorem 4.4 is written $ma + na = (m + n)a$. (Note the "abuse of notation" here. When we wrote $ma + na$, the "+" refers to the binary operation on D. When we wrote $m + n$, the "+" refers to the "normal" addition of integers.) Theorem 4.5 is written $n(a + b) = na + nb$. The reader is asked to write Theorem 4.6 using this additive notation.

Recall that if we are working with a ring, then we have two operations on our set. We usually use addition to denote one of them and multiplication to denote the other, as in Definition 3.12. In this situation, the notation a^n is used for the multiplication operation, and the notation na is used for the addition operation.

Finally, consider the (as yet undeveloped) set of real numbers. If $a \in \mathbf{R}$ and $n \in \mathbf{N}$, then the expression na could be interpreted in two ways. One interpretation is as we just discussed; the addition of a with itself n times. The other interpretation is the multiplication of the real number n with the real number a. It should come as no surprise to you that we later show that both interpretations result in the same answer.

Generalized Binary Operations

Next, we generalize what we just accomplished. Let D be a nonempty set. As before, let us consider a binary operation on D that is commutative and associative. For notational convenience, we usually represent this operation by addition for the remainder of this section. Of course, we could use whatever notation we like to denote this binary operation. Choice of notation does not change basic results.

Consider any $n \in \mathbf{N}$ and any $(d_1, d_2, \ldots, d_n) \in D^n$. We want to define what is meant by $d_1 + d_2 + \cdots + d_n$; that is, the sum of $d_1, d_2, \ldots, d_n$ (using additive notation). In the Exercises, the reader is asked to show that this has already been accomplished for the case in which each d_i is the same element, say d, for each $i \in \mathbf{I}_n$. In this special case, the reader is asked in the Exercises to show that the desired sum turns out to be nd.

For the general case, we use the Recursion Theorem to show the existence of a sequence of functions $\{s_n\}_{n=1}^{\infty}$, in which each $s_n: D^n \to D$. For each $(d_1, d_2, \ldots, d_n) \in D^n$, $s_n(d_1, d_2, \ldots, d_n)$ will denote $d_1 + d_2 + \cdots + d_n$. Then it will be shown that this "general sum" is commutative and associative. The proof of commutativity is not hard. However, the proof of associativity is a little complicated. The associativity proof has been designated as optional.

Definition 4.5

Let D be a nonempty set, and suppose there is a binary operation on D that is associative and commutative that we denote by addition. We define $\{s_n\}_{n=1}^{\infty}$ by the Recursion Theorem as follows, in which each $s_n: D^n \to D$:

1) $s_1(d) = d$ for each $d \in D$.

2) $s_{n+1}(d_1, d_2, \ldots, d_n, d_{n+1}) = s_n(d_1, d_2, \ldots, d_n) + d_{n+1}$ for each $n \in \mathbf{N}$ and each $(d_1, d_2, \ldots, d_n, d_{n+1}) \in D^{n+1}$.

To show how simple this definition really is, consider some $(d_1, d_2, d_3, d_4) \in D^4$. Suppose that we want to find $s_4(d_1, d_2, d_3, d_4)$. Property 1 of Definition 4.5 says that $s_1(d_1) = d_1$. This is the "sum" having only one element. Now, Property 2 says that $s_2(d_1, d_2) = s_1(d_1) + d_2 = d_1 + d_2$. Continuing to use Property 2, we have $s_3(d_1, d_2, d_3) = s_2(d_1, d_2) + d_3 = (d_1 + d_2) + d_3$. Finally, $s_4(d_1, d_2, d_3, d_4) = s_3(d_1, d_2, d_3) + d_4 = ((d_1 + d_2) + d_3) + d_4$. We have to keep the parentheses for the moment, because we have not yet discussed the generalized associativity law.

Now, as previously mentioned, the function g of Theorem 4.1 (the Recursion Theorem) is not usually explicitly mentioned in most advanced mathematics texts. However, it is not always obvious as to what this function g should be. Because the author of this book is compassionate, and this is a "bridging text," we will demonstrate how to define this function for Definition 4.5.

Hence, we need to define some function g so that $s_{n+1} = g(s_n)$ for each $n \in \mathbf{N}$. Recall that each s_n is to be a function with domain D^n and range D.

Therefore, however we define g, its domain has to be a set of functions. Consequently, it seems reasonable to define the domain of g to be the set of all functions with domain D^n and range D for all positive integers n. Let us call this domain Ψ. Hence, $\Psi = \{f: D^n \to D | n \in \mathbf{N}\}$. (Another way to write this is $\Psi = \cup_{n=1}^{\infty} \Psi_n$ in which $\Psi_n = \{f | f: D^n \to D\}$ for each $n \in \mathbf{N}$.) Hence, to conform to the Recursion Theorem, the range of g will also be Ψ. Hence, the sequence $\{s_n\}_{n=1}^{\infty}$ that we are about to define will have domain $\mathbf{N}$ (of course) and range Ψ.

To define $g: \Psi \to \Psi$, we must specify $g(f)$ for each $f \in \Psi$. This makes for some awkward notation because given that we must have $g(f) \in \Psi$, this means that $g(f): D^m \to D$ for some positive integer m (to be chosen below). We shall write the image of each $(d_1, d_2, \ldots, d_m)$ in D^m with respect to $g(f)$ as $\langle g(f) \rangle (d_1, d_2, \ldots, d_m)$. The author almost apologizes for this annoying notation. Please note that $g(f)$ is **not** the composition of g and f.

With this (temporary) notation, we now formally define $g: \Psi \to \Psi$ as follows: Given an $f \in \Psi$, there must exist a unique integer $n \in \mathbf{N}$ such that $f: D^n \to D$. We define $g(f): D^{n+1} \to D$ by $\langle g(f) \rangle (d_1, d_2, \ldots, d_n, d_{n+1}) = f(d_1, d_2, \ldots, d_n) + d_{n+1}$. Hence, we may now formally invoke the Recursion Theorem to state that there exists a unique sequence $\{s_n\}_{n=1}^{\infty}$ with range Ψ, with $s_n: D^n \to D$ for each $n \in \mathbf{N}$ such that

1) $s_1(d) = d$ for each $d \in D$; and

2) $s_{n+1} = g(s_n)$ for each $n \in \mathbf{N}$.

In the exercises, the reader is asked to complete the proof that, in fact, $s_{n+1}(d_1, d_2, \ldots, d_n, d_{n+1}) = s_n(d_1, d_2, \ldots, d_n) + d_{n+1}$ for each $n \in \mathbf{N}$ and each $(d_1, d_2, \ldots, d_n, d_{n+1}) \in D^{n+1}$.

It often will be the case in advanced mathematics that we will have to count how many integers there are in a list of integers. Theorem 4.7 specifies the number of positive integers that are in the set $\{i \in \mathbf{N} | k \leq i \leq n\}$, where $k, n \in \mathbf{N}$ with $k \leq n$. The reader may want to review the use of the notation **card** defined in Section 3.4. Theorem 4.7 below is very important in its own right. Our reason for introducing it here is that we shall need it for additional results that we want to prove about binary operations.

Theorem 4.7   Suppose $k, n \in \mathbf{N}$ with $k \leq n$. Then $\mathbf{card}(\{i \in \mathbf{N} | k \leq i \leq n\}) = (n - k) + 1$.

Proof: Induction on n. To show the result holds for 1, we must show that if $k \in \mathbf{N}$ with $k \leq 1$, then $\mathbf{card}(\{i \in \mathbf{N} | k \leq i \leq 1\}) = (1 - k) + 1$. However, it is clear that in this case we must have $k = 1$, so $(1 - k) + 1 = (1 - 1) + 1 = 0 + 1 = 1$. Also, the set $\{i \in \mathbf{N} | 1 \leq i \leq 1\}$ in this case is the set $\{1\}$, which is clearly of size 1. Hence, the result holds for 1.

We now make the inductive step and assume that the result holds for some $n \in \mathbf{N}$. We will show that this forces the result to also hold for $n + 1$. To show that it holds for $n + 1$, suppose that $k \in \mathbf{N}$ with $k \leq n + 1$. We must show that $\mathbf{card}(\{i \in \mathbf{N} | k \leq i \leq n + 1\}) = ((n + 1) - k) + 1$. Of course, $((n + 1) - k) + 1 = n - k + 2$. Now, $\{i \in \mathbf{N} | k \leq i \leq n + 1\} = \{i \in \mathbf{N} | k \leq i \leq n\} \cup \{n + 1\}$. Now, by Exercise 5 of Section 3.4, $\mathbf{card}(\{i \in \mathbf{N} | k \leq i \leq n + 1\}) = \mathbf{card}(\{i \in \mathbf{N} | k \leq i \leq n\}) + \mathbf{card}(\{n + 1\})$. By the induction hypothesis, we have $\mathbf{card}(\{i \in \mathbf{N} | k \leq i \leq n\}) = (n - k) + 1$. The result clearly follows because $\mathbf{card}(\{n + 1\}) = 1$.

Example 4.7

Theorem 4.7 indicates that $\mathbf{card}(\{i \in \mathbf{N} | 3 \leq i \leq 7\}) = (7 - 3) + 1 = 5$. This is easy to verify because the set $\{i \in \mathbf{N} | 3 \leq i \leq 7\}$ is simply the set $\{3, 4, 5, 6, 7\}$, and it is clear that this set is of size 5.

The reader is asked to show (Exercise 6) that the conclusion of Theorem 4.7 also holds if $\mathbf{N}$ is replaced with $\mathbf{Z}$.

As an important application of Theorem 4.7, suppose D is a nonempty set, and consider some $(d_1, d_2, \ldots, d_n) \in D^n$, in which $n \in \mathbf{N}$ with $n > 1$. Then, for any $k \in \mathbf{N}$ with $k \leq n$, Theorem 4.7 indicates that $(d_k, d_{k+1}, \ldots, d_n)$ is a $n - k + 1$-tuple. That is, $(d_k, d_{k+1}, \ldots, d_n) \in D^{n-k+1}$. [Of course, by the expression $n - k + 1$ we mean $(n - k) + 1$.]

Now consider any $n \in \mathbf{N}$ with $n > 1$, and suppose $m \in \mathbf{N}$ with $m < n$. The next theorem indicates that the "sum" $s_n(d_1, d_2, \ldots, d_n)$ can be obtained by "adding" $s_m(d_1, d_2, \ldots, d_m)$ to $s_{n-m}(d_{m+1}, d_{m+2}, \ldots, d_n)$. We note that this makes sense because $(d_{m+1}, d_{m+2}, \ldots, d_n)$ is an $n - (m + 1) + 1$-tuple; that is, it is an $n - m$-tuple.

Theorem 4.8

Suppose that D and $\{s_n\}_{n=1}^{\infty}$ are as defined in Definition 4.5. Then for each $n \in \mathbf{N}$ with $n > 1$, and each $m \in \mathbf{N}$ with $m < n$, we have

$$s_n(d_1, d_2, \ldots, d_n) = s_m(d_1, d_2, \ldots, d_m) + s_{n-m}(d_{m+1}, d_{m+2}, \ldots, d_n)$$

for each $(d_1, d_2, \ldots, d_n) \in D^n$.

Proof: We perform an induction on n starting with 2. For $n = 2$, the theorem claims that $s_2(d_1, d_2) = s_1(d_1) + s_1(d_2)$. The reader should be able to see immediately that this true.

Next, we take the inductive step and assume that the result holds for some $n \in \mathbf{N}$ with $n > 1$. Our goal is show that it holds for $n + 1$. To do so, consider any $m \in \mathbf{N}$ with $m < n + 1$. We must show that

$$s_{n+1}(d_1, d_2, \ldots, d_n, d_{n+1})$$
$$= s_m(d_1, d_2, \ldots, d_m) + s_{n+1-m}(d_{m+1}, d_{m+2}, \ldots, d_n, d_{n+1}).$$

Now, if $m = n$, we note that $s_{n+1-m}(d_{m+1}, d_{m+2}, \ldots, d_n, d_{n+1})$ is simply $s_1(d_{m+1}) = d_{m+1}$, so $s_m(d_1, d_2, \ldots, d_m) + s_{n+1-m}(d_{m+1}, d_{m+2}, \ldots, d_n, d_{n+1}) = s_m(d_1, d_2, \ldots, d_m) + d_{m+1}$. However, by the recursive definition of s_{m+1}, this is simply $s_{m+1}(d_1, d_2, \ldots, d_m, d_{m+1})$, which is the same as $s_{n+1}(d_1, d_2, \ldots, d_n, d_{n+1})$ because $m = n$.

For the remainder of the proof, we examine the case in which $m < n$. Now,

$$s_m(d_1, d_2, \ldots, d_m) + s_{n+1-m}(d_{m+1}, d_{m+2}, \ldots, d_n, d_{n+1})$$

$$= s_m(d_1, d_2, \ldots, d_m) + (s_{n-m}(d_{m+1}, d_{m+2}, \ldots, d_n) + d_{n+1})$$
$$\text{(definition of } s_{n+1-m})$$

$$= (s_m(d_1, d_2, \ldots, d_m) + s_{n-m}(d_{m+1}, d_{m+2}, \ldots, d_n)) + d_{n+1}$$
$$\text{(associativity)}$$

$$= s_n(d_1, d_2, \ldots, d_n) + d_{n+1} \quad \text{(inductive step)}$$

$$= s_{n+1}(d_1, d_2, \ldots, d_n, d_{n+1}) \quad \text{(definition of } s_{n+1}).$$

Some comments concerning notation will be very useful. Recall that $s_n(d_1, d_2, \ldots, d_n)$ represents what we frequently write as $d_1 + d_2 + \cdots + d_n$ using the "additive" notation for the binary operation. Using the latter notation, we can express the conclusion of Theorem 4.8 as

$$d_1 + d_2 + \cdots + d_n = (d_1 + d_2 + \cdots + d_m) + (d_{m+1} + d_{m+2} + \cdots + d_n).$$

No doubt, the result of the theorem is more familiar to the reader with this notation.

There is another notation that is frequently employed with "additive" notation to represent the given binary operation. The reader may be familiar with the use of the capital Greek letter sigma, Σ, to represent "sums." Employing this, we can write $d_1 + d_2 + \cdots + d_n$ as $\sum_{i=1}^n d_i$. Of course, the letter i could be replaced by any letter (except for n, of course). For example, $\sum_{k=1}^n d_k$ means the same thing. Note that using the *sigma* notation, Definition 4.5 says:

1) $\sum_{k=1}^1 d_k = d_1$; and
2) $\sum_{k=1}^{n+1} d_k = \sum_{k=1}^n d_k + d_{n+1}$.

The reader is asked to get used to the *sigma* notation. This notation will now be extended. Let D be as in Definition 4.5. Suppose $m, n \in \mathbf{Z}$ with $m \leq n$. Let $f: \{m, m+1, \ldots, n\} \to D$. Then, by $\sum_{k=m}^n f(k)$, we mean $s_{n-m+1}(f(m), f(m+1), \ldots, f(n))$. (Note that Exercise 6 guarantees that this makes sense.) Of course, we can also write this as $f(m) + f(m+1) + \cdots + f(n)$. In this case, the integer m is often called the **lower limit of summation**, and the integer n is often called the **upper limit of summation**. The letter k, which can be replaced by any letter (except m and n in this case), is often called the **index of summation**. Replaceable letters in an expression are often referred to as **dummy variables**.

Using the *sigma* notation, the conclusion of Theorem 4.8 can be written as $\sum_{i=1}^{n} d_i = \sum_{i=1}^{m} d_i + \sum_{i=m+1}^{n} d_i$. (In the determination of $\sum_{i=m+1}^{n} d_i$, what is the function $f: \{m + 1, m + 2, \ldots, n\} \to D$ in this case? It is simply $f(i) = d_i$ for each $i \in \{m + 1, m + 2, \ldots, n\}$. Usually, the function f is not explicitly defined; it is understood to be there.)

In practice, $\sum_{k=m}^{n} f(k)$ is often calculated as follows: First, the dummy variable k is set to be m, resulting in $f(m)$. Next, the dummy variable is set to $m + 1$, and the sum $f(m) + f(m + 1)$ is determined. This process continues until the dummy variable is set to n, and the final sum is determined.

Example 4.8

Write out the sums for

1) $\sum_{k=2}^{5} k$;
2) $\sum_{s=-1}^{2} s^2$;
3) $\sum_{i=-3}^{0} 2$; and
4) $\sum_{n=1}^{3} (n + 1)^2$.

Solution:

1) $\sum_{k=2}^{5} k = 2 + 3 + 4 + 5 = 14$.
2) $\sum_{s=-1}^{2} s^2 = (-1)^2 + 0^2 + 1^2 + 2^2 = 1 + 0 + 1 + 4 = 6$.
3) $\sum_{i=-3}^{0} 2 = 2 + 2 + 2 + 2 = 8$.
4) $\sum_{n=1}^{3} (n + 1)^2 = 2^2 + 3^2 + 4^2 = 4 + 9 + 16 = 29$.

Example 4.9

Suppose the set D is as in Definition 4.5, and suppose we have a finite sequence $\{d_n\}_{n=1}^{10}$ of elements of D. Write out $\sum_{n=1}^{5} d_{2n-1}$.

Solution: $\sum_{n=1}^{5} d_{2n-1} = d_1 + d_3 + d_5 + d_7 + d_9$.

Now that we feel comfortable with the *sigma* notation, we are ready for the next theorem. This theorem says that the sum of the terms of the sum of two finite sequences can be obtained by first summing the terms of each sequence and then adding these two results.

Theorem 4.9

Suppose that D is defined as in Definition 4.5. Then, using the "additive notation," for each $n \in \mathbf{N}$ and each $(d_1, d_2, \ldots, d_n), (e_1, e_2, \ldots, e_n) \in D^n$, we have

$$\sum_{i=1}^{n} (d_i + e_i) = \sum_{i=1}^{n} d_i + \sum_{i=1}^{n} e_i.$$

Proof: We perform an induction on n. The result is trivial for 1. Now, make the induction hypothesis that the result holds for some $n \in \mathbf{N}$. The goal is now to show that $\sum_{i=1}^{n+1}(d_i + e_i) = \sum_{i=1}^{n+1} d_i + \sum_{i=1}^{n+1} e_i$ for any $(d_1, d_2, \ldots, d_n, d_{n+1}), (e_1, e_2, \ldots, e_n, e_{n+1}) \in D^{n+1}$. Now,

$$\sum_{i=1}^{n+1}(d_i + e_i) = \sum_{i=1}^{n}(d_i + e_i) + (d_{n+1} + e_{n+1}) \qquad (\text{definition of } \sum_{i=1}^{n+1}(d_i + e_i))$$

$$= \left(\sum_{i=1}^{n} d_i + \sum_{i=1}^{n} e_i\right) + (d_{n+1} + e_{n+1}) \qquad (\text{induction hypothesis})$$

$$= \left(\sum_{i=1}^{n} d_i + d_{n+1}\right) + \left(\sum_{i=1}^{n} e_i + e_{n+1}\right)$$

$$\text{(associativity and commutativity)}$$

$$= \sum_{i=1}^{n+1}(d_i + e_i)$$

$$\text{(definition of } \sum_{i=1}^{n+1} d_i \text{ and } \sum_{i=1}^{n+1} e_i).$$

The next theorem generalizes the distributive law for rings. The induction proof is left to the reader.

Theorem 4.10   **(Generalized Distributive Law)**

Suppose that D is a ring, with "addition" and "multiplication" represented as in Definition 3.12. Then for each $n \in \mathbf{N}$, each $(d_1, d_2, \ldots, d_n) \in D^n$, and each $a \in D$, we have $a \sum_{k=1}^{n} d_k = \sum_{k=1}^{n} ad_k$. That is, $a(d_1 + d_2 + \cdots + d_n) = ad_1 + ad_2 + \cdots + ad_n$.

As we have pointed out before, there is nothing sacred about representing a binary operation using "additive" notation. All the results of this section hold no matter what notation is used to represent the binary operation on D (provided the binary operation is associative and commutative). There is a special notation that is often used when the binary operation is represented using "multiplicative" notation. The letter *sigma* is replaced with Π, the capital Greek letter *pi*. Hence, $\Pi_{k=m}^{n} f(k)$ is used to denote $f(m) \cdot f(m + 1) \cdots \cdots f(n)$. Of course, in systems such as the integers and the real numbers, there are two binary operations of interest; one is expressed using "additive" notation and the other is expressed using "multiplicative" notation. Hence, both the *sigma* and the *pi* notation is used. Note that using the *pi* notation, Definition 4.5 says

1) $\Pi_{k=1}^{1} \, d_k = d_1$; and
2) $\Pi_{k=1}^{n+1} \, d_k = (\Pi_{k=1}^{n} \, d_k) \, d_{n+1}$.

Using this notation, the conclusion of Theorem 4.8 can be written as

$$\prod_{i=1}^{n} d_i = \left(\prod_{i=1}^{m} d_i \right) \left(\prod_{i=m+1}^{n} d_i \right).$$

The conclusion of Theorem 4.9 can be written as

$$\prod_{i=1}^{n} d_i e_i = \left(\prod_{i=1}^{n} d_i \right) \left(\prod_{i=1}^{n} e_i \right).$$

Finally (for this subsection), we formally define the **factorial function**, even though we already did this in Example 4.6 of Section 4.1.

Definition 4.6

$0!$ is defined to be 1, and $n!$ is defined to be $\Pi_{k=1}^{n} \, k$ for each $n \in \mathbf{N}$.

Generalized Commutative and Associative Laws

As before, let us consider a commutative and associative binary operation on a nonempty set D. For notational convenience, we will continue to represent this operation by "addition."

Consider a "sum" $s_n(d_1, d_2, \ldots, d_n)$ of elements of D, as defined in Definition 4.5. The next goal is to show the generalized commutative law. Intuitively, this law states that no matter how we rearrange the order of these terms, we get the same "sum." For example, this law will imply that $s_4(d_1, d_2, d_3, d_4)$ is the same thing as $s_4(d_2, d_3, d_4, d_1)$. That is, the law will imply that $d_1 + d_2 + d_3 + d_4 = d_2 + d_3 + d_4 + d_1$. However, we first need a rigorous way to define what is meant by "rearranging the order." This is provided by Definition 4.7.

Definition 4.7

Let S be a nonempty set. A **permutation** of S is a bijection $f \colon S \to S$.

Now consider our "friend" $\mathbf{I}_n$, for some $n \in \mathbf{N}$, and consider a permutation of $\mathbf{I}_n$. Frequently, the Greek letter π is used to represent permutations of $\mathbf{I}_n$, although this is by no means required.

| **Example 4.10** | List all of the permutations of $I_3 = \{1, 2, 3\}$. |

Solution: Let us denote these permutations by $\pi_1, \pi_2, \ldots, \pi_6$ as defined by:

$$\pi_1(1) = 1, \pi_1(2) = 2, \text{ and } \pi_1(3) = 3;$$
$$\pi_2(1) = 1, \pi_2(2) = 3, \text{ and } \pi_2(3) = 2;$$
$$\pi_3(1) = 2, \pi_3(2) = 1, \text{ and } \pi_3(3) = 3;$$
$$\pi_4(1) = 2, \pi_4(2) = 3, \text{ and } \pi_4(3) = 1;$$
$$\pi_5(1) = 3, \pi_5(2) = 1, \text{ and } \pi_5(3) = 2; \text{ and}$$
$$\pi_6(1) = 3, \pi_6(2) = 2, \text{ and } \pi_6(3) = 1.$$

Note that Example 4.10 conforms to our intuitive concept that listing all permutations of I_n has the same meaning as listing all the different ways that members of the set can be arranged. It is easy to see what we mean when we say that we get the "sum" $s_n(d_1, d_2, \ldots, d_n)$ no matter how we may arrange the terms. By this we mean that $s_n(d_1, d_2, \ldots, d_n)$ is the same thing as $s_n(d_{\pi(1)}, d_{\pi(2)}, \ldots, d_{\pi(n)})$ for each $n \in \mathbf{N}$, each n-tuple $(d_1, d_2, \ldots, d_n) \in D^n$, and each permutation π of I_n. Also, we go ahead and use the *sigma* notation because we now have it at our disposal.

| **Theorem 4.11** | **(Generalized Commutative Law)** |

Let D be a nonempty set, and suppose there is a binary operation on D that is associative and commutative that we denote by "addition." Then, for each $n \in \mathbf{N}$, each n-tuple $(d_1, d_2, \ldots, d_n) \in D^n$, and each permutation π of I_n, we have

$$\sum_{k=1}^{n} d_{\pi(k)} = \sum_{k=1}^{n} d_k.$$

Proof: Part of the proof will be left to the reader. We perform an induction on n. It clearly holds for 1 because the only permutation of I_1 is the identity mapping on I_1.

We now make the induction hypothesis that the result holds for some $n \in \mathbf{N}$. Let $(d_1, d_2, \ldots, d_n, d_{n+1}) \in D^{n+1}$, and suppose π is a permutation of I_{n+1}. The goal is to show that $\sum_{k=1}^{n+1} d_{\pi(k)} = \sum_{k=1}^{n+1} d_k$.

Because π is a permutation of I_{n+1}, there exists a unique $m \in I_{n+1}$ such that $\pi(m) = n + 1$. We consider three cases: the case in which $m = 1$, the case in which $1 < m < n + 1$, and the case in which $m = n + 1$.

Case 1: $m = 1$. Hence, it is clear that the restriction of π to $I_{n+1} - \{1\}$ is a bijection with range I_n. Define $\phi: I_n \to I_n$ by $\phi(k) = \pi(k + 1)$ for all $k \in I_n$. Then it is clear that ϕ is a permutation of I_n. Now,

$$\sum_{k=2}^{n+1} d_{\pi(k)} = \sum_{k=1}^{n} d_{\pi(k+1)} \quad \text{(same terms of the "sum")}$$

$$= \sum_{k=1}^{n} d_{\phi(k)} \quad \text{(definition of } \phi)$$

$$= \sum_{k=1}^{n} d_k \quad \text{(induction hypothesis)}.$$

Now,

$$\sum_{k=1}^{n+1} d_{\pi(k)} = \sum_{k=1}^{1} d_{\pi(k)} + \sum_{k=2}^{n+1} d_{\pi(k)} \quad \text{(Theorem 4.8)}$$

$$= d_{\pi(1)} + \sum_{k=2}^{n+1} d_{\pi(k)} \quad \left(\text{definition of } \sum_{k=1}^{1} d_{\pi(1)}\right)$$

$$= d_{n+1} + \sum_{k=2}^{n+1} d_{\pi(k)} \quad \text{(Case 1 assumption)}$$

$$= d_{n+1} + \sum_{k=1}^{n} d_k \quad \text{(above result)}$$

$$= \sum_{k=1}^{n} d_k + d_{n+1} \quad \text{(commutativity)}$$

$$= \sum_{k=1}^{n+1} d_k \quad \text{(Definition 4.5)}.$$

Case 2: $1 < m < n + 1$. Hence, it is clear that that the restriction of π to $\mathbf{I}_{n+1} - \{m\}$ is a bijection with range $\mathbf{I}_n$.
Define $\phi: \mathbf{I}_n \to \mathbf{I}_n$ by

$$\phi(k) = \begin{cases} \pi(k) & \text{for } k \in \{1, 2, \cdots, m-1\} \\ \pi(k+1) & \text{for } k \in \{m, m+1, \cdots, n\} \end{cases}.$$

Then it is clear that ϕ is a permutation of $\mathbf{I}_n$. The remainder of Case 2 is left to the reader.

Case 3: $m = n + 1$. Hence, it is clear that the restriction of π to $\mathbf{I}_n$ is a permutation of $\mathbf{I}_n$. Hence,

$$\sum_{k=1}^{n+1} d_{\pi(k)} = \sum_{k=1}^{n} d_{\pi(k)} + d_{\pi(n+1)} \quad \text{(Definition 4.5)}$$

$$= \sum_{k=1}^{n} d_{\pi(k)} + d_{n+1} \quad \text{(Case 3 assumption)}$$

$$= \sum_{k=1}^{n} d_k + d_{n+1} \quad \text{(induction hypothesis)}$$

$$= \sum_{k=1}^{n+1} d_k \quad \text{(Definition 4.5)}.$$

This concludes the induction argument. We have shown that the order in which we add the elements always results in the same sum.

———

Next, we discuss the **generalized associative law**. This turns out to be a little more difficult, so we have made the formal part of the discussion optional. The basic idea is discussed here. We want to show that no matter how we arrange parentheses within a sum, we get the same answer. The problem comes in quantifying what is meant by "no matter how we arrange parentheses." The following argument is taken from Burril (1967).

Given any $n \in \mathbf{N}$ and each n-tuple $(d_1, d_2, \ldots, d_n) \in D^n$, the basic idea is to use the Generalized Recursion Theorem to create a set, which will be denoted $S_n(d_1, d_2, \ldots, d_n)$. Here, we are using capital S instead of lowercase s. This set will contain all of the various ways that "sums" can be written using all of the components of $(d_1, d_2, \ldots, d_n)$ in order, in which parentheses are rearranged differently.

For example, $S_1(d_1)$ will simply be the set $\{d_1\}$. $S_2(d_1, d_2)$ will be the set $\{d_1 + d_2\}$. $S_3(d_1, d_2, d_3)$ will be the set of "sums" $\{(d_1 + d_2) + d_3, d_1 + (d_2 + d_3)\}$. Of course, you see that because the binary operation is associative, the two "sums" in $S_3(d_1, d_2, d_3)$ are the same. In general, it will be shown that all of the sums of $S_n(d_1, d_2, \ldots, d_n)$ are the same; they are all equal to $s_n(d_1, d_2, \ldots, d_n)$ of Definition 4.5 (i.e., $\sum_{i=1}^{n} d_i$).

Intuitively using the Generalized Recursion Theorem (Theorem 4.3), $S_n(d_1, d_2, \ldots, d_n)$ (for $n > 1$) is the set of all sums $u = v + w$ in which $v \in S_m(d_1, d_2, \ldots, d_m)$ for $m < n$, and $w \in S_{n-m}(d_{m+1}, d_{m+2}, \ldots, d_n)$. [Note from Theorem 4.7 that $(d_{m+1}, d_{m+2}, \ldots, d_n)$ is an $n - m$-tuple.]

Example 4.11

To demonstrate this, consider $S_4(d_1, d_2, d_3, d_4)$. In this example, $n = 4$. Now, start with $m = 1$. Now, the only member of $S_1(d_1)$ is d_1. There are two members of $S_3(d_2, d_3, d_4)$; they are $(d_2 + d_3) + d_4$ and $d_2 + (d_3 + d_4)$. Hence, $S_1(d_1)$ and $S_3(d_2, d_3, d_4)$ contribute the following members to $S_4(d_1, d_2, d_3, d_4)$: $d_1 + ((d_2 + d_3) + d_4)$ and $d_1 + (d_2 + (d_3 + d_4))$.

Next set $m = 2$. The only member of $S_2(d_1, d_2)$ is $d_1 + d_2$. Similarly, the only member of $S_2(d_3, d_4)$ is $d_3 + d_4$. Hence, the sets $S_2(d_1, d_2)$ and $S_2(d_3, d_4)$ contribute only $(d_1 + d_2) + (d_3 + d_4)$ to $S_4(d_1, d_2, d_3, d_4)$.

Finally, set $m = 3$. The members of $S_3(d_1, d_2, d_3)$ are $(d_1 + d_2) + d_3$ and $d_1 + (d_2 + d_3)$. The only member of $S_1(d_4)$ is d_4. Hence, the sets $S_3(d_1, d_2, d_3)$ and $S_1(d_4)$ contribute $((d_1 + d_2) + d_3) + d_4$ and $(d_1 + (d_2 + d_3)) + d_4$ to $S_4(d_1, d_2, d_3, d_4)$. This is illustrated in Figure 4.1. The contributions to $S_4(d_1, d_2, d_3, d_4)$ of each of the combinations of sets is indicated by an arrow.

As we move in the direction of more formality, the Generalized Recursion Theorem is used (as shown below) to construct a sequence $\{S_n\}_{n=1}^{\infty}$. Each S_n

Figure 4.1
Illustrating Example
4.11

$$S_1(d_1) \text{ and } S_3(d_2,d_3,d_4) \quad \rightarrow \quad d_1 + ((d_2 + d_3) + d_4) \text{ and}$$
$$d_1 + (d_2 + (d_3 + d_4))$$

$$S_2(d_1,d_2) \text{ and } S_2(d_3,d_4) \quad \rightarrow \quad (d_1 + d_2) + (d_3 + d_4)$$

$$S_3(d_1,d_2,d_3) \text{ and } S_1(d_4) \quad \rightarrow \quad ((d_1 + d_2) + d_3) + d_4 \text{ and}$$
$$(d_1 + (d_2 + d_3)) + d_4$$

is to be a function with domain D^n. What is used for the range of each S_n? Recall that we "want" $S_n(d_1, d_2, \ldots, d_n)$ to be a set of elements of D for each $(d_1, d_2, \ldots, d_n) \in D^n$. Therefore, to be safe, it is reasonable to let the range of each S_n be the set of all subsets of D, which we shall denote by $\mathbf{P}(D)$. (The capital $\mathbf{P}$ stands for *power set*.) Of course, it will turn out that each $S_n(d_1, d_2, \ldots, d_n)$ will be a set consisting of only one element, that element being $\sum_{i=1}^n d_i$. This is the essence of the generalized associative law. This law is stated precisely in the next subsection. The next subsection has been designated as optional because it is somewhat dry and complicated (although not beyond the studious student's grasp).

Generalized Associative Law [Optional]

As before, let D be a nonempty set, and we consider a binary operation on D that is commutative and associative, which we represent by "addition." Define the set V by $V = \{\phi: D^n \rightarrow \mathbf{P}(D) | n \in \mathbf{N}\}$. Also, define $c \in V$ by $c: D^1 \rightarrow \mathbf{P}(D)$ with $c(d) = \{d\}$ for each $d \in D^1$. (Of course, Definition 4.3 mandates that D^1 is just D.)

The goal is to use the Generalized Recursion Theorem to show the existence of a sequence $\{S_n\}_{n=1}^\infty$ with range V. Recall that for any $(d_1, d_2, \ldots, d_n) \in D^n$, we "want" each $S_n(d_1, d_2, \ldots, d_n)$ to be a set that contains all of the various ways that "sums" can be written using all of the components of $(d_1, d_2, \ldots, d_n)$ in order, in which parentheses are rearranged differently. Note that this will require that each $S_n \in V$.

To properly invoke the Generalized Recursion Theorem, we will define an appropriate function $g: \cup_{n=1}^\infty V^n \rightarrow V$. What must we do to specify g? Consider any $h \in \cup_{n=1}^\infty V^n$. Then, by definition, $h \in V^n$ for exactly one $n \in \mathbf{N}$. That is, $h = (h_1, h_2, \ldots, h_n)$ in which each $h_i \in V$. Hence, for each such h_i, there is an $m_i \in \mathbf{N}$ (which depends on h) so that $h_i: D^{m_i} \rightarrow \mathbf{P}(D)$.

We will now specify $g(h): D^{n+1} \rightarrow \mathbf{P}(D)$. We will express the image of each $(d_1, d_2, \ldots, d_n, d_{n+1}) \in D^{n+1}$ with respect to $g(h)$ as

$\langle g(h)\rangle(d_1, d_2, \ldots, d_n, d_{n+1})$. First, we examine the case in which $m_i = i$ for each $i \in \mathbf{I}_n$. In this case, we specify $\langle g(h)\rangle(d_1, d_2, \ldots, d_n, d_{n+1})$ to be the subset of D defined by $\cup_{k=1}^{n}\{a + b | a \in h_k(d_1, d_2, \ldots, d_k)$ and $b \in h_{n-k+1}(d_{k+1}, \ldots, d_{n+1})\}$. Now, we really do not care about the case in which $m_i \neq i$ for some $i \in \mathbf{I}_n$. However, we must define $\langle g(h)\rangle(d_1, d_2, \ldots, d_n, d_{n+1})$ for completeness. In this case, we define $\langle g(h)\rangle(d_1, d_2, \ldots, d_n, d_{n+1})$ to be the empty set, which is a valid member of $\mathbf{P}(D)$.

Definition 4.8

Suppose that D, V, c, and g are as discussed above. Then we define the sequence $\{S_n\}_{n=1}^{\infty}$ via the Generalized Recursion Theorem as follows:
i) $S_1 = c$; and
ii) $S_{n+1} = g(S_1, S_2, \ldots, S_n)$ for each $n \in \mathbf{N}$.

Now, we have $S_n \in V$ for each $n \in \mathbf{N}$ (by definition). However, as Lemma 4.1 indicates, we can state more than this.

Lemma 4.1

Suppose that D, V, c g, and $\{S_n\}_{n=1}^{\infty}$ are as in Definition 4.8. Then $S_n : D^n \to \mathbf{P}(D)$ for each $n \in \mathbf{N}$.

Proof: We do an induction on n. Now, $S_1 = c$, and $c : D^1 \to \mathbf{P}(D)$. Hence, the result holds for 1. Suppose (our induction hypothesis) that it holds for some $n \in \mathbf{N}$. We must show that $S_{n+1} : D^{n+1} \to \mathbf{P}(D)$. Now, $S_{n+1} = g(S_1, S_2, \ldots, S_n)$, so the result is immediate by the definition of g.

Now, from the definition of g and from Lemma 4.1, it is clear that for each $n \in \mathbf{N}$, we have $S_{n+1}(d_1, d_2, \ldots, d_{n+1}) = \cup_{k=1}^{n}\{a + b | a \in S_k(d_1, d_2, \ldots, d_k)$ and $b \in S_{n-k+1}(d_{k+1}, \ldots, d_{n+1})\}$ for each $(d_1, d_2, \ldots, d_n, d_{n+1}) \in D^{n+1}$. (How did we use Lemma 4.1?) Hence, we now know that each $S_{n+1}(d_1, d_2, \ldots, d_{n+1})$ is a set of "sums" of elements of D, as discussed at the end of the previous subsection. We are now ready for the main result, which simply states that the only member of $S_n(d_1, d_2, \ldots, d_n)$ is $s_n (d_1, d_2, \ldots, d_n)$ (i.e., $\sum_{i=1}^{n} d_i$).

Theorem 4.12   (Generalized Associative Law)

Let $\{S_n\}_{n=1}^{\infty}$ be as in Definition 4.8, and let $\{s_n\}_{n=1}^{\infty}$ be as in Definition 4.5. Then for each $(d_1, d_2, \ldots, d_n) \in D^n$, $s_n(d_1, d_2, \ldots, d_n)$ is the only member of $S_n(d_1, d_2, \ldots, d_n)$.

Proof: We use the **Second Principle of Mathematical Induction** (Theorem 3.23). For any $d \in D$, $S_1(d) = \{d\} = \{s_1(d)\}$, so the result holds for 1. Assume the induction hypothesis for some $n \in \mathbf{N}$. For the Second Principle of Mathematical Induction, this hypothesis is that the result holds for any $k \in \mathbf{N}$ with $k \le n$. Let $(d_1, d_2, \ldots, d_{n+1}) \in D^{n+1}$. The goal is to show that $s_{n+1}(d_1, d_2, \ldots, d_{n+1})$ is the only member of $S_{n+1}(d_1, d_2, \ldots, d_{n+1})$.

Now, from the discussion immediately following Lemma 4.1, we know that $S_{n+1}(d_1, d_2, \ldots, d_{n+1}) = \cup_{k=1}^{n}\{a + b | a \in S_k(d_1, d_2, \ldots, d_k)$ and $b \in S_{n-k+1}(d_{k+1}, \ldots, d_{n+1})\}$. However, by the induction hypothesis, $S_k(d_1, d_2, \ldots, d_k) = \{s_k(d_1, d_2, \ldots, d_k)\}$ and $S_{n-k+1}(d_{k+1}, \ldots, d_{n+1}) = \{s_{n-k+1}(d_{k+1}, \ldots, d_{n+1})\}$ for each $k \in \mathbf{I}_n$. Now, by Theorem 4.8, $s_k(d_1, d_2, \ldots, d_k) + s_{n-k+1}(d_{k+1}, \ldots, d_{n+1}) = s_{n+1}(d_1, d_2, \ldots, d_{n+1})$ for each $k \in \mathbf{I}_n$. This says that $S_{n+1}(d_1, d_2, \ldots, d_{n+1}) = \cup_{k=1}^{n}\{s_{n+1}(d_1, d_2, \ldots, d_{n+1})\} = \{s_{n+1}(d_1, d_2, \ldots, d_{n+1})\}$. That is, $s_{n+1}(d_1, d_2, \ldots, d_{n+1})$ is the only member of $S_{n+1}(d_1, d_2, \ldots, d_{n+1})$.

■

Note how much work was necessary to show something that is intuitively obvious!

EXERCISES

1. Demonstrate the existence of a function $g: D \to D$ for Definition 4.4 so that the Recursion Theorem is properly invoked.

2. Prove Theorem 4.6.

3. Write Theorem 4.6 using the "additive" notation.

4. Suppose the conditions of Definition 4.5 hold, and suppose $d \in D$. Show that $s_n(d_1, d_2, \ldots, d_n) = nd$, provided each $d_i = d$. Note that this substantiates our claim that Definition 4.4 (using "additive" notation) is a special case of Definition 4.5 for binary operations that are commutative and associative.

5. Complete the proof that the function $g: \Psi \to \Psi$ defined after Definition 4.5 results in Property 2 of Definition 4.5 holding.

6. Suppose $k, n \in \mathbf{Z}$ with $k \le n$. Show that $\mathbf{card}(\{i \in \mathbf{Z} | k \le i \le n\}) = (n - k) + 1$.

7. Simplify $\sum_{i=-3}^{1}(2 - i)$ and $\Pi_{i=-3}^{1}(2 - i)$.

8. Prove Theorem 4.10.

9. Show that the number of distinct permutations of $\mathbf{I}_n$ is $n!$, for each $n \in \mathbf{N}$.

10. Complete the remainder of Case 2 in the proof of Theorem 4.11.

Section 4.3   ## Introduction to Number Theory

Now that we have resolved some theoretical issues concerning integers (such as the generalized commutative and associative laws), we can do some very interesting things with these integers. There is a whole branch of mathematics known as **number theory**. Number theory is basically concerned with properties of integers such as divisibility and primeness. The subject is very vast, and we only sample a small portion, just enough to prepare the reader for more advanced courses. The reader may also be surprised to hear that there are many unsolved problems in number theory that are easy to state, and one of them is mentioned. After all, mathematics is very much about discovery. The last subsection of this section, which is optional, discusses the justification of the grade-school algorithm for adding integers.

Divisibility

At a very basic level, number theory is concerned with divisibility. The student instinctively knows that when one integer is divided into another, sometimes the remainder is zero, other times not. For example, when the integer 2 is divided into the integer 6, the result is a quotient of 3 and the remainder is 0. On the other hand, when the integer 2 is divided into the integer 7, the result is a quotient of 3 and a remainder of 1. When the integer a is divided into the integer b, and a remainder of 0 results, we say that a *divides* b. We can also say that b is *divisible* by a. Definition 4.9 makes this precise.

Definition 4.9

Let $a, b \in \mathbf{Z}$ with $a \neq 0$. We say that a **divides** b provided there is a $c \in \mathbf{Z}$ such that $b = ac$. We can also express this by saying b is **divisible** by a. This can also be expressed by saying that a is a **divisor** of b. Finally, it can also be expressed by saying that a is a **factor** of b. The notation $a|b$ may also be used to say that a divides b.

Example 4.12

We see that $-7|21$ because there is an integer c such that $21 = (-7)c$. Specifically, $c = 3$ works. On the other hand, it is not the case that $3|-7$ because we cannot express -7 as 3 times some integer. (Think about how you can prove this latter statement. Hint: Use inequalities.)

Example 4.13

Let $a \in \mathbf{Z}$ with $a \neq 0$. Then $a|0$ because $0 = a \cdot 0$. Hence, every nonzero integer divides 0.

Next are two very simple but important theorems about divisibility. The proof of Theorem 4.14 is left to the reader.

Theorem 4.13   Let a, b, $c \in \mathbf{Z}$ with $a \neq 0$ and $b \neq 0$. Suppose $a|b$ and $b|c$. Then $a|c$.

Proof: We must show that there is some integer, say z, such that $c = az$. Now, because $a|b$, there is some integer x such that $b = ax$. Because $b|c$, there is some integer y such that $c = by$. Hence, $c = by = (ax)y = a(xy)$. Hence, setting z to be xy works.

Theorem 4.14   Let a, b, m, $n \in \mathbf{Z}$ with $a \neq 0$. Suppose $a|b$ and $a|c$. Then $a|(mb + nc)$.

Example 4.14   We know that $3|6$ and $3|9$. Then, by Theorem 4.14, we know that $3|((-2) \cdot 6 + 10 \cdot 9)$. That is, we know that $3|78$.

The next theorem is a very important result known as the **Division Algorithm**. This theorem states that whenever we divide an integer a by an integer b with $b > 0$, the result is a unique quotient q and unique remainder r with the stipulation that $0 \leq r < b$. That is, there are integers q and r with $0 \leq r < b$ such that

1) $a = bq + r$; and
2) for any integers q^* and r^*, if $0 \leq r^* < b$ and $a = bq^* + r^*$, then $q^* = q$ and $r^* = r$.

Property 2 expresses the uniqueness of the integers q and r. Before formally stating and proving the theorem, we provide some examples.

Example 4.15   Let us contemplate the division of 23 by 5. In this case, $a = 23$ and $b = 5$. Because we are dividing by 5, we fully expect to get a remainder r that lies in the set $\{0, 1, 2, 3, 4\}$. That is, we expect to have $0 \leq r < b$. Indeed, if we do the division, we get $23 = 5 \cdot 4 + 3$; that is, we get a quotient of 4 and a remainder of 3. On the other hand, if we divide 30 by 5, we get a quotient of 6 and a remainder of 0; that is, $30 = 5 \cdot 6 + 0$. In fact, notice that because $5|30$, of course we get a remainder of 0.

Example 4.16   The situation is a little less intuitive if we divide into a negative number (i.e., if a is negative). Consider the division of -23 by 5. It can be seen that $-23 = 5 \cdot (-5) + 2$. Note that we "overshoot" on the quotient and make it up with the remainder. By playing around with this, the reader should see that there is no other possible quotient and remainder, provided that we insist that the remainder lie in the set $\{0, 1, 2, 3, 4\}$.

We are now ready to state and prove the Division Algorithm. The proof makes excellent use of the fact that every set of integers bounded below has a least element.

Theorem 4.15 **(Division Algorithm)**

Let $a, b \in \mathbf{Z}$ with $b > 0$. Then there are unique integers q and r such that $a = bq + r$ with $0 \le r < b$. The number q is called the quotient, and the number r is called the remainder.

Proof: First we will prove the existence of integers q and r such that $a = bq + r$ and $0 \le r < b$. Then we show that these integers are unique.

Define the set A by $A = \{a - bx \,|\, x \in \mathbf{Z}\}$. Define the set B by $B = \{y \in A \,|\, y \ge 0\}$. That is, B is the set of nonnegative elements of A. We would like to work with the least element of B, but first we have to make sure that B is not the empty set. This is equivalent to proving that the inequality $a - bx \ge 0$ holds for some integer x. Now, if $a \ge 0$, then setting x to be 0 "works." What if $a < 0$? Then, the reader should "play around" with the inequality until it becomes clear that setting x to be a "works." In either case, we have shown that B is nonempty.

Let r denote the least element of B. Because $r \in A$, we must have $a - bq = r$ for some $q \in \mathbf{Z}$. Then, of course, we must have $a = bq + r$. To complete the existence portion of the proof, we need only show that $0 \le r < b$. We already know that $r \ge 0$ due to the fact that all elements of B are nonnegative, and $r \in B$. Now (recalling that $b > 0$), note that $r - b < r$. Because r is the least element of B and $r - b < r$, $r - b$ cannot be a member of B. However, $r - b$ is a member of A because $r - b = a - bq - b = a - b(q + 1)$. By the definition of B, something that is in A but not in B must be negative. Hence, $r - b < 0$. That is, $r < b$. We have shown that $0 \le r < b$, so this concludes the existence portion of the proof.

Finally, we must show uniqueness. Suppose q^* and r^* are integers such that $0 \le r^* < b$ with $a = bq^* + r^*$. The goal is to prove that $q^* = q$ and $r^* = r$. Now, subtracting the equation $a = bq^* + r^*$ from the equation $a = bq + r$ gives the equation $0 = b(q - q^*) + (r - r^*)$. This can be written as $r^* - r = b(q - q^*)$. However, recall that $0 \le r < b$ and $0 \le r^* < b$. Working with these two inequalities, the reader should be able to produce the inequality $-b < r^* - r < b$. We now break into cases.

Case 1: $r^* - r > 0$. Because $r^* - r = b(q - q^*)$, we have $q - q^* > 0$, so $r^* - r = b(q - q^*) \ge b$. But this contradicts the statement $-b < r^* - r < b$, so this case is not possible.

Case 2: $r^* - r < 0$. We can arrive at a similar contradiction as in Case 1.

The only remaining possibility is that $r^* - r = 0$ (i.e., that $r^* = r$). Hence, from the equation $r^* - r = b(q - q^*)$, we also immediately conclude that $q^* = q$.

─────────

Two comments concerning the Division Algorithm are merited. The first comment is that the student should see that the positive integer b divides the integer a iff the remainder from the Division Algorithm is 0.

The second comment concerns the Division Algorithm in the case in which the divisor, b, is 2. Let $a \in \mathbf{Z}$. The Division Algorithm tells us that there exist unique integers q and r such that $a = 2q + r$ with $0 \leq r < 2$. Hence, $r = 0$ or $r = 1$. In the case in which $r = 0$, we call the integer a **even**. In the case in which $r = 1$, we call the integer a **odd**. Note that the uniqueness of the remainder of the Division Algorithm is our guarantee that an integer is either even or odd, but not both. (Remember that we used this in Section 1.4, and we referred to it as "Informal Fact I"? What a relief that this is now a proven fact!) Hence, the integer a is even iff it can be expressed as $a = 2q$ for some $q \in \mathbf{Z}$, and it is odd iff it can be expressed as $a = 2q + 1$ for some $q \in \mathbf{Z}$.

Representation of Integers

Recall that one of the main reasons for humans' initial interest in integers is that integers provide a means of counting. To use integers most effectively to count, it is important to have a mechanism for expressing integers in an efficient way that makes it easy to manipulate them and combine them with other integers. In this subsection, we develop a means of representing integers.

We use the decimal "base 10" representation in everyday life. That is, we express integers in terms of powers of 10. For example, when we write the integer 4306, what we really mean is $(4 \times 10^3) + (3 \times 10^2) + (0 \times 10^1) + (6 \times 10^0)$. (We have not officially defined what is meant by raising an integer to the power 0, so we hereby declare that for any $a \in \mathbf{Z}$ with $a \neq 0$, by a^0 we mean the integer 1. This should not come as a surprise to anyone.)

There is nothing necessary about using the number 10 as a base. It is likely that 10 is used because members of our species tend to have 10 figures. The Babylonians used base 60, and the Mayans used base 20. Electronic computers make use of base 2, base 8, and base 16. (Base 2 is called *binary*, and base 16 is called *hexadecimal* or just *hex*.) It turns out that any integer b with $b \geq 2$ can be used as a base. That is, given any $n \in \mathbf{N}$, there exists a nonnegative integer k and integers $a_0, a_1, \ldots, a_k$, with each a_i satisfying $0 \leq a_i \leq b - 1$ such that $n = a_k b^k + a_{k-1} b^{k-1} + \cdots + a_1 b + a_0$, with $a_k > 0$. (The reason why we stipulate that $a_k > 0$ is because we are not usually interested in padding the expression of the integer with leading zeroes.) We refer to this as the **base b expansion of n.** We further claim that there is only one base b expansion of n; that is, the expansion is unique. (Note that for the base 10 expansion, the stipulation that each a_i satisfies $0 \leq a_i \leq b - 1$ says that each a_i is greater than or equal to 0, but less than or equal to 9.) We can use the notation $(a_k a_{k-1} \cdots a_1 a_0)_b$ to refer to the base b expansion of n, rather than writing out $n = a_k b^k + a_{k-1} b^{k-1} + \cdots + a_1 b + a_0$. (If b is 10, we can write the number "normally.")

We are about to prove the above result. However, before doing so, it is helpful to prove two lemmas that are used in the proof. The first one is quite obvious.

Lemma 4.2   Let $a, b \in \mathbf{Z}$ with $b > 0$. If $a \geq 0$, then the quotient q obtained from the Division Algorithm (in which $a = bq + r$) is nonnegative.

Proof: From the Division Algorithm, there are integers q and r such that $a = bq + r$ and $0 \leq r < b$. The goal is to prove that $a \geq 0 \Rightarrow q \geq 0$. It suffices (of course) to prove the contrapositive; that is, we show that $q < 0 \Rightarrow a < 0$. If $q < 0$, then $q < -1$. Because $b > 0$, we have $bq \leq -b$. Hence, $bq + r \leq -b + r$. Because $r < b$, we have $-b + r < 0$, so $bq + r < 0$, but $bq + r = a$, so we have reached the conclusion that $a < 0$.

The idea behind the next lemma is as follows: Suppose that we have a sequence of nonnegative (i.e., not negative) integers such that whenever a term of the sequence is positive, the next term is strictly less than this term. Then we must eventually arrive at a term of the sequence that is zero. This is because either the first term is 0, in which case we are done, or else it is positive, causing the next term to be less. So, eventually, we run out of potential positive integers, and we arrive at 0.

Lemma 4.3   Suppose $\{q_m\}_{m=1}^{\infty}$ is a sequence of nonnegative integers with the property that for each $m \in \mathbf{N}$, if $q_m > 0$, then $q_{m+1} < q_m$. Then there exists an $i \in \mathbf{N}$ such that $q_i = 0$.

Proof: Let $A = \{q_m \mid m \in \mathbf{N}\}$. Because each member of this set is nonnegative (i.e., ≥ 0), A must have a least element; let us call it w. Because $w \in A$, we must have $w = q_i$ for some $i \in \mathbf{N}$. Suppose (to reach a contradiction) we have $q_i > 0$. Then, we must have $q_{i+1} < q_i$, which contradicts the fact that q_i is the least element of A. Thus, $q_i = 0$.

We are now ready to prove that if $b \in \mathbf{N}$ with $b \geq 2$, then for each $n \in \mathbf{N}$, n has a unique base b expansion. For easier readability, this result is broken into two theorems. Theorem 4.16 states that n has a base b expansion, and Theorem 4.17 states that this expansion is unique. The proofs of these theorems given here are based on the proofs given in Rosen (2000) (with some slight modifications).

Theorem 4.16

Let $b \in \mathbf{N}$ with $b \geq 2$. Given any $n \in \mathbf{N}$, there exists a nonnegative integer k and integers $a_0, a_1, \ldots, a_k$, with each a_j satisfying $0 \leq a_j \leq b - 1$ such that $n = a_k b^k + a_{k-1} b^{k-1} + \cdots + a_1 b + a_0$, with $a_k > 0$.

Proof: We use the Recursion Theorem to define a sequence $\{(q_m, a_m)\}_{m=0}^{\infty}$ of ordered pairs. By the Division Algorithm, there are integers q_0 and a_0 with $0 \leq a_0 \leq b - 1$, such that $n = bq_0 + a_0$. Given (q_m, a_m) for $m \in \mathbf{N} \cup \{0\}$, by the Division Algorithm, there are integers q_{m+1} and a_{m+1} with $0 \leq a_{m-1} \leq b - 1$, such that $q_m = bq_{m+1} + a_{m+1}$. (Note that as promised, we are now invoking the Recursion Theorem informally. If you feel uncomfortable with this, then by all means fill in the gaps.)

Now, by Lemma 4.2, $\{q_m\}_{m=0}^{\infty}$ is a sequence of nonnegative integers. (Why?) Suppose that $q_m > 0$ for some $m \in \mathbf{N} \cup \{0\}$. Then $q_m = bq_{m+1} + a_{m+1}$, so $q_{m+1} < q_m$. (Why?) Hence, by Lemma 4.3, there exists a $i \in \mathbf{N} \cup \{0\}$ such that $q_i = 0$. (The reason why we conclude $i \in \mathbf{N} \cup \{0\}$ instead of just $i \in \mathbf{N}$ is that the sequence $\{q_m\}_{m=0}^{\infty}$ starts at 0.) So, it is possible for i to be 0. We now choose $k \in \mathbf{N} \cup \{0\}$ to be the smallest $m \in \mathbf{N} \cup \{0\}$ such that $q_m = 0$. Hence, $q_k = 0$. If $k = 0$, then because $n = bq_0 + a_0$, we have $n = b \cdot 0 + a_0 = a_0$, and in fact we are done! For the remainder of the proof, we consider the case in which $k > 0$. Plugging in $k - 1$ for m in the recursion statement $q_m = bq_{m+1} + a_{m+1}$ results in $q_{k-1} = bq_k + a_k$. Hence, $q_{k-1} = b \cdot 0 + a_k$, so $q_{k-1} = a_k$. Now, by the definition of k, $q_{k-1} > 0$ (i.e., $a_k > 0$).

Using induction, the reader can verify that $n = b^m q_{m-1} + \sum_{j=1}^{m} a_{m-j} b^{m-j}$ for each $m \in \mathbf{N}$. (Note that the last term of this sum is $a_0 b^0 = a_0 \cdot 1 = a_0$. Plugging in k for m in this equation results in $n = b^k q_{k-1} + \sum_{j=1}^{k} a_{k-j} b^{k-j}$. However, we have already noted that $q_{k-1} = a_k$. Hence, $n = a_k b^k + \sum_{j=1}^{k} a_{k-j} b^{k-j}$. That is, $n = a_k b^k + a_{k-1} b^{k-1} + \cdots + a_1 b + a_0$. We have already shown that $a_k > 0$.

──────

There are some important points to be made concerning the proof of Theorem 4.16. The first point has to do with the "slick" formula $n = b^m q_{m-1} + \sum_{j=1}^{m} a_{m-j} b^{m-j}$ that the reader was asked to establish via induction. Before performing an induction, it is often useful for the student to see if an intuitive explanation can be found to help motivate the induction. Often, such motivation is left out of the "slick" proofs of advanced mathematics, so the student should get used to searching for the intuition.

To accomplish this with the above-mentioned induction, start with the equation $n = bq_0 + a_0$ and the recursive statement $q_m = bq_{m+1} + a_{m+1}$.

Setting m to be 0 results in $q_0 = bq_1 + a_1$. But now "plugging in" this value of q_0 into the equation $n = bq_0 + a_0$ results in $n = b(bq_1 + a_1) + a_0$, which simplifies to $n = b^2q_1 + a_1b + a_0$. Setting m to be 1 in the recursive statement $q_m = bq_{m+1} + a_{m+1}$ results in $q_1 = bq_2 + a_2$. Now we can "plug" this value of q_1 into the equation $n = b^2q_1 + a_1b + a_0$. We then simplify this. Continuing this process, we see that we motivate the "slick" formula $n = b^mq_{m-1} + \sum_{j=1}^{m} a_{m-j}b^{m-j}$.

The second point concerning the proof of Theorem 4.16 has to do with the distinction between existence proofs and proofs that demonstrate how to construct the object whose existence is being proven. By an **existence proof**, it is meant a proof that demonstrates that some object must exist without indicating how to specifically identify or construct the object. This is contrasted with Theorem 4.16, which not only proves that a base b expansion of a positive integer n exists, it also indicates how to construct this expansion. Such a proof can be called a **constructive proof**.

To illustrate this, recall that we start with $n = bq_0 + a_0$. Now, if $q_0 = 0$, we are done. The base b expansion of n is given by $n = (a_0)_b$. Otherwise, we generate the following equations:

$$n = bq_0 + a_0.$$
$$q_0 = bq_1 + a_1.$$
$$q_1 = bq_2 + a_2.$$
$$\vdots$$
$$q_{k-1} = bq_k + a_k.$$

(Because $q_k = 0$, the last equation can be written $q_{k-1} = a_k$). Note from the proof that the base b expansion can be read directly from this. That is, $n = (a_ka_{k-1} \cdots a_1a_0)_b$.

Example 4.17	Find the base 3 expansion of 101.

Solution: We systematically apply the Division Algorithm as in the proof of Theorem 4.16:

$$101 = 3 \cdot 33 + 2.$$
$$33 = 3 \cdot 11 + 0.$$
$$11 = 3 \cdot 3 + 2.$$
$$3 = 3 \cdot 1 + 0.$$
$$1 = 3 \cdot 0 + 1.$$

Hence, $101 = (10202)_3$. To check this, simply calculate $(1 \times 3^4) + (0 \times 3^3) + (2 \times 3^2) + (0 \times 3) + 2$. This comes to 101, as it should.

■

We now present Theorem 4.17, which states that the base b expansion of a positive integer is unique.

Theorem 4.17   Let $b \in \mathbf{N}$ with $b \geq 2$. Given any $n \in \mathbf{N}$, there is a unique base b expansion of n.

Proof: It has already been shown that there is at least one base b expansion of n. Suppose (to reach a contradiction) that we have two distinct base b expansions of n. Now, if these two expansions have a different number of terms, we can add leading zero terms to the shorter one so that the number of terms of the expansions are the same. Having done so, let us represent the two expansions of n as $n = a_m b^m + a_{m-1} b^{m-1} + \cdots + a_1 b + a_0$, and $n = c_m b^m + c_{m-1} b^{m-1} + \cdots + c_1 b + c_0$ for some $m \in \mathbf{N}$, in which each a_i satisfies $0 \leq a_i \leq b - 1$ and each c_i satisfies $0 \leq c_i \leq b - 1$. (Because we may have added leading zero terms, we do not make the claim that $a_m > 0$ and $c_m > 0$.)

Subtracting one of these expansions from the other results in $(a_m - c_m) b^m + (a_{m-1} - c_{m-1}) b^{m-1} + \cdots + (a_1 - c_1) b + (a_0 - c_0) = 0$. Because we have made the assumption that these two expansions are not identical, there must be a smallest integer j, with $0 \leq j \leq m$, such that $a_j \neq c_j$. Hence, we have $(a_m - c_m) b^m + (a_{m-1} - c_{m-1}) b^{m-1} + \cdots + (a_{j+1} - c_{j+1}) b^{j+1} + (a_j - c_j) b^j = 0$. Factoring out b^j gives $b^j \lfloor (a_m - c_m) b^{m-j} + (a_{m-1} - c_{m-1}) b^{m-j-1} + \cdots + (a_{j+1} - c_{j+1}) b + (a_j - c_j) \rfloor = 0$. Hence, $(a_m - c_m) b^{m-j} + (a_{m-1} - c_{m-1}) b^{m-j-1} + \cdots + (a_{j+1} - c_{j+1}) b + (a_j - c_j) = 0$.

Now, solving for $a_j - c_j$ results in $a_j - c_j = (c_m - a_m) b^{m-j} + (c_{m-1} - a_{m-1}) b^{m-j-1} + \cdots + (c_{j+1} - a_{j+1}) b$. We see from this that $b | (a_j - c_j)$, so that $a_j - c_j = bw$ for some integer w. Recall that $0 \leq a_j < b$ and $0 \leq c_j < b$. Working with these two inequalities, the reader should be able to produce the inequality $-b < a_j - c_j < b$. Now, (just as in Case 1 and Case 2 at the end of the proof of Theorem 4.15) the reader should be able to see that we cannot have $a_j - c_j > 0$, and we cannot have $a_j - c_j < 0$. Hence, we must have $a_j = c_j$. However, recall that the integer j was defined so that $a_j \neq c_j$. We have arrived at the desired contradiction.

Primes and the Fundamental Theorem of Arithmetic

The student is likely to have had some exposure previously to the concept of **prime numbers**. In any case, the definition of a prime number is provided next.

Definition 4.10

Let $n \in \mathbf{N}$ with $n > 1$. Then n is called **prime** provided the only positive divisors of n are 1 and n. Otherwise n is called **composite**.

Example 4.18

The positive integer 5 is prime because the only positive divisors of 5 are 1 and 5. The positive integer 6 is composite because it is divisible by 2; that is, 6 has positive divisors other than 1 and 6. The integer 2 is the only even prime. The integer 1 is classified as neither prime nor composite.

Of paramount importance to mathematics is the **Fundamental Theorem of Arithmetic**. (You can tell by its name that it is important.) This theorem states that every positive integer $n > 1$ is either prime, in which case it cannot be written as a product of any other primes, or n is composite, in which case it can be written as a product of primes; and this product is unique (ignoring the order of their listing). This theorem has extensive applications both within number theory and in other subjects. For example, the theorem has applications to the subject of cryptology. We will prove the theorem by first showing that every positive integer $n > 1$ is either prime or can be written as a product of primes. Then the uniqueness portion will be proved. Both proofs make use of the Second Principle of Mathematical Induction (Theorem 3.23).

Lemma 4.4

Let $n \in \mathbf{N}$ with $n > 1$. Then either n is prime, or else n can be written as a product of primes.

Proof: Clearly the result holds for 2. Now, to make the induction hypothesis, let $n \in \mathbf{N}$ with $n > 1$ and suppose that the result holds true for all $k \in \mathbf{N}$ with $1 < k \leq n$. The goal is to show that the result holds for $n + 1$. Now, if $n + 1$ is prime, we are done, so let us examine the case in which it is composite. Then $n + 1$ has a positive integer divisor u such that $1 < u < n + 1$. Note that $u \leq n$ (of course). Because $u | n + 1$, there must be a positive integer v such that $n + 1 = uv$. It is clear that we also have $1 < v \leq n$. (Why?) Hence, by the induction hypothesis, both u and v are a product of primes. Therefore, the product uv is also. That is, $n + 1$ is a product of primes.

The upcoming uniqueness proof is a little harder. To make things a little easier, when we write a product of primes such as $n = p_1 p_2 \cdots p_k$, we often will place the primes in order so that $p_1 \leq p_2 \leq \cdots \leq p_k$. (Technically, an induction argument can be used to prove that any list of integers can be "reordered" into a list of integers that are "in order." See Exercise 9 for more precision.) Now we can state and prove the Fundamental Theorem of Arithmetic. The proof is based on the proof given in Jacobson (1974).

Theorem 4.18

(Fundamental Theorem of Arithmetic)

Let $n \in \mathbf{N}$ with $n > 1$. Then there exist a unique positive integer s and unique primes $p_1, p_2, \ldots, p_s$ satisfying both $p_1 \leq p_2 \leq \cdots \leq p_s$ and $n = p_1 p_2 \cdots p_s$.

Proof: We will again make use of the Second Principle of Mathematical Induction. The result clearly holds for 2. Now, to make the induction hypothesis, let $n \in \mathbf{N}$ with $n > 1$, and suppose that the result holds true for all $k \in \mathbf{N}$ with $1 < k \leq n$. The goal is to show that the result holds for $n + 1$.

By Lemma 4.4 and Exercise 9 of this section, there exists a positive integer s and primes $p_1, p_2, \cdots, p_s$ satisfying both $p_1 \leq p_2 \leq \cdots \leq p_s$ and $n + 1 = p_1 p_2 \cdots p_s$. To prove uniqueness, suppose there is a positive integer t and primes $q_1, q_2, \ldots, q_t$ satisfying both $q_1 \leq q_2 \leq \cdots \leq q_t$ and $n + 1 = q_1 q_2 \cdots q_t$. Then the goal is to show that $s = t$ and that $q_i = p_i$ for each $i \in \mathbf{I}_s$. We will first show that the theorem holds provided that $q_1 = p_1$. Then, we will show that if $q_1 \neq p_1$, then we arrive at a contradiction so that we must have $q_1 = p_1$ after all.

Suppose that $q_1 = p_1$. If either s or t is 1, it is clear that we are done, so consider the case where they both exceed 1. Now, we have $p_1 p_2 \cdots p_s = p_1 q_2 \cdots q_t$. We can cancel p_1 to get $p_2 \cdots p_s = q_2 \cdots q_t$. Note that $p_2 \cdots p_s < n + 1$ (because $n + 1 = p_1 p_2 \cdots p_s$). Hence, we apply the induction hypothesis to $p_2 \cdots p_s$, and we get the desired result.

Suppose $q_1 \neq p_1$. Then either $p_1 < q_1$, or else $q_1 < p_1$. We analyze the case in which $p_1 < q_1$. The case in which $q_1 < p_1$ is very analagous, and the reader should be able to easily fill it in after seeing the case in which $p_1 < q_1$. Recall that $n + 1 = q_1 q_2 \cdots q_t$. Because $p_1 < q_1$, we have $p_1 q_2 \cdots q_t < q_1 q_2 \cdots q_t$. Hence, $p_1 q_2 \cdots q_t < n$. Also note that $t > 1$ because otherwise we would have $p_1 p_2 \cdots p_s = q_1$, which is impossible. (Why? Because this would imply that $p_1 | q_1$, which in turn would imply that $q_1 = p_1$ because the only positive divisors of the prime q_1 are 1 and q_1.)

Now, define m by $m = n + 1 - p_1 q_2 \cdots q_t$. Hence, $m = (q_1 q_2 \cdots q_t) - (p_1 q_2 \cdots q_t) = (q_1 - p_1)(q_2 \cdots q_t)$. Hence, because $q_1 - p_1 < q_1$, we have $m < n + 1$. Also note that because $t > 1$, we must also have $m > 1$. (Why?) Recall that $n + 1$ is also given by $n + 1 = p_1 p_2 \cdots p_s$. Therefore, another way to write m is $m = p_1 p_2 \cdots p_s - p_1 q_2 \cdots q_t = p_1[(p_2 \cdots p_s) - (q_2 \cdots q_t)]$. (Note that this makes sense because we also must have $s > 1$.)

Because the two ways of writing m must be equal to each other, we have $m = p_1[(p_2 \cdots p_s) - (q_2 \cdots q_t)] = (q_1 - p_1)(q_2 \cdots q_t)$. Also recall that

$m < n + 1$. Because $1 < m < n + 1$, we can apply the induction hypothesis to m. From the left-hand side way of writing m, we conclude that one of the primes in the factorization of m into primes is p_1. Therefore, looking at the right-hand side way of writing m, we conclude that p_1 must divide $(q_1 - p_1)(q_2 \cdots q_t)$. Now, let $(w_1 w_2 \cdots w_v)(q_2 \cdots q_t)$ be the prime factorization of $(q_1 - p_1)(q_2 \cdots q_t)$. (We know that the primes $q_2, \ldots, q_t$ must be in this factorization because each of these primes is a factor, and we are using the induction hypothesis that m can be written uniquely as a product of primes in increasing order.) Therefore, because we have already concluded that p_1 is one of these primes, p_1 must be one of the primes of the listing $q_2, \ldots, q_t$ or one of the primes in the listing $w_1, \ldots, w_v$ (or both). However, we know that $p_1 < q_1 \leq q_2 \leq \cdots q_t$. Therefore, p_1 cannot be one of the primes of the listing $q_2, \cdots, q_t$. Hence, p_1 must be one of the primes in the listing $w_1, \cdots, w_v$. Because the product of the primes in this list is $(q_1 - p_1)$, we have $p_1 | (q_1 - p_1)$. Clearly, $p_1 | p_1$. Hence, by Theorem 4.14, $p_1 | q_1$. (Why?) But, q_1, being a prime, has no positive divisors other than 1 and q_1. Hence, we have arrived at our desired contradiction.

Example 4.19   The integer 48 can be written as a product of primes as $2 \cdot 2 \cdot 2 \cdot 2 \cdot 3$. In practice, we often collect common terms together using exponents, so we would write this as $2^4 \cdot 3$.

We have a little more to say about number theory in upcoming sections. In the next section, we present the classic proof that there are infinitely many primes. We could do this now, but we wait for the formal definition (in the next section) of an infinite set. We also briefly discuss the concept of greatest common divisors in Chapter 5 within the context of rediscussing the proof that the real number $\sqrt{2}$ is irrational.

Finally, it has already been mentioned briefly that there are many unsolved problems in number theory. One of the most famous of these is **Goldbach's Conjecture**, named after the mathematician Christian Goldbach. This conjecture, which dates from the 18th century, states that *every even integer greater than 2 can be written as a sum of two (not necessarily distinct) primes*. For example, 4 can be written as the sum of 2 and 2, and 8 can be written as the sum of 3 and 5. Some integers (such as 10) can be written as a sum of two primes in more than one way. If Goldbach's Conjecture is true, then every even integer greater than 2 can be written as a sum of two primes in at least one way (i.e., possibly more).

Even though this has been verified for many even integers, no proof has been discovered that it holds for *all* even integers greater than 2. However, the conjecture is believed to be true, and of course no counterexample has been discovered. Perhaps you, the reader, inspired by this book, will discover a proof some day (provided it is still unsolved when you read it).

A Revisit to Grade School [Optional]

Finally, we top off this section with a little fun. Remember the grade-school algorithm for adding positive integers? For example, let us add 548 and 471. We do this as follows:

$$\begin{array}{r} \mathbf{11} \\ 548 \\ \underline{471} \\ 1019 \end{array}$$

The bolded digits above the number 548 are the **carries**. To do this arithmetic, first we add the digits 8 and 1 to get the digit 9. Next we add 4 and 7 to get 11. However, instead of writing 11, we only write 1 because $11 > 10$, and we *carry* a 1 over to the left. Then we add this additional 1 to the sum of 5 and 4. This gives us the number 10. Actually, we just write 0 and do another *carry* to the left to get the last digit, which is 1.

We are now in a position to show why this works! This argument is based on the discussion in Rosen (2000) with some added details. We will provide the algorithm for finding the digits of the sum of positive integers a and b in any positive integer base ≥ 2, say base r. Now, suppose that both a and b have n digits, with $a = (a_{n-1} \cdots a_1 a_0)_r$ and $b = (b_{n-1} \cdots b_1 b_0)_r$. (We can pad a or b with initial digits of zeroes if necessary, so that a and b both have n digits.) The goal is to find the digits of the base r expansion of $a + b$.

By the Division Algorithm, there are integers C_0 and s_0 such that $a_0 + b_0 = C_0 r + s_0$, in which $0 \leq s_0 \leq r - 1$. Now, because $0 \leq a_0 \leq r - 1$ and $0 \leq b_0 \leq r - 1$, we must have $0 \leq a_0 + b_0 \leq 2r - 2$. The reader should verify that this shows that $C_0 \in \{0, 1\}$. (This makes sense because C_0 is the *carry* over to the next digit.)

We can proceed via recursion to get integers C_i and s_i for each $i \in \{1, 2, \ldots, n - 1\}$ so that $a_i + b_i + C_{i-1} = C_i r + s_i$, $0 \leq s_i \leq r - 1$, and $C_i \in \{0, 1\}$ for each $i \in \{1, 2, \ldots, n - 1\}$. (For each such i, the number C_{i-1} is the *carry* from the previous addition.) We then define s_n to be C_{n-1}. (This is the last carry.)

A comment concerning the recursion may be useful here. The student may note that the Recursion Theorem has been (informally) invoked to obtain a *finite sequence*, $\{(C_i, s_i)\}_{i=0}^{n-1}$, instead of a sequence. How is this possible? The answer is that a sequence can be defined by formally invoking the Recursion Theorem, but then a finite sequence can be formed from this sequence by "discarding" the terms that we do not care about. (The student is encouraged to do this once, just to see how it is done.)

Now, it can be shown that $a + b = (s_n s_{n-1} \cdots s_0)_r$. One way to do this is to perform an induction on the number of digits n. To show that it holds for 1-digit numbers, consider any $a = (a_0)_r$ and $b = (b_0)_r$. Then using the above algorithm, we have $a_0 + b_0 = C_0 r + s_0$ with $0 \leq s_0 \leq r - 1$ and $C_0 \in \{0, 1\}$. Also, the algorithm specifies that $s_1 = C_0$. Now, we must show that $a + b = (s_1 s_0)_r$. Now, consider the case in which $C_0 = 0$. Then $a_0 + b_0 = s_0$, and $s_1 = 0$. It is clear that the algorithm works in this case. Suppose that $C_0 = 1$. Then $a_0 + b_0 = r + s_0$, and $s_1 = 1$. It is also clear in this case that the algorithm works, as $a + b = (1 s_0)_r$. Hence, it works for 1.

We now make the induction hypothesis that the algorithm works for the sum of any two n-digits positive integers for some $n \in \mathbf{N}$. The goal is to prove that it holds for the sum of any two $n + 1$-digit positive integers. Toward that end, suppose $a = (a_n a_{n-1} \cdots a_1 a_0)_r$ and $b = (b_n b_{n-1} \cdots b_1 b_0)_r$. Note that $a + b = (a_n r^n + b_n r^n) + (a_{n-1} \cdots a_1 a_0)_r + (b_{n-1} \cdots b_1 b_0)_r$. Now the induction hypothesis is applied to $(a_{n-1} \cdots a_1 a_0)_r + (b_{n-1} \cdots b_1 b_0)_r$. The details are left to the reader.

For a quick demonstration that the method works, consider the sum of 548 and 471 discussed above. Applying the algorithm gives:

$$8 + 1 = 0 \cdot 10 + 9, \text{ so } C_0 = 0 \text{ and } s_0 = 9.$$
$$4 + 7 + 0 = 1 \cdot 10 + 1, \text{ so } C_1 = 1 \text{ and } s_1 = 1.$$
$$5 + 4 + 1 = 1 \cdot 10 + 0, \text{ so } C_2 = 1 \text{ and } s_2 = 0.$$
$$s_3 = C_2 = 1.$$

Hence, the algorithm claims that $548 + 471 = (s_3 s_2 s_1 s_0)_{10} = 1019$, and we know that this is true.

EXERCISES

1. Prove Theorem 4.14.

2. Suppose m and n are nonzero integers such that $m|n$ and $n|m$. What can you conclude about m and n?

3. Find the quotient and remainder using the Division Algorithm for 1) $a = 1$ and $b = 7$; 2) $a = -1$ and $b = 7$; and 3) $a = 21$ and $b = 4$.

4. Extend the Division Algorithm to allow negative divisors as follows: Show that if $a, b \in \mathbf{Z}$ with $b < 0$, then there exist unique integers q and r such that $a = bq + r$ and $0 \leq r < -b$. (Hint: Use the Division Algorithm; you do not need to reinvent the wheel.)

5. When using hexadecimal representation (base 16), we need to use 15 symbols. Of course, the symbols 0 to 9 represent themselves. It is traditional to use the letters A, B, C, D, E, and F to represent 10, 11, 12, 13, 14, and 15, respectively. Using this notation, find the hexadecimal representation of the base-10–expressed number 2740.

6. Find the binary (base 2) expansion of the base-10–expressed number 37.

7. Suppose that the base b expansion of a positive integer n is given by $n = (a_k a_{k-1} \cdots a_1 a_0)_b$. What is the base b expansion of $b^m n$, where m is a positive integer?

8. Suppose that the base 10 expansion of a positive integer n is given by $n = (a_k a_{k-1} \cdots a_1 a_0)_{10}$. Show that n is even iff $2|a_0$. Note that this shows that an in-

teger is even iff the last digit of the base 10 expansion is even. (Hint: Use the beginning of the proof of Theorem 4.16 along with the definition of being even.)

9. Show that given any n-tuple $(a_1, a_2, \cdots, a_n)$ of integers, there is a permutation π of $\mathbf{I}_n$ so that $a_{\pi(1)} \leq a_{\pi(2)} \leq \cdots \leq a_{\pi(n)}$.

10. Find the unique factorization of the number 90 into primes.

11. Find integers a and b and a nonzero integer m such that m does not divide a and m does not divide b, but $m|ab$.

12. Suppose that a and b are integers, and p is a prime such that $p|ab$. Show that $p|a$ or $p|b$.

13. Show that the following holds for each $n \in \mathbf{N}$: Given any integers $a_1, a_2, \cdots, a_n$, if p is any prime such that $p|(a_1 a_2 \cdots a_n)$, then $p|a_i$ for at least one $i \in \mathbf{I}_n$. (Hint: Use Exercise 12 and induction.)

14. (For readers of the optional subsection "A Revisit to Grade School") Find the binary (base 2) expansion of the sum of the binary-expressed integers $(10101)_2$ and $(1101)_2$.

15. (For readers of the optional subsection "A Revisit to Grade School") Formally use the Recursion Theorem to construct the finite sequence $\{(C_i, s_i)\}_{i=0}^{n-1}$.

16. (For readers of the optional subsection "A Revisit to Grade School") Complete the induction argument that proves that $a + b = (s_n s_{n-1} \cdots s_0)_r$.

17. (For readers of the optional subsection "A Revisit to Grade School") **CLASS PROJECT**: Try to come up with formal algorithms to find 1) the base r expansion of the subtraction $a - b$ of positive integers (with $a > b$), and 2) the base r expansion of the multiplication ab of positive integers (which is a little harder), and prove that these algorithms work. You can also try to find the base r expansion of the quotient of the Division Algorithm, but it will be easier if you wait until the discussion of real numbers.

Section 4.4   Finite and Infinite Sets

A discussion of counting began in Section 3.4. The discussion is continued and extended here and in the next section. The extension leads to some all-important concepts involved with **infinite sets**. The reader will begin to get a sampling of the mysteries of the infinite in this section and the next section, while learning material that is absolutely vital for advanced mathematics. Technically, the theory of infinite sets goes beyond applications of integers. However, it makes sense to include this theory in this chapter because it can be viewed as an extension of the concept of basic counting. Also, there is a strong tie between infinite sets and number theory. One very famous result we prove is that there are infinitely many primes.

Finite Sets

In this subsection, we derive the well-known properties of finite sets. Recall from Section 3.4 that if S is any nonempty set, then if S is of size n (with $n \in \mathbf{N}$), we can say S has cardinality n, and we write this as **card**$(S) = n$. The definition of cardinality is now extended to the empty set, which will be assigned a cardinality of 0. Why did we not do this in Section 3.4? Recall that in Section 3.4, we did not yet have the integer 0. We did not have access to the number 0 until we constructed the full set of integers. We also now extend the definition of a set being of finite cardinality to apply to the empty set.

Definition 4.11

The cardinality of the empty set is defined to be 0. That is, $\mathbf{card}(\varnothing) = 0$. Now, for any set S, we say that the **cardinality of S is finite** (or S is of **finite cardinality**) provided that there is an $n \in \mathbf{N} \cup \{0\}$ such that $\mathbf{card}(S) = n$.

Hence, a set S is of finite cardinality if it is the empty set, or else if $S \approx \mathbf{I}_n$ for some $n \in \mathbf{N}$. (Recall the definition of $\approx$ from Definition 3.7.) It is easy to define what is meant by a **finite** set. We will begin a discussion of sets that are not finite, called **infinite** sets, in the next subsection. Both kinds of sets are defined below.

Definition 4.12

A set S is said to be a **finite** set provided S is of finite cardinality. Otherwise, S is said to be an **infinite** set. If S is a finite set, we can say S is finite. If S is an infinite set, we can say S is infinite.

The proof of Theorem 4.19 below is mostly a summary of some results that were discussed in Section 3.4. Also, the theorem makes intuitive sense.

Theorem 4.19   Suppose A and B are sets. Then the following hold:

1) Suppose A is finite and $B \subseteq A$. Then B is finite, and $\mathbf{card}(B) \leq \mathbf{card}(A)$.

2) Suppose A is finite and $B \subseteq A$. Then $\mathbf{card}(B) < \mathbf{card}(A)$ iff $B \subset A$.

3) Suppose A and B are finite with A nonempty such that $\mathbf{card}(A) < \mathbf{card}(B)$. Then if $f: A \to B$, f is not onto. Also if $g: B \to A$, g is not one-to-one.

4) Suppose A and B are finite and nonempty. Then $A \approx B$ iff $\mathbf{card}(A) = \mathbf{card}(B)$.

5) Suppose A and B are finite and disjoint. Then $A \cup B$ is finite, and $\mathbf{card}(A \cup B) = \mathbf{card}(A) + \mathbf{card}(B)$.

6) Suppose A is finite and $B \subseteq A$. Then $A - B$ is finite, and $\mathbf{card}(A - B) = \mathbf{card}(A) - \mathbf{card}(B)$.

Proof:

1) If $B = \emptyset$, then B is finite with $\mathbf{card}(B) = 0$. Hence, $\mathbf{card}(B) \leq \mathbf{card}(A)$ because $\mathbf{card}(A) \geq 0$. The result is completely obvious if $A = \emptyset$. Consider the case in which neither A nor B is the empty set. Then the result follows directly from Theorem 3.29.

2) To show that $(\mathbf{card}(B) < \mathbf{card}(A)) \Rightarrow (B \subset A)$, suppose $\mathbf{card}(B) < \mathbf{card}(A)$. Now we are given that $B \subseteq A$. Were B to equal A, of course we would have $\mathbf{card}(B) = \mathbf{card}(A)$, which we do not have. Therefore, B must be a proper subset of A. That is, $B \subset A$. Next, to prove the converse, suppose $B \subset A$. Note that $A \neq \emptyset$ because the empty set does not have any proper subsets. Hence, $\mathbf{card}(A) > 0$. Thus, if $B = \emptyset$, clearly $0 = \mathbf{card}(B) < \mathbf{card}(A)$. If $B \neq \emptyset$, then the conclusion that $\mathbf{card}(B) < \mathbf{card}(A)$ is just a restatement of Theorem 3.29.

3) It is clear that B is nonempty since $\mathbf{card}(B) > 0$. The conclusion that f is not onto follows directly from Part 2) of Theorem 3.26. The conclusion that g is not one-to-one follows directly from Part 2) of Theorem 3.28.

4) This is left to the reader.

5) The result is obvious if $A = \emptyset$ or $B = \emptyset$. If $A \neq \emptyset$ and $B \neq \emptyset$, then the result is just a restatement of Exercise 5 of Section 3.4.

6) Note that B is finite by Part 1. $A - B$ is also finite by Part 1 because $A - B \subseteq A$. The result is obvious if $A = \emptyset$ or $A = B$. Otherwise, the result is just a restatement of Exercise 6 of Section 3.4.

A corollary to Theorem 4.19 is presented next. For the proof, recall (from Definition 3.10) that if we are given a function f with domain S, then $\mathbf{im}(f)$ is defined to be the set $\{f(x) | x \in S\}$.

Corollary 4.1

Suppose that A and B are finite sets, both with cardinality $n \in \mathbf{N}$. Then the following hold:

1) Every function $f: A \rightarrow B$ that is one-to-one is also onto.

2) Every function $g: A \rightarrow B$ that is onto is also one-to-one.

Proof: The proof of Part 1 is proved here, and the proof of Part 2 is left to the reader. To prove Part 1, suppose $f: A \rightarrow B$ is one-to-one. Assume (to reach a contradiction) that f is not onto. Then the set $\mathbf{im}(f)$ is a proper subset of B (i.e., $\mathbf{im}(f) \subset B$). Hence, by Part 2 of Theorem 4.19,

$\mathbf{card}(\mathbf{im}(f)) < n$. Now, consider the function $h\colon A \to \mathbf{im}(f)$ defined by $h(a) = f(a)$ for each $a \in A$. Note that h is the same as f, except that the range of h has been restricted so that h is a bijection. (This is where we use the given fact that f is one-to-one.). Hence, $A \approx \mathbf{im}(f)$. By Part 4 of Theorem 4.19, $\mathbf{card}(A) = \mathbf{card}(\mathbf{im}(f))$. However, it is given that $\mathbf{card}(A) = n$. This contradicts our earlier conclusion that $\mathbf{card}(\mathbf{im}(f)) < n$. Hence, f must be onto.

The next theorem is an extension of Part 5 of Theorem 4.19. The reader is asked to recall the definition of pairwise disjoint sets (Defintion 2.9). The proof of this theorem is left to the reader. In words, the theorem can be expressed by saying the cardinality of a finite union of pairwise disjoint finite sets is the sum of the cardinalities of these sets.

Theorem 4.20 Suppose that $\{A_k\}_{k=1}^{n}$ is a finite sequence of pairwise disjoint finite sets. Then $\bigcup_{k=1}^{n} A_k$ is finite, and $\mathbf{card}(\bigcup_{k=1}^{n} A_k) = \sum_{k=1}^{n} \mathbf{card}(A_k)$.

What if finite sets A and B are not disjoint? Is $A \cup B$ finite? Our intuition tells us strongly that it must be.

Theorem 4.21 Suppose that A and B are finite sets. Then $A \cup B$ is finite, and $\mathbf{card}(A \cup B) = \mathbf{card}(A - B) + \mathbf{card}(B - A) + \mathbf{card}(A \cap B)$.

Proof: Note that $A \cup B$ can be written as $A \cup B = (A - B) \cup (B - A) \cup (A \cap B)$. Because both $A - B$ and $A \cap B$ are subsets of the finite set A, these sets are finite. (Of course, $A \cap B$ is also a subset of B.) Similary, $B - A$ is finite because it is a subset of the finite set B. Also note that the family of sets $\{A - B, B - A, A \cap B\}$ is pairwise disjoint. The result follows immediatedly from Theorem 4.20.

The proof of Theorem 4.21 demonstrates the point made previously that it is often important in mathematics to break a union of sets into a union of pairwise disjoint sets. See the remarks following Definition 2.9. The sets are displayed in Figure 4.2. A is represented by the circle on the left, and B is represented by the circle on the right.

Example 4.20 Define the sets A and B by $A = \{-5, -3, 0, 1, 2\}$ and $B = \{-5, -3, 6, 9, 12, 100\}$. Then $A \cup B = \{-5, -3, 0, 1, 2, 6, 9, 12, 100\}$,

Figure 4.2
Venn diagram for
Theorem 4.21

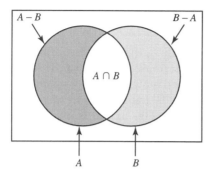

so **card**$(A \cup B) = 9$. $A - B = \{0, 1, 2\}$, so **card**$(A - B) = 3$. $B - A = \{6, 9, 12, 100\}$, so **card**$(B - A) = 4$. $A \cap B = \{-5, -3\}$, so **card**$(A \cap B) = 2$. Hence, **card**$(A \cup B) = $ **card**$(A - B) + $ **card**$(B - A) + $ **card**$(A \cap B)$, as required by Theorem 4.21.

◼

Theorem 4.22 indicates another way to express the cardinality of $A \cup B$. Theorem 4.23 states that a finite union of finite sets is finite. The proofs of both theorems are left to the reader.

Theorem 4.22   Suppose that A and B are finite sets. Then

$$\mathbf{card}(A \cup B) = \mathbf{card}(A) + \mathbf{card}(B) - \mathbf{card}(A \cap B).$$

Theorem 4.23   Suppose that $\{A_k\}_{k=1}^n$ is a finite sequence of finite sets. Then $\cup_{k=1}^n A_k$ is finite.

Before discussing infinite sets, we present one more (obvious) fact about finite sets.

Theorem 4.24   Suppose that A and B are nonempty sets, and suppose A is finite with cardinality n. Suppose $f: A \to B$. Then **im**(f) is finite, with **card**$(\mathbf{im}(f)) \leq n$.

Proof: First, note that $\mathbf{im}(f) = \cup_{a \in A}\{f(a)\}$. Hence, **im**$(f)$ is finite, by Theorem 4.23. Now, define $g: A \to \mathbf{im}(f)$ by $g(a) = f(a)$ for each $a \in A$. Then g is onto. Hence, **card**$(\mathbf{im}(f)) \leq n$, by Part 3 of Theorem 4.19. (Why?)

Introduction to Infinite Sets

We see from Definition 4.12 that a set S is infinite provided it is nonempty and provided that the statement $S \approx \mathbf{I}_n$ is false for each $n \in \mathbf{N}$. The reader should be able to immediately see from Exercise 7 of Section 3.4 that $\mathbf{N}$ is infinite. By the way, this fact is the reason why the Axiom of Infinity (Section 2.7) is aptly named. This axiom is used to prove the existence of a Peano space, and for any Peano space $(P, s, 1)$, the set P is infinite. Hence, the Axiom of Infinity is used to prove that an infinite set exists!

We begin by presenting a famous theorem and proof. The proof that there are infinitely many primes is attributed to Euclid. This celebrated proof is a classic example of proof by contradiction.

Theorem 4.25   There are infinitely many primes.

Proof: Let S denote the set of primes. Suppose (to reach a contradiction) that S is a finite set. Clearly S is nonempty because $2 \in S$. Hence, by definition, there is an $n \in \mathbf{N}$ such that $S \approx \mathbf{I}_n$. That is, there are precisely n primes. Let us list them as $p_1, p_2, \ldots, p_n$. Now, let $Q = (p_1 p_2 \cdots p_n) + 1$; that is, Q is the sum of the product of these n primes with 1. By the Fundamental Theorem of Arithmetic (Theorem 4.18), Q is either a prime or else it is a product of primes. In either case, there exists a prime that divides Q. Let us call such a prime r, so that $r|Q$. However, r must be one of the primes of the list of primes $p_1, p_2, \ldots, p_n$ because we have made the assumption that this list contains all of the primes. Hence, we must have $r|(p_1 p_2 \cdots p_n)$. Hence, by Theorem 4.14, we must have $r|(Q - (p_1 p_2 \cdots p_n))$. (Why?) However, $Q - (p_1 p_2 \cdots p_n) = 1$, so we have reached the conclusion that $r|1$. Of course, it is impossible for a prime to divide 1, so we have arrived at a contradiction. Thus, there does not exist an $n \in \mathbf{N}$ such that $S \approx \mathbf{I}_n$. That is, the set of primes is an infinite set.

For the record, we list the next two theorems, which are intuitively obvious.

Theorem 4.26   Suppose that A is an infinite set, and B is a nonempty finite set. Then

1) If $f: A \to B$, then f is not one-to-one.
2) If $g: B \to A$, then g is not onto.

Proof: We prove Part 2, and leave the proof of Part 1 for the reader. Suppose $g: A \to B$. Now, suppose (to reach a contradiction) that g is onto. Then, we have $\mathbf{im}(g) = A$ (by the definition of *onto*). By Theorem 4.24,

$\mathbf{im}(g)$ is finite. However, this contradicts the given statement that A is infinite. Thus, g is not onto.

Theorem 4.27   Suppose that A is an infinite set. Then

1) $A \cup B$ is infinite for any set B.
2) If B is a finite set, then $A - B$ is infinite.

Proof:

1) Let B be any set. Suppose (to reach a contradiction) that $A \cup B$ is finite. Note that $A \subseteq A \cup B$. Hence, we conclude that A is finite by Part 1 of Theorem 4.19, and this is a contradiction. Hence, $A \cup B$ is infinite.
2) Let B be any finite set. Suppose (to reach a contradiction) that $A - B$ is finite. Clearly, $A \cup B = B \cup (A - B)$, so we conclude that $A \cup B$ is finite by Theorem 4.21. However, this contradicts Part 1 above. Hence $A - B$ is infinite.

Next, we begin to discuss some of the unintuitive properties of infinite sets. This discussion will continue in the next section. Consider finite nonempty sets A and B. Recall that there is a one-to-one correspondence (i.e., a bijection) between these sets iff $\mathbf{card}(A) = \mathbf{card}(B)$ (i.e., iff these sets have the same size). Intuitively, we can regard a one-to-one correspondence as a pairing of elements of B with elements of A such that every element of B gets assigned to one and only one element of A, and vice versa. This is illustrated in Figure 4.3.

Figure 4.3
One-to-one
correspondence
between finite sets

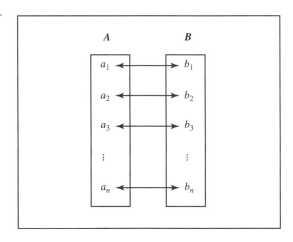

Now, suppose A and B are infinite sets. We can still regard a one-to-one correspondence as a pairing in the same manner as in the case of finite sets. However, the concept of *size* is not the same as with finite sets. As an example, consider the mapping $f: \mathbf{Z} \to \mathbf{N}$ defined by

$$f(m) = \begin{cases} 1 - 2m & if & m \in \mathbf{Z} - \mathbf{N} \\ 2m & if & m \in \mathbf{N} \end{cases}.$$

This mapping will show that there is a one-to-one correspondence between $\mathbf{Z}$ and $\mathbf{N}$. A little reflection will indicate that the top portion of the definition of f generates all of the odd positive integers, and the bottom portion generates all of the even integers, so the mapping is onto. It is also not hard to see that the mapping is one-to-one; that is, f is a one-to-one correspondence between $\mathbf{Z}$ and $\mathbf{N}$. Just to be extra careful, we will formally prove this.

To show that f is onto, suppose $n \in \mathbf{N}$. We must show that there is an $m \in \mathbf{Z}$ so that $f(m) = n$. Now, if n is odd, it can be written as $n = 2k + 1$ for some $k \in \mathbf{N} \cup \{0\}$. Hence, $-k \in \mathbf{Z} - \mathbf{N}$. Also, $f(-k) = 1 - 2(-k) = 1 + 2k = n$, so setting m to be $-k$ "works." What if n is even? Then it can be written as $n = 2k$ for some $k \in \mathbf{N}$. Hence, $f(k) = 2k = n$, so setting m to be k "works." We have shown that f is onto.

To show that f is one-to-one, suppose $f(m_1) = f(m_2)$ for some $m_1, m_2 \in \mathbf{Z}$. The goal is to show that $m_1 = m_2$. This is easy to see if m_1 and m_2 are both in $\mathbf{Z} - \mathbf{N}$, or if they both are in $\mathbf{N}$. Suppose that one of them, say m_1, is in $\mathbf{Z} - \mathbf{N}$, and the other is in $\mathbf{N}$? Then we must have $1 - 2m_1 = 2m_2$. However, this is not possible because $2m_2$ is even and $1 - 2m_1$ is odd. Hence, f is one-to-one.

Thus, we have produced a one-to-one correspondence between $\mathbf{Z}$ and $\mathbf{N}$. In that sense, these two sets can be regarded as having the same number of elements because the elements of these sets can be paired off such that every element of $\mathbf{Z}$ is assigned exactly one element of $\mathbf{N}$, and every element of $\mathbf{N}$ is assigned to exactly one element of $\mathbf{Z}$.

However, in another sense they do not have the same number of elements. We tend to regard $\mathbf{Z}$ as much "bigger" than $\mathbf{N}$. Indeed, $\mathbf{N} \subset \mathbf{Z}$. We may be tempted to reconcile these views by noting that in the above assignment of elements of $\mathbf{Z}$ to elements of $\mathbf{N}$, we never have a "shortage" of elements of $\mathbf{N}$ because there is an infinite supply of them. However, as we will see in the next section, this is an insufficient argument. Rather, infinite sets are classified by whether or not it is possible to define bijections to other given infinite sets. This is discussed in the next section.

We conclude this section with a brief discussion of a famous theorem. We have just seen that if A and B are infinite sets, it is not always so easy to use intuition to determine if $A \approx B$. The infamous Cantor-Schröder-Bernstein Theorem states that if there are one-to-functions $f: A \to B$ and $g: B \to A$, then indeed a bijection $h: A \to B$ can be found, so that that $A \approx B$. The proof of this theorem is not conceptually hard. However, it is long and somewhat unintuitive. For that reason, the proof has been placed in an Appendix of this text (Appendix A). You are encouraged to read it because it makes good use of many of the concepts that we have covered so far in this text.

| Theorem 4.28 | **(Cantor-Schröder-Bernstein)** |

Suppose that A and B are nonempty sets. Suppose there are functions $f: A \rightarrow B$ and $g: B \rightarrow A$ that are both one-to-one. Then $A \approx B$.

Note that the sets A and B are allowed to be finite in the statement of the theorem. However, in this case, the conclusion follows from our work with finite sets.

EXERCISES

1. Prove Part 4 of Theorem 4.19.
2. Prove Part 2 of Corollary 4.1.
3. Prove Theorem 4.20.
4. Prove Theorem 4.22. (Hint: Use Theorem 4.21 and Part 5 of Theorem 4.19.)
5. Prove Theorem 4.23.
6. Suppose that $\{A_i\}_{i=1}^n$ is a finite sequence of finite sets. Show that $\mathbf{card}(\cup_{i=1}^n A_i) \leq \sum_{i=1}^n \mathbf{card}(A_i)$.
7. Prove Part 1 of Theorem 4.26.
8. Find a bijection between the set $\mathbf{N}$ and the set of even members of $\mathbf{N}$.
9. Prove Theorem 4.28 in the case where both sets A and B are finite.
10. Use the Cantor-Shröder-Bernstein Theorem to show that $\mathbf{Z}$ is countable.

| Section 4.5 | # Infinite Sets: Countable and Uncountable |

In this section we continue examining some of the mysteries of infinite sets. Two major classifications of infinite sets will be discussed: **countable sets** and **uncountable sets**. We will also briefly introduce a topic in set theory: the Axiom of Choice. This axiom will be used to prove some properties of infinite sets. These proofs are in Appendix B.

Countable Sets

It was mentioned at the end of Section 4.4 that there are different types of infinite sets. The most basic type of classification of an infinite set is whether or not it is **countable**.

Definition 4.13

A set S is said to be a **countable** set provided $\mathbf{N} \approx S$.

Hence, S is countable provided there exists a one-to-one correspondence between $\mathbf{N}$ and S. With this definition, note that because we showed $\mathbf{Z} \approx \mathbf{N}$ in the last section, we conclude that $\mathbf{Z}$ is countable. (Recall that for sets A and B, $A \approx B$ iff $B \approx A$.) You may be interested to hear that the set $\mathbf{Q}$ of rational numbers is countable (as we show in Chapter 5) and that the set $\mathbf{R}$ of real numbers is not countable (as we show in Chapter 7).

If a set S is countable, its elements can be listed as the terms of a sequence. That is, there exists a sequence $\{s_n\}_{n=1}^{\infty}$, with $s_i \neq s_j$ whenever $i \neq j$, such that $S = \bigcup_{n=1}^{\infty}\{s_n\}$. Of course, the expression $S = \bigcup_{n=1}^{\infty}\{s_n\}$ can also be written as $S = \{s_1, s_2, \ldots\}$. (Instead of always writing out the cumbersome expression "$s_i \neq s_j$ whenever $i \neq j$," we often will indicate this by saying that we have a listing of *distinct* elements. We can also indicate this by simply saying that each s_i is distinct.) It should be pointed out that some authors use the term *denumerable* to mean what we have defined as countable, and they use the term *countable* to mean finite or denumerable.

The next goal is to prove that the union of a countable set with a finite set is a countable set. It will be much easier to accomplish this by first proving the following lemma.

Lemma 4.5

Suppose that A is countable and B is finite, and A and B are disjoint. Then $A \cup B$ is countable.

Proof: If $B = \varnothing$, the result is trivial. Consider the case in which B is nonempty, so that $B \approx \mathbf{I}_n$ for some $n \in \mathbf{N}$. Then we can write B as $B = \{b_1, b_2, \ldots, b_n\}$, in which each b_i is distinct. Because A is countable, A can be written as $A = \{a_1, a_2, \ldots\}$, in which each a_i is distinct. Furthermore, because A and B are disjoint sets, each element of B is distinct from each element of A.

Now, define $f \colon \mathbf{N} \to A \cup B$ by

$$f(m) = \begin{cases} b_m & \text{if} & m \in \mathbf{I}_n \\ a_{m-n} & \text{if} & m \in \mathbf{N} - \mathbf{I}_n \end{cases}.$$

It is easy to see that f is a bijection. Hence, $A \cup B$ is countable.

The reader should not be intimidated by the compactness of notation in the definition of the function f in the proof of Lemma 4.5. All that we are saying is that the elements of $A \cup B$ can be listed by listing the elements of B first and then listing the elements of A. That is, $A \cup B = \{b_1, b_2, \ldots, b_n, a_1, a_2, \ldots\}$. Hence, $f(n + 1) = a_1$, $f(n + 2) + a_2$, and so on.

The proof of the next theorem is left to the reader.

Theorem 4.29 Suppose that A is countable and B is finite. Then both $A \cup B$ and $A - B$ are countable.

The next theorem is very important. It states that any infinite subset of a countable set is also countable. Before doing this, it will be useful to introduce a definition and a lemma.

Definition 4.14

Suppose $\{a_n\}_{n=1}^{\infty}$ is a sequence of positive integers. Then we say that $\{a_n\}_{n=1}^{\infty}$ **is strictly increasing** provided $a_{n+1} > a_n$ for each $n \in \mathbf{N}$.

Simply put, the sequence $\{a_n\}_{n=1}^{\infty}$ of positive integers is strictly increasing if each term is larger than the previous term. We have more to say about this concept in a more general setting later in the text.

Example 4.21 Consider the sequence $\{2n\}_{n=1}^{\infty}$. Of course, this is the sequence of even positive integers. In this case, the statement $a_{n+1} > a_n$ means $2(n + 1) > 2n$, which is clearly true for each $n \in \mathbf{N}$. Hence, $\{2n\}_{n=1}^{\infty}$ is strictly increasing. The reader should also easily see that $\{n\}_{n=1}^{\infty}$, which is the sequence of all positive integers, is also strictly increasing.

Lemma 4.6 Suppose $\{a_n\}_{n=1}^{\infty}$ is a strictly increasing sequence of positive integers. Then $a_n \geq n$ for each $n \in \mathbf{N}$.

Proof: We perform a proof by induction. Clearly $a_1 \geq 1$ because no positive integer is less than 1. We now make the induction hypothesis that $a_n \geq n$ for some $n \in \mathbf{N}$. The goal is to show that $a_{n+1} \geq n + 1$. Now, $a_{n+1} > a_n$ (by the definition of strictly increasing), and $a_n \geq n$ by the induction hypothesis. Hence, $a_{n+1} > n$. Thus, $a_{n+1} \geq n + 1$.

Theorem 4.30   Suppose that B is countable, A is infinite, and $A \subseteq B$. Then A is countable.

Proof: Because B is countable, it can be written as $B = \{b_1, b_2, \ldots\}$, in which each b_i is distinct. We will use the generalized Recursion Theorem to prove the existence of a sequence $\{k_n\}_{n=1}^{\infty}$ of positive integers as follows: Choose k_1 to be the smallest positive integer j such that $b_j \in A$. (Note that we are guaranteed the existence of k_1 because A is nonempty, and $A \subseteq B$. Now, suppose we have chosen $k_1, k_2, \ldots, k_n$ for some $n \in \mathbf{N}$, such that $k_1 < k_2 < \cdots < k_n$. Note that the set $A - \{b_{k_1}, b_{k_2}, \ldots, b_{k_n}\}$ is nonempty because A is infinite. We choose k_{n+1} to be that smallest positive integer $j > k_n$ such that $b_j \in A - \{b_{k_1}, b_{k_2}, \ldots, b_{k_n}\}$. Note that $\{k_n\}_{n=1}^{\infty}$ is a strictly increasing sequence of positive integers.

We will now show that $A = \{b_{k_1}, b_{k_2}, \ldots\}$. Suppose (to reach a contradiction) there exists an $a \in A$ with $a \notin \{b_{k_1}, b_{k_2}, \ldots\}$. Because $A \subseteq B$, we must have $a = b_i$ for some $i \in \mathbf{N}$. Now, by Lemma 4.6, $k_i \geq i$. Thus, the set $\{j \in \mathbf{N} | k_j \geq i\}$ is nonempty, so it must have a least element, which we denote by w.

We first examine the case in which $w = 1$. Then we must have $1 \leq i \leq k_1$. But this implies that $k_1 = i$ because k_1 has been defined to be the smallest positive integer j such that $b_j \in A$, but $b_i \in A$ (because $a = b_i$). Hence, $b_i = b_{k_1} \in \{b_{k_1}, b_{k_2}, \ldots\}$, but this contradicts our assumption that $a \notin \{b_{k_1}, b_{k_2}, \ldots\}$. Hence, this case is not impossible.

Hence, we must have $w > 1$. Note that the definition of w forces us to conclude that $k_{w-1} < i \leq k_w$. (Why?) However, recall that k_w is the smallest positive integer j with the properties that $j > k_{w-1}$ and $b_j \in A - \{b_{k_1}, b_{k_2}, \ldots, b_{k_{w-1}}\}$. Hence, $k_w \leq i$. Now, because we have both $i \leq k_w$ and $k_w \leq i$, this forces the conclusion that $k_w = i$, and again we arrive at a contradiction.

Note that the proof of Theorem 4.30 invokes the generalized Recursion Theorem "informally." In the Exercises, the student is asked to invoke it formally and show that it works.

Example 4.22   Consider the set S of all primes. By Theorem 4.25, S is infinite. Because $S \subseteq \mathbf{N}$, Theorem 4.30 indicates that S is countable.

Theorem 4.31   Suppose that A and B are countable. Then $A - B$ is either finite or countable.

Proof: Note that $A - B \subseteq A$. Now, if $A - B$ is finite, we are done. If $A - B$ is infinite, then it is an infinite subset of a countable set, so $A - B$ is countable by Theorem 4.30.

We would like to prove that the union of two countable sets is a countable set. As before, it will be much easier to accomplish this by first proving it is true if the sets are disjoint.

Lemma 4.7

Suppose that A and B are countable disjoint sets. Then $A \cup B$ is countable.

Proof: We can write A as $A = \{a_1, a_2, \ldots\}$, in which each a_i is distinct. We can write B as $B = \{b_1, b_2, \ldots\}$, in which each b_i is distinct. Furthermore, because A and B are disjoint sets, each element of B is distinct from each element of A. Now we define $f: A \cup B \to \mathbf{N}$ by

$$f(x) = \begin{cases} 2i - 1 & if \quad x = a_i \\ 2i & if \quad x = b_i \end{cases}.$$

It is easy to see that f is a bijection. Hence, $A \cup B$ is countable.

The reader should have an easy time seeing that the inverse mapping of f in the proof of Lemma 4.7 is simply the listing of the elements of $A \cup B$ as $a_1, b_1, a_2, b_2, \ldots$. Now the reader should be able to supply the proof of the next two theorems.

Theorem 4.32

Suppose that A and B are countable sets. Then $A \cup B$ is countable.

Theorem 4.33

Suppose that $\{A_k\}_{k=1}^{n}$ is a finite sequence of countable sets. Then $\cup_{k=1}^{n} A_k$ is countable.

A very important theorem is presented next. This theorem is a good example of how results about infinite sets can be surprising. Suppose A and B are any countable sets. Then the theorem states that $A \times B$ is countable. What is so surprising about this is that intuitively there are so many more elements in $A \times B$ than there are in A or B. Yet, the elements of $A \times B$ can be put into one-to-one correspondence with $\mathbf{N}$. To show the power of the methods that you now have at your disposal, two proofs are provided. The first two proofs make excellent use of the Fundamental Theorem of Arithmetic. Then a third proof is outlined. This third proof is important, not only for historical reasons, but because it also offers some intuition as to what is going on.

Theorem 4.34

Suppose that A and B are countable sets. Then $A \times B$ is countable.

Proof 1: Because A and B are countable, we can write A as $A = \{a_1, a_2, \ldots\}$, in which each a_i is distinct, and we can write B as $B = \{b_1, b_2, \ldots\}$, in which each b_i is distinct. We define $f\colon A \times B \to \mathbf{N}$ by $f(a_m, b_n) = 2^m 3^n$ for each $m, n \in \mathbf{N}$. We note that from the uniqueness portion of the Fundamental Theorem of Arithmetic, this function is one-to-one. Now, define $g\colon \mathbf{N} \to A \times B$ by $g(n) = (a_n, b_1)$ for each $n \in \mathbf{N}$. This mapping is clearly also one-to-one. Hence, by Theorem 4.28, $\mathbf{N} \approx A \times B$, so $A \times B$ is countable.

Proof 2: Because A and B are countable, we can write A as $A = \{a_1, a_2, \ldots\}$, in which each a_i is distinct, and we can write B as $B = \{b_1, b_2, \ldots\}$, in which each b_i is distinct. We define $f\colon A \times B \to \mathbf{N}$ by $f(a_m, b_n) = 2^m 3^n$ for each $m, n \in \mathbf{N}$. We note that from the uniqueness portion of the Fundamental Theorem of Arithmetic, this function is one-to-one. Now, let $S = \mathbf{im}(f)$. It is clear that S is an infinite set. (Why?) Noting that $S \subseteq \mathbf{N}$, Theorem 4.30 mandates that S is countable, so $\mathbf{N} \approx S$. Now, define g to be identical with f, with the exception that g has range S. Hence, g is a bijection, so $A \times B \approx S$. It is now clear that $\mathbf{N} \approx A \times B$, so $A \times B$ is countable.

Outline of Proof 3: We now show how to intuitively construct a direct bijection from $A \times B$ to $\mathbf{N}$. Consider Figure 4.4.

In the diagram, we represent the element (a_m, b_n) of $A \times B$ as simply (m, n), and we place them in tabular form. Hence, the element (a_m, b_n) is

Figure 4.4
Tabular intuitive "proof" of Theorem 4.34

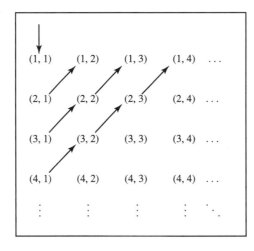

placed in the mth row and nth column. The basic idea is to "count" the elements of $A \times B$ by moving diagonally in the direction of the arrows in the figure. Hence, we first list $(1, 1)$. Then we jump down to count $(2, 1)$ and $(1, 2)$. Then we jump down to $(3, 1)$ and keep going in this manner. It is intuitively obvious that every element eventually gets listed. Why cannot we just complete one row before going to the next row? Because we will never finish the first row!

Note that by this method, we are really listing all pairs (m, n) that satisfy (m, n) such that $m + n = k$ for a fixed k before doing the same for $k + 1$. For example, we first list all pairs (m, n) satisfying $m + n = 2$. This gives us the pair $(1, 1)$ only. Next, we list all pairs (m, n) satisfying $m + n = 3$. This gives us the pairs $(2, 1)$ and $(1, 2)$. This process continues.

It turns out that the above mapping can be expressed by $f(a_m, b_n) = [\frac{(m + n - 1)(m + n - 2)}{2}] + n$. The reader is invited (but not required) to show directly that this mapping is a bijection. If you try to do this, we suggest that you make use of Example 2.33, in which we show via induction that $1 + 2 + \cdots + k = \frac{k(k + 1)}{2}$ for each positive integer k. (Note: The astute student may object to our right to use this induction yet because we have not developed rational numbers and therefore cannot divide by 2. However, because we can easily show that $k(k + 1)$ must be even for any integer k, we can easily get around this "problem." How?]

We have more to say about countable sets, both later in this section and in upcoming sections. Recall the "argument" in Section 4.4 that we contemplated while trying to explain why $\mathbf{Z}$ is countable, even though it appears to have many more members than $\mathbf{N}$. We considered the possibility that $\mathbf{Z}$ is countable because we have an inexhaustible supply of members of $\mathbf{N}$ to pair with members of $\mathbf{Z}$. However, this argument does not sit well with the existence of uncountable sets, and we will shortly prove that uncountable sets exist.

Uncountable Sets

Recall that an infinite set is called uncountable provided it cannot be put into one-to-one correspondence with $\mathbf{N}$. Hence, intuitively, we view such a set as "very large." First some basic theorems about uncountable sets are proved. Then a classic example is presented. The proofs of Theorems 4.35 and 4.37 below are left to the reader.

Theorem 4.35   Suppose that A is uncountable, and B is any set. Then $A \cup B$ is uncountable.

Theorem 4.36   Suppose that A is uncountable, and B is finite. Then $A - B$ is uncountable.

Proof: By Part 2 of Theorem 4.27, $A - B$ is infinite. Hence, it is either countable or uncountable. Suppose (to reach a contradiction), $A - B$ is countable. Now, we can express A as $A = (A - B) \cup (A \cap B)$. Because $A \cap B \subseteq B$, $A \cap B$ is finite by Part 1 of Theorem 4.19. Thus, A has been written as a union of a countable set and a finite set, so A is countable by Theorem 4.29, which is a contradiction. Hence, $A - B$ is uncountable.

Theorem 4.37   Suppose that A is uncountable, and B is countable. Then $A - B$ is uncountable.

Now it may have come to the reader's attention that we have not yet demonstrated the existence of an uncountable set. To address this situation, we discuss the power set. We have already used the power set on several occasions, but we have not yet provided it with a definition number.

Definition 4.15

Suppose S is a set. The **power set** of S, written $\mathbf{P}(S)$, is the set of all subsets of S.

Now, if S is a finite set, it is obvious that there is not a one-to-one correspondence between S and $\mathbf{P}(S)$. For example, if $S = \{1, 2\}$, then $S = \{\varnothing, \{1\}, \{2\}, \{1, 2\}\}$. Hence, **card**$(S) = 2$ and **card**$(\mathbf{P}(S)) = 4$. In the Exercises, the reader is asked to show that if S is finite, then **card**$(\mathbf{P}(S)) = 2^{\mathbf{card}(S)}$. Theorem 4.38 settles the issue in general. The proof of this classical theorem is a superb example of proof by contradiction.

Theorem 4.38   Let S be any set. Then there does not exist a one-to-one correspondence between S and $\mathbf{P}(S)$.

Proof: The result is trivial if $S = \varnothing$, so consider the case in which S is not the empty set. Now, suppose (to reach a contradiction) that there exists a bijection $f: S \to \mathbf{P}(S)$. In particular, f must be onto. Now, note that for each $s \in S$, $f(s)$ is a subset of S, so we could have $s \in f(s)$, or we could have $s \notin f(s)$. Define the set T by $T = \{s \in S \,|\, s \notin f(s)\}$. Now, T is a subset of S, so $T \in \mathbf{P}(S)$.

Because we have assumed that f is onto, there must be an $a \in S$ such that $f(a) = T$. Now either $a \in T$ or else $a \notin T$. Suppose $a \in T$. Then by the definition of T, we must have $a \notin f(a)$; that is $a \notin T$. What has happened here? We have just shown that $a \in T \Rightarrow a \notin T$, and this clearly is not possible. Hence, we cannot have $a \in T$, so we must have $a \notin T$. However, by the defintion of T, this means that $a \in f(a)$; that is, $a \in T$. So now we have shown that $a \notin T \Rightarrow a \in T$, and this is also not possible. Therefore, we have arrived at a contradiction. Hence, we conclude that f is not onto. That is, there does not exist a one-to-one correspondence between S and $\mathbf{P}(S)$.

We have an immediate corollary from this that identifies an uncountable set. The corollary tells us that the set of subsets of $\mathbf{N}$ is not countable. Later, it will be proven that the set $\mathbf{R}$ of real numbers is uncountable.

Corollary 4.2

The set $\mathbf{P}(\mathbf{N})$ is uncountable.

Proof: For $\mathbf{P}(\mathbf{N})$ to be countable, we must have $\mathbf{N} \approx \mathbf{P}(\mathbf{N})$. However, by Theorem 4.38, there does not exist a one-to-one correspondence between $\mathbf{N}$ and $\mathbf{P}(\mathbf{N})$, so we cannot have $\mathbf{N} \approx \mathbf{P}(\mathbf{N})$. Therefore, $\mathbf{P}(\mathbf{N})$ is uncountable.

The significance of Theorem 4.38 goes beyond just proving the existence of an uncountable set. It also demonstrates that if we "start" with an infinite set, then we can keep getting "higher levels" of infinity by continuing to look at the set of subsets of the set of subsets, and so on. There is a whole theory of infinite sets, and we only scratch the surface in this text.

The Axiom of Choice

Recall from the beginning of Chapter 2 that axioms can be specified for set theory. This is a very involved process that can be postponed by the budding mathematician until later courses. Note, however, that we did explicitly mention the Axiom of Infinity (Section 2.7).

There is another axiom of set theory that deserves mention: the **Axiom of Choice**. Although it is very simple to state this axiom, it has been the subject of some controversy in the mathematics world. Some mathematicians felt that it should be provable from the other axioms of set theory. However, in 1963 the mathematician Paul Cohen proved that this axiom could not be derived from the others. Most mathematicians make use of it without realizing it. However, because branches of set theory have been developed that do not use the Axiom of Choice, it has become a courtesy for many mathematicians to mention it explicitly when it is used in a proof. Also, it is often used in a proof only as a last resort.

Axiom of Choice

Suppose F is a nonempty family of nonempty sets. Then there exists a function $f: F \rightarrow \cup_{A \in F} A$ such that $f(A) \in A$ for each $A \in F$.

Although this looks complicated at first glance, all it really says is that given a family of nonempty sets, there exists a function that "chooses" an element from each member of the family. The function f is often called a **choice function**. One reason that it caused concern for mathematicians is that it provides no guidance on how this choice is made.

It turns out (for reasons beyond the scope of this text) that the axiom is not needed if the family of sets is a finite family. It also is not needed if the family of sets has a structure that allows one to specifically choose a specific well-defined member of each set in the family. Example 4.24 provides an example of when the axiom is needed.

Example 4.23   Consider an infinite family F of nonempty sets in which $A \subseteq \mathbf{N}$ for each $A \in F$. Even though F is an infinite family, we can still select a choice function without using the Axiom of Choice. Specifically, it is known that each $A \in F$ has a least element (because every nonempty subset of positive integers has a least element). Let us refer to this least element of $A \in F$ as b_A. Then we can define $f: F \rightarrow \cup_{A \in F} A$ by $f(A) = b_A$ for each $A \in F$.

Example 4.24   Consider an infinite family F of nonempty sets in which $A \subseteq \mathbf{R}$ for each $A \in F$. Now, an arbitrary subset of the real numbers might not have a least or greatest element. If we do not have additional information about these sets, there is no obvious way to choose an element from each set. So unless one can be discovered, the Axiom of Choice must be invoked to assert the existence of a function that "chooses" an element from each set in the family.

More information about the Axiom of Choice can be found in Goldrei (1996). The reason for it being mentioned in this section is that there are some theorems about infinite sets that the student will encounter in higher mathematics that require the use of the Axiom of Choice in their proofs. Three such important theorems are presented here. The complete proofs are in Appendix B.

Theorem 4.39 Let S be any infinite set. Then there exists a countable subset of S.

Theorem 4.40 Let S be any infinite set. Then there exists an $A \subset S$ such that $S \approx A$.

Theorem 4.41 Suppose we have the countable family F of countable sets. Then $\cup_{A \in F} A$ is countable.

Theorem 4.41 can be summarized by saying the union of a coutable family of countable sets is countable. Finally, with regard to Theorem 4.40, it should be noted that the Axiom of Choice is not needed to prove that a *countable* set S has a proper subset A such that $S \approx A$.

Exercises

1. Prove Theorem 4.29.
2. Formally invoke the generalized Recursion Theorem for the proof of Theorem 4.30, and show that this invocation produces the desired result. (Hint: Define the function g of Theorem 4.3 to be such that $g: \cup_{n=1}^{\infty} \mathbf{N}^n \to \mathbf{N}$.)
3. Prove Theorem 4.32. (Hint: Use Theorem 4.31 and Lemma 4.7.)
4. Prove Theorem 4.33.
5. Suppose that A is a countable set. Show that A^n is countable for each $n \in \mathbf{N}$.
6. Prove Theorem 4.35.
7. Prove Theorem 4.37.
8. Show that if S is any finite set, then $\mathbf{card}(\mathbf{P}(S)) = 2^{\mathbf{card}(S)}$. (Hint: Use induction.)
9. Show that if S and T are nonempty sets with $S \approx T$, then $\mathbf{P}(S) \approx \mathbf{P}(T)$.
10. Suppose S is any countable set. Show that there exists an $A \subset S$ such that $S \approx A$.

Chapter 5

Fields and the Rational Numbers

N ext on the agenda is the development of the rational numbers. You likely will find that this construction does not require as much work as was needed for the construction of the integers. Just as we developed the integers within the context of the algebraic concept of a *ring*, so we will develop the rational numbers within the context of the algebraic concept of what is called a *field*. The process of proving results in a general context often will be continued because this makes it easier for us in the next chapter when the full set of real numbers is developed. Properties of fields are discussed in Section 5.1. The postulates for rational numbers are discussed in Section 5.2, as well as initial results about rational numbers. In Section 5.2, we also demonstrate that a set actually exists that obeys these postulates.

There are occasions in this chapter in which we are working with a Peano space and integer space that are subsets of a particular set of interest. Additionally, we work with several integer spaces simultaneously. Hence, we do not use the notation **N** and **Z** again until we specify them in the next chapter.

Introduction to Fields

Before reading this section, the student may want to review the material on rings and ordered integral domains.

Definition and Basic Properties of Fields

Definition 5.1

A commutative ring $(F, +, \cdot, 0, 1)$ is called a **field** if for each $a \in F$ with $a \neq 0$, there is an element $b \in F$ such that $ab = 1$. The element b is said to be a **multiplicative inverse** of a.

Instead of expressing the field as $(F, +, \cdot, 0, 1)$, we often "abuse notation" and say that F is the field. Also, note that if we have $ab = 1$ for some $a, b \in F$, then we also have $ba = 1$ because multiplication is commutative.

Theorem 5.1   Suppose $(F, +, \cdot, 0, 1)$ is a field, and suppose $a \in F$ with $a \neq 0$. Then a has a unique multiplicative inverse.

Proof: Suppose the elements $b, c \in F$ are multiplicative inverses of a. Then $ab = 1 = ac$. Hence, $ab - ac = 0$, so $a(b - c) = 0$. Now, we also have $ba = 1$. Note that by multiplying both sides of the equation $a(b - c) = 0$ by b gives us $b(a(b - c)) = b \cdot 0 = 0$. By associativity, we have $(ba)(b - c) = 0$, so $1 \cdot (b - c) = 0$. Hence, $b = c$.

Due to Theorem 5.1, we can speak of **the** multiplicative inverse of an $a \in F$ with $a \neq 0$ because there is only one such inverse. The multiplicative inverse of a is normally written as a^{-1}. Also, by reviewing Definition 2.31, the reader should be able to see that the nonzero elements of a field form a *group* with respect to the binary operation of multiplication.

Also, note that in the definition of a ring (Definition 3.12), we require that the binary operation of addition is not identical to the binary operation of multiplication. This forces the conclusion that $0 \neq 1$ (so technically the requirement in Definition 3.12 that *0 and 1 are distinct* is unnecessary). To show this, suppose (to reach a contradiction) that $0 = 1$. Then, for any element a of the ring, multiplying this equation by a gives $a \cdot 0 = a \cdot 1$. This reduces to $0 = a$. Hence, 0 is the only element of the ring. However, if this is the case, then the operation of addition is identical to the binary operation of multiplication because we know that $0 \cdot 0 = 0 + 0$. This contradicts our given fact that addition is not identical to multiplication. Therefore, $0 \neq 1$. Because this is true for any ring, it is also true for any field. It should be noted that many authors allow for a ring or field in which $0 = 1$, and they refer to this as the **trivial ring** and the **trivial field**.

Example 5.1

Consider the ring containing two elements of Example 3.4 in Section 3.5. The reader should be able to easily verify that this is a field and that $b^{-1} = b$.

Example 5.2

It will soon be clear that **Q** and **R** are fields with respect to the "usual" operations of addition and multiplication. However, **Z** is not a field. Note that the only elements of the ring **Z** that have multiplicative inverses are 1 and -1.

An additional example is provided in Exercise 17.

Theorem 5.2

Suppose F is a field, and $a \in F$ with a nonzero. Then $(a^{-1})^{-1} = a$.

Proof: $(a^{-1})^{-1}$ is defined to be that $b \in F$ such that $a^{-1}b = 1$. By Theorem 5.1, this b is unique. However, note that $a^{-1}a = 1$. Hence, we must have $b = a$. That is, we must have $(a^{-1})^{-1} = a$.

Note that Theorem 5.2 says that the multiplicative inverse of the multiplicative inverse of a is a. The proofs of the next two theorems are easy and are left to the reader.

Theorem 5.3

Suppose F is a field, and $a, b \in F$ with a and b nonzero. Then $(ab)^{-1} = a^{-1}b^{-1}$.

Theorem 5.4

Every field is an integral domain.

Next, we shall introduce a notation that should be instantly recognizable to the reader.

Definition 5.2

Suppose F is a field, and $a, b \in F$ with b nonzero. Then we can write ab^{-1} as $\dfrac{a}{b}$.

Note that if $a \in F$ with $a \neq 0$, then $a^{-1} = 1 \cdot a^{-1} = \dfrac{1}{a}$. Now, the proofs of the following two theorems are easy, so not all of the steps are listed. The student should be able to easily fill in any missing steps. Also, some parts of the proofs are left to the reader.

Theorem 5.5   Suppose F is a field, and $a, b, c, d \in F$ with b and d nonzero. Then:

1) $\dfrac{ad}{bd} = \dfrac{a}{b}$.
2) $\left(\dfrac{a}{b}\right)\left(\dfrac{c}{d}\right) = \dfrac{ac}{bd}$.
3) $\dfrac{a}{b} + \dfrac{c}{b} = \dfrac{a+c}{b}$.
4) $\dfrac{a}{b} + \dfrac{c}{d} = \dfrac{ad+bc}{bd}$.

Proof:

1) $\dfrac{ad}{bd} = (ad)((bd)^{-1}) = adb^{-1}d^{-1} = ab^{-1} = \dfrac{a}{b}$.
2) Left to the reader.
3) $\dfrac{a}{b} + \dfrac{c}{b} = ab^{-1} + cb^{-1} = (a+c)b^{-1} = \dfrac{a+c}{b}$.
4) Left to the reader.

Theorem 5.6   Suppose F is a field, and $a, b, c, d \in F$ with b, c, and d nonzero. Then

1) $\left(\dfrac{a}{b}\right)/\left(\dfrac{c}{d}\right) = \dfrac{ad}{bc}$.
2) $\left(\dfrac{a}{b}\right)/c = \dfrac{a}{bc}$.
3) $1/\left(\dfrac{1}{c}\right) = c$.
4) $\left(\dfrac{b}{c}\right)^{-1} = \dfrac{b^{-1}}{c^{-1}} = \dfrac{c}{b}$.

Proof:

1) $\left(\dfrac{a}{b}\right)/\left(\dfrac{c}{d}\right) = (ab^{-1})((cd^{-1})^{-1}) = ab^{-1}c^{-1}d = (ad)((bc)^{-1}) = \dfrac{ad}{bc}$.
2) Left to the reader.
3) Left to the reader.
4) $\left(\dfrac{b}{c}\right)^{-1} = 1/\left(\dfrac{b}{c}\right) = \left(\dfrac{1}{1}\right)/\left(\dfrac{b}{c}\right) = \dfrac{c}{b} = cb^{-1} = b^{-1}((c^{-1})^{-1}) = \dfrac{b^{-1}}{c^{-1}}$.

The following two definitions will prove useful to us. They should not come as a great surprise, provided that the student understands the concepts of ring isomorphisms and subrings (Section 3.5).

Definition 5.3

Suppose F and G are fields, and $f: F \to G$ is a ring isomorphism. Then f is also called a **field isomorphism**.

Definition 5.4

Suppose F is a field, and H is a nonempty subset of F, such that H is itself a field with respect to the restrictions of addition and multiplication to $H \times H$. Then H is called a **subfield** of F.

Note that every field F has at least one subfield: F itself. Also, note that the following must hold for the nonempty set H to be a subfield of a field F:

1) H must be closed with respect to addition and multiplication (See Definition 3.15).

2) H must contain 0 and 1.

3) For each $a \in H$, we have $-a \in H$.

4) For each nonzero $a \in H$, we must have $a^{-1} \in H$.

Example 5.3 As we stated in Example 3.9 in Section 3.5, $\mathbf{Q}$ is a subring of $\mathbf{R}$ (i.e., this will be the case after we construct these sets). Because $\mathbf{Q}$ is a field in its own right (with respect to addition and multiplication on $\mathbf{R}$), it is clear that $\mathbf{Q}$ is a subfield of $\mathbf{R}$. ∎

Before moving on to the next topic, the student should be aware that the theory of fields is extensive, and we are only scratching the surface here. Specifically, from this point on we are focusing more and more on those aspects of abstract algebra that will be most useful for constructing the real number system. The student will learn much more about fields in a course in abstract algebra. Upon completing this book, the student will be well prepared to take such a course.

Ordered Fields

We know from Theorem 5.4 that any given field is an integral domain. If this given field is also an ordered integral domain, it is called an **ordered field**.

Definition 5.5

Suppose F is a field. Then F is called an **ordered field** if it is also an ordered integral domain.

Theorem 5.7

Suppose F is an ordered field with linear order $\leq$. Suppose $x, y \in F$ with $0 < x < y$. Then $0 < \frac{1}{y} < \frac{1}{x}$.

Proof: We first show that $0 < \frac{1}{y}$. Suppose (to reach a contradiction) that $\frac{1}{y} < 0$. Then $0 < -\frac{1}{y}$. Because we have both $0 < -\frac{1}{y}$ and $0 < y$, we have $0 < (-\frac{1}{y})y$. That is, $0 < -1$, so $1 < 0$. However, this is impossible because we know that $0 < 1$ for ordered integral domains. Hence, we must have $0 < \frac{1}{y}$. Now, we show that $\frac{1}{y} < \frac{1}{x}$. Starting with the inequality $x < y$, multiply both sides by the positive value $\frac{1}{y}$. This gives $\frac{x}{y} < 1$. Noting that we also have $0 < \frac{1}{x}$ (for the same reason that we have $0 < \frac{1}{y}$), we multiply the inequality $\frac{x}{y} < 1$ by $\frac{1}{x}$ to get $\frac{1}{y} < \frac{1}{x}$.

■

Next, we discuss **absolute values**. The reader is likely to have encountered absolute values (for real numbers) in previous mathematics training. In a sense, the study of absolute values may be regarded as the beginning of the portion of mathematics known as **analysis**. Analysis is the realm of distance, limits, continuity, integration, and so on. Most of our results on absolute value apply to integral domains. However, we shall keep ordered fields as our structure of interest because we will be dealing with them as we construct the real numbers.

Definition 5.6

Suppose F is an ordered field with linear order $\leq$. Suppose $a \in F$. Then the **absolute value** of a is denoted by $|a|$ and is defined as follows:

$$|a| = \begin{cases} a & if \quad a \geq 0 \\ -a & if \quad a < 0 \end{cases}.$$

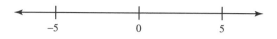

Figure 5.1
Illustrating absolute
value of real
numbers

Now, it is clear that $|a| \geq 0$ for any $a \in F$. When F is the field of real numbers (or any subset of the field of real numbers), the absolute value conforms to our concept of **distance** from the origin along a number line. In Figure 5.1, both the real number 5 and -5 are a distance of 5 from the origin denoted by 0. That is, $|5| = 5 = |-5|$.

Also, when $a, b \in \mathbf{R}$, $|a - b|$ conforms to our concept of the distance between a and b. This can be seen by drawing a number line and looking at all cases in terms of how a and b are situated with respect to each other and to the origin 0. For example, in Figure 5.2, we have $a < 0 < b$. From the figure, it can be seen that the distance from a to b is the distance from a to 0 plus the distance from 0 to b. Now, the distance from a to 0 is $-a$, which is positive because a is negative. The distance from 0 to b is b because b is positive. These distances are indicated in the figure by the line segments above the number line. Hence, the distance from a to b is $-a + b = b - a$. Now, $a - b < 0$, so $|a - b| = -(a - b) = b - a$, which is the distance from a to b. Other cases will "work" similarly.

The proof of the following theorem is left to the reader.

Theorem 5.8 Suppose F is an ordered field with linear order $\leq$. Suppose $a, b \in F$. Then

1) $|-a| = |a|$.
2) $a \leq |a|$.
3) $-a \leq |a|$.
4) $|b - a| = |a - b|$.

Of course, the reader has probably seen these results of Theorem 5.8 before. The reader has also seen the results in Theorems 5.9 through 5.12 as well. Now the reader can see how these results follow from earlier concepts.

Figure 5.2
Illustrating the
distance between two
real numbers

Theorem 5.9   Suppose F is an ordered field with linear order $\leq$. Suppose $a, b \in F$. Then

1) $|ab| = |a||b|$.
2) If $b \neq 0$ then $\left|\frac{a}{b}\right| = \frac{|a|}{|b|}$.
3) $|a + b| \leq |a| + |b|$.

Proof:

1) Case 1: $a > 0$ and $b > 0$. Then $|a| = a$ and $|b| = b$. Also, $ab > 0$, so $|ab| = ab$. Hence, $|ab| = ab = |a||b|$.

    Case 2: $a < 0$ and $b < 0$. Then $|a| = -a$ and $|b| = -b$. Also, $ab > 0$, so $|ab| = ab$. Hence, $|ab| = ab = (-a)(-b) = |a||b|$.

    Case 3: $a > 0$ and $b < 0$. Then $|a| = a$ and $|b| = -b$. Also, $ab < 0$, so $|ab| = -(ab)$. Hence, $|ab| = -(ab) = a(-b) = |a||b|$.

    Case 4: $a < 0$ and $b > 0$. This is the same as Case 3 with the roles of a and b reversed.

    Case 5: $a = 0$ or $b = 0$. The result is obvious.

2) Left to the reader.

3) By Theorem 5.8, we have $a \leq |a|$ and $b \leq |b|$. Hence $a + b \leq |a| + |b|$. Also by Theorem 5.8, we have $-a \leq |a|$ and $-b \leq |b|$. Hence, $-(a + b) \leq |a| + |b|$. Now, $|a + b|$ is equal to $a + b$ or is equal to $-(a + b)$ (by the definition of absolute value). In either case, it is $\leq |a| + |b|$.

The next theorem is so important to analysis that it has its own name: **the triangle inequality**. It is called this because with respect to the mathematical theory of *vector analysis*, it states that the length of a given side of a triangle cannot exceed the sum of the lengths of the other two sides.

Theorem 5.10   **(Triangle Inequality)**

Suppose F is an ordered field with linear order $\leq$. Suppose $a, b, c \in F$. Then $|a - c| \leq |a - b| + |b - c|$.

Proof: $a - c = (a - b) + (b - c)$. Hence by Part 3 of Theorem 5.9, we have $|a - c| = |(a - b) + (b - c)| \leq |a - b| + |b - c|$.

It should be noted that Part 3 of Theorem 5.9 is also frequently referred to as the triangle inequality. The proof of the next theorem is left to the reader.

Theorem 5.11   Suppose F is an ordered field with linear order $\leq$. Suppose $a, b \in F$. Then $\big||a| - |b|\big| \leq |a - b|$.

Finally, the next theorem is also important to analysis.

Theorem 5.12   Suppose F is an ordered field with linear order $\leq$. Suppose $a, b \in F$ with $b > 0$. Then

1) $\{x \in F \mid |x| < b\} = \{x \in F \mid -b < x < b\}$.
2) $\{x \in F \mid |x - a| < b\} = \{x \in F \mid a - b < x < a + b\}$.

Proof:

1) To show that $\{x \in F \mid |x| < b\} \subseteq \{x \in F \mid -b < x < b\}$, suppose $x \in F$ with $|x| < b$. Now, if $x \geq 0$, then $|x| = x$, so $x < b$. Because $b > 0$, we also have $-b < 0 \leq x$, so $-b < x$. Hence, $-b < x < b$. The other possibility is that $x < 0$, so that $|x| = -x$. Hence, $-x < b$, so $-b < x$. Now $x < 0 < b$. Hence, again we have $-b < x < b$. Thus, in either case we have $x \in \{x \in F \mid -b < x < b\}$. Thus, $\{x \in F \mid |x| < b\} \subseteq \{x \in F \mid -b < x < b\}$.

   To show that $\{x \in F \mid -b < x < b\} \subseteq \{x \in F \mid |x| < b\}$, suppose $x \in F$ with $-b < x < b$. If $x \geq 0$, then $|x| = x < b$. If $x < 0$, then $|x| = -x$. Now, because $-b < x$, we have $-x < b$. Hence, $|x| = -x < b$. We have shown that in either case, we have $|x| < b$. Thus, $x \in \{x \in F \mid |x| < b\}$. We have shown that $\{x \in F \mid -b < x < b\} \subseteq \{x \in F \mid |x| < b\}$. The result follows immediately.

2) Applying Part 1 to $|x - a|$, we get $\{x \in F \mid |x - a| < b\} = \{x \in F \mid -b < x - a < b\}$. The result now follows immediately from the fact that for each $x \in F$, we have $-b < x - a < b$ iff $a - b < x < a + b$.

Note that for the set of real numbers in which absolute value represents distance, Part 2 of Theorem 5.12 says that the set of points x whose distance from a is less than b is the set of points x that satisfy the inequality $a - b < x < a + b$.

Next, we prove the all-important fact that every ordered field contains a subring that is an integer space (Theorem 5.13). This result proves very useful in Chapter 6. In fact, we can prove a more general result: Every ordered integral domain contains a subring that is an integer space. Because it is just as easy to prove the more general result, we do this. We use two lemmas to help us. The first one is presented immediately following.

Lemma 5.1 Suppose R is a ring, and G is a nonempty set of subrings of R. Then $\bigcap_{A \in G} A$ is a subring of R.

Proof: Because $0, 1 \in A$ for each $A \in G$, we have $0, 1 \in \bigcap_{A \in G} A$. Suppose $a \in \bigcap_{A \in G} A$. Then $a \in A$ for each $A \in G$. Hence, $-a \in A$ for each $A \in G$. Thus, $-a \in \bigcap_{A \in G} A$. Finally, suppose $b, c \in \bigcap_{A \in G} A$. Then $b, c \in A$ for each $A \in G$. Hence, $b + c \in A$ and $bc \in A$ for each $A \in G$. Hence, $b + c \in \bigcap_{A \in G} A$ and $bc \in \bigcap_{A \in G} A$. We have shown that $\bigcap_{A \in G} A$ contains 0 and 1, is closed under addition and multiplication, and contains the additive inverse of each of its members. Therefore, $\bigcap_{A \in G} A$ is a subring of R.

Before continuing, it will be helpful if we discuss some notation. Suppose that D is any nonempty set with a binary operation on D that is associative and commutative, and furthermore suppose that this binary operation is represented by "addition." Suppose $a \in D$ and $n \in \mathbf{N}$. Recall (just after Theorem 4.6 in Section 4.2) that we employed the notation na to mean the addition of a with itself n times. Throughout Chapter 4, we only needed to work with one integer space. Therefore, for notational convenience we called this integer space $\mathbf{Z}$; we knew that it did not matter which integer space was chosen for the purposes of the work in Chapter 4, and that we would eventually choose *the* set $\mathbf{Z}$. Of course, we know that however we choose *the* set $\mathbf{Z}$, our choice for $\mathbf{N}$ will be determined; it will be $\mathbf{N} = \{n \in \mathbf{Z} | n \geq 1\}$. This choice for $\mathbf{N}$ is "determined" by our desire to have $\mathbf{N} \subset \mathbf{Z}$.

However, for the proof of Theorem 5.13 below, we will need to work with a Peano space for the purpose of "using" na (mentioned above) at the same time that we are demonstrating that a particular set is indeed an integer space. Therefore, we shall use a Peano space that we call P to do this counting. (Of course, it does not matter which Peano space is chosen.) At the same time, we need to work with an ordered integral domain (that we shall call D). Note that each of P and D have their own unit element, their own addition and subtraction, and so on. We will use the symbols $0'$, $1'$, $\oplus$, $\otimes$, and $\leq$ to represent the zero, the unit element, addition, multiplication, and the linear order, respectively, for D. We will reserve the "usual" symbols for P. (Of course, there is no zero element for P.) We will use the minus sign to indicate additive inverses for elements of D. For each $a \in D$, by a^2 we mean $a \otimes a$. (We have already pointed out that in many advanced mathematics books, symbols often are used for several purposes simultaneously, and we have already done this on occasion. However, we do not do this in explanations for which the simultaneous use of symbols for different purposes could lead to confusion.)

We need a result that we discussed in the third paragraph following Theorem 4.6. With the above-mentioned notation, this result is as follows:

$$ma \oplus na = (m + n)a \text{ for all } a \in D \text{ and } m, n \in P.$$

This result will be used in the proof of Lemma 5.2 and Theorem 5.13 without formal reference. Lemma 5.2 below will be used in the proof of Theorem 5.13.

Lemma 5.2

Let D be an ordered integral domain and P a Peano space. Then (invoking the above-mentioned notation), we have $ma \otimes na = (mn)a^2$ for all $a \in D$ and $m, n \in P$.

Proof: Let $a \in D$. We perform an induction on m. For $m = 1$, we need to show that $a \otimes na = na^2$. However, this is just an application of the generalized distributive law (Theorem 4.10), so it clearly holds.

Next, we take the inductive step and assume that $ma \otimes na = (mn)a^2$ for a given $m \in P$ for each $n \in P$. The goal is to show that $((m + 1)a) \otimes na = ((m + 1)n)a^2$. We leave out the justification for the below steps; make sure that you can fill them in.

$$\begin{aligned}
((m + 1)a) \otimes na &= (ma \oplus a) \otimes (na) \\
&= ((ma) \otimes (na)) \oplus (a \otimes (na)) \\
&= ((mn)a^2) \oplus (na^2) \\
&= (mn + n)a^2 \\
&= ((m + 1)n)a^2.
\end{aligned}$$

Do not let the "fancy" notation of Lemma 5.2 fool you into thinking there is anything complicated about it. All that we are saying is $(a \oplus a \oplus \cdots \oplus a) \otimes (a \oplus a \oplus \cdots \oplus a) = a^2 \oplus a^2 \oplus \cdots \oplus a^2$, in which the first sum has m terms, the second sum has n terms, and the third sum has mn terms.

The idea behind the proof of upcoming Theorem 5.13 is really quite simple. Let G be the set of all subrings of the ordered integral domain D, and consider $\cap_{X \in G} X$, which is the intersection of all subrings of D. We know from Lemma 5.1 that $\cap_{X \in G} X$ is itself a subring of D. Because $1' \in \cap_{X \in G} X$, we must have $1' \oplus 1' \in \cap_{X \in G} X$. Then, we must have $1' \oplus 1' \oplus 1' \in \cap_{X \in G} A$, and we continue with this, so that we get all sums obtained by continuing to add $1'$. Because we are working with an ordered integral domain, each such sum results in a different value from every other such sum. If we then include the additive inverse of each of these elements, and also include $0'$, we get an integer space.

In fact, this concept is considered so self-evident that many advanced mathematics texts that include this result do not even bother with a formal proof. However, because this is a "gateway" book, we include some details of how pieces of mathematical structures fit together that others consider obvious. Hence, we "pamper" the student a little. Parts of the proof of Theorem 5.13 are left to the reader to fill in.

Theorem 5.13   Suppose D is an ordered integral domain, and G is the set of subrings of D. Then $\cap_{X \in G} X$ is an integer space (with respect to the addition and multiplication on D).

Proof: Define the set A by $A = \{n1' \mid n \in P\}$ (where P is a Peano space). Define the set B by $B = \{-(n1') \mid n \in P\}$. Finally, define the set I by $I = A \cup B \cup \{0'\}$. First, we show that I is an integer space. Then it is shown that $I = \cap_{X \in G} X$.

To show that I is an integer space, we first show that it is an ordered integral domain. To show that I is an ordered integral domain, it suffices to show that I is a subring of D. (Why?)

Now, it is clear that $0'$, $1' \in I$, and that I contains the additive inverse of each of its members. Hence, to complete the proof that I is a subring of D, it suffices to show that I is closed under multiplication and addition.

To show closure under multiplication, suppose $x, y \in I$. We want to show that $x \otimes y \in I$.

Case 1: $x, y \in A$. Then $x = m1'$ and $y = n1'$ for some $m, n \in P$. Now, by Lemma 5.2, $m1' \otimes n1' = (mn)(1')^2 = (mn)1' \in I$.

Case 2: $x, y \in B$. Then $x = -(m1')$ and $y = -(n1')$ for some $m, n \in P$. Hence, $(-(m1')) \otimes (-(n1')) = (mn)1' \in I$. (All we did was use Theorem 3.37 and Case 1 above.)

Case 3: $x \in A$ and $y \in B$. Left to the reader.

Case 4: $x \in B$ and $y \in A$. This is the same as Case 3 with x and y "switched."

Case 5: $x = 0'$ or $y = 0'$. Then $x \otimes y = 0' \in I$.

In all cases, $x \otimes y \in I$.

Next, we show closure under addition. That is, we show that $x \oplus y \in I$.

Case 1: $x, y \in A$. Then $x = m1'$ and $y = n1'$ for some $m, n \in P$. Hence, $m1' \oplus n1' = (m + n)1' \in I$.

Case 2: $x, y \in B$. Left to the reader.

Case 3: $x \in A$ and $y \in B$. Then $x = m1'$ and $y = -(n1')$ for some $m, n \in P$.
Subcase 3a: $m > n$. Then $m = n + k$ for some $k \in P$. Hence, $x \oplus y = m1' \oplus -(n1') = (n + k)1' \oplus -(n1') = n1' \oplus k1' \oplus -(n1') = k1' \in I$.
Subcase 3b: $m < n$. Left to the reader.
Subcase 3c: $m = n$. Then $x \oplus y = 0' \in I$.

Case 4: $x \in B$ and $y \in A$. This is the same as Case 3 with x and y "switched."

Case 5: $x = 0'$ or $y = 0'$. This is obvious.

In all cases, $x \oplus y \in I$.

This completes the proof that I is an ordered integral domain that is a subring of D. To complete the proof that I is an integer space, it will be shown that every nonempty subset of I that is bounded below (by an element of I, of course) has a minimum.

To do this, suppose C is a nonempty subset of I, and C is bounded below by some member of I.

Case 1: C contains negative elements. Then there exists $-(n1') \in C$ for some $n \in P$. Now, because C is bounded below by a member of I, there is a $b \in P$ so that $-(b1') \leqq -(n1') \in C$. The reader is now asked to show that $n \leq b$. (Hint: You need to use the fact that $\leqq$ is a linear order.) Define the set S by $S = \{k \in P | k \leq b \ and -(k1') \in C\}$. Note that S is nonempty because $n \in S$. Then S has a maximum because S is bounded above. Let us denote this largest element by L. Then $-(L1')$ is the maximum of C. (Why?)

Case 2: C does not contain negative elements. Left to the reader.

Finally, we show that $I = \cap_{X \in G} X$. First, note that the set G of all subrings of D is nonempty because D is a subring of itself. Hence, by Lemma 5.1, $\cap_{X \in G} X$ is a subring of D. Recall that A was defined as $A = \{n1' | n \in P\}$. An easy induction (do it!) shows that $A \subseteq \cap_{X \in G} X$. Hence, it is clear that $I \subseteq \cap_{X \in G} X$ (because $\cap_{X \in G} X$ contains $0'$ and contains the additive inverse of each of its elements). We have already shown that I is a subring of D, so that $I \in G$. That is, I is one of the sets that is being "intersected" in the intersection $\cap_{X \in G} X$. Hence, $\cap_{X \in G} X \subseteq I$. Therefore, we must have $I = \cap_{X \in G} X$. This concludes the proof.

Before ending this section, we make one final point. Consider the ordered integral domain D of Theorem 5.13, and keep the notation used in the proof for a moment. We showed the existence of a subring I of D that is an integer space. Let Y denote the positive members of D. Then Y is itself a Peano space. Now, let $a \in D$. Therefore, for each $n \in Y$, we have the element $na \in D$. We also have the element $n \otimes a \in D$, which makes sense because $a, n \in D$. An easy induction argument (do it!) shows that $na = n \otimes a$ for each $n \in Y$.

Let us briefly recap what we have accomplished in this section. We defined the concepts of a field and an ordered field. We then derived some results that hold for all fields and some results that hold for all ordered fields. Because it will turn out that the set of real numbers is an ordered field, these results will automatically apply to the real numbers. Finally, we showed that every ordered integral domain contains an integer space as a subring.

EXERCISES

1. Prove Theorem 5.3.
2. Prove Theorem 5.4.
3. Prove Parts 2 and 4 of Theorem 5.5.

4. Prove Parts 2 and 3 of Theorem 5.6.

5. Suppose that F is a field, and that a, b, $c \in F$ with a nonzero. Consider the *linear* equation $ax + b = c$, in which x is a variable. Show that this equation has a unique solution for x, and identify this solution. Is it enough just to "solve" the equation for x?

6. Suppose that F is a field. Show that for each finite sequence $\{x_i\}_{i=1}^n$ of nonzero elements of F, we have $(\Pi_{i=1}^n x_i)^{-1} = \Pi_{i=1}^n(x_i^{-1})$.

7. Show that if F is a field such that 0 has a multiplicative inverse, then F is the *trivial field* (and hence is not a field at all according to our definitions).

8. Suppose F is a field and G is a ring, and $f: F \to G$ is a ring isomorphism. Show that G is also a field.

9. Suppose F and G are fields, and $f: F \to G$ is a field isomorphism. Suppose $a \in F$ with a nonzero. Show that $f(a^{-1}) = (f(a))^{-1}$. (By a^{-1}, we mean the multiplicative inverse of a with respect to the multiplication on F. By $(f(a))^{-1}$, we mean the multiplicative inverse of $f(a)$ with respect to the multiplication on G.)

10. Prove Theorem 5.8.

11. Prove Part 2 of Theorem 5.9.

12. Suppose F is an ordered field with linear order $\leq$. Suppose $a \in F$. Show that $|a^n| = |a|^n$ for each positive integer n.

13. Prove Theorem 5.11. (Hint: Use the triangle inequality.)

14. Suppose F is a field, and G is a nonempty set of subfields of F. Show that $\bigcap_{A \in G} A$ is a subfield of F.

15. Fill in the missing "justifications" in the proof of Lemma 5.2.

16. Fill in all missing pieces of the proof of Theorem 5.13.

17. Using your prior knowledge of real numbers, define the set S by $S = \{a + b\sqrt{2} \mid a, b \in \mathbf{R}\}$. Show that S is a subfield of $\mathbf{R}$.

Section 5.2   The Rational Numbers and the Field of Quotients

In this section, we first list the postulates obeyed by rational numbers, and some initial properties are derived. We revisit the proof that there does not exist a rational number whose square is 2. It is then shown that any set obeying these postulates is countable. Finally, the existence of a set obeying the postulates is demonstrated. In demonstrating the existence of such a set, we construct what is often called the **field of quotients** from an integer space.

Rational Spaces

In this subsection we work with an ordered field F. We use the "usual" symbols for addition, multiplication, order, and so on. Recall from Theorem

5.13 that F contains a subring I that is an integer space. Recall that $I = \cap_{X \in G} X$, where G is the set of subrings of F. (Actually, Theorem 5.13 says that every ordered integral domain contains a subring that is an integer space, but recall that every ordered field is also an ordered integral domain.)

Definition 5.7

Suppose F is an ordered field. Let $I = \cap_{X \in G} X$, where G is the set of subrings of F. We shall call F a **rational space**, provided that for each $q \in F$, there are $m, n \in I$ with $n \neq 0$ such that $q = \frac{m}{n}$.

The properties of Definition 5.7 are what we mean by the postulates for the rational numbers. As presented in the definition, it looks like only one postulate. However, many of the postulates are "buried" in the statement that the set F of Definition 5.7 is an ordered field. Note that this definition "captures" the essence of what we associate with rational numbers. They are ratios (e.g., quotients of integers). Also note that if $q \in F$ with $q = \frac{m}{n}$ in which $m, n \in I$ and $n \neq 0$, we can also express q as a ratio of integers given by $q = \frac{km}{kn}$ for any nonzero $k \in I$. (For example, $q = \frac{1}{3}$ can be written as $q = \frac{2 \cdot 1}{2 \cdot 3}$; i.e., $q = \frac{2}{6}$.) You will probably not find the term *rational space* in the mathematics literature, but it serves us well here. The following lemma is useful.

Lemma 5.3

Suppose F is a rational space. Let $I = \cap_{X \in G} X$, where G is the set of subrings of F. Then for each $q \in F$, there are $m, n \in I$ with $n > 0$ such that $q = \frac{m}{n}$.

Proof: Let $q \in F$. Then there are $a, b \in I$ with $b \neq 0$ such that $q = \frac{a}{b}$. If $b > 0$, then we are done. If $b < 0$, note that we can write q as $q = \frac{-a}{-b}$, and that $-b > 0$.

To make clear that it does not matter which rational space we use, our first major order of business is to demonstrate that if F_1 and F_2 are rational spaces, then there is a field isomorphism $h: F_1 \rightarrow F_2$. That is, any two rational spaces are isomorphic. We use the "usual" notation to represent the zero element, the unit element, addition, multiplication, additive and multiplicative inverses, and the "usual" linear order for both rational spaces. The context makes clear what is meant. The reader should pay close attention to the proof of Theorem 5.14 because a very important point is made in this proof. Part of the proof is left to the reader.

Theorem 5.14   Suppose that F_1 and F_2 are rational spaces. Then there is a field isomorphism $h: F_1 \to F_2$ such that $\forall q, r \in F_1$, we have $q \leq r$ iff $h(q) \leq h(r)$.

Proof: We first give a brief "road map" of the proof. We perform the following steps:

1) Define a mapping $h: F_1 \to F_2$, and show that it is well defined.
2) Show that h is one-to-one.
3) Show that h is onto.
4) Show that $h(qr) = h(q)h(r)$ for each $q, r \in F_1$. [The reader is asked to show that $h(q + r) = h(q) + h(r)$.]
5) Show that $q \leq r$ iff $h(q) \leq h(r)$ for each $q, r \in F_1$.

Let us get started. Let $I_1 = \cap_{X \in G} X$, where G is the set of subrings of F_1, and let $I_2 = \cap_{X \in H} X$, where H is the set of subrings of F_2. Then I_1 and I_2 are integers spaces. Recall from Section 3.6 that there is a ring isomorphism $g: I_1 \to I_2$ such that $\forall m, n \in I_1$, we have $m \leq n$ iff $g(m) \leq g(n)$.

Consider any $q \in F_1$. Then there are $m, n \in I_1$ with $n \neq 0$ such that $q = \dfrac{m}{n}$. We would like to define $h: F_1 \to F_2$ by $h(q) = \dfrac{g(m)}{g(n)}$. We are assured that $g(n) \neq 0$ (because g is one-to-one with $g(0) = 0$ and $n \neq 0$). However, we are still faced with a potential problem. There are many ways to express q as a ratio of integers (i.e., members of I_1). Yet, we have specified h in terms of one specific representation of q, namely m in the numerator and n in the denominator. We indicate this potential problem by saying that we have not yet shown that h is **well defined** (i.e., that the definition makes sense and is completely and unambiguously specified). The issue of demonstrating that a definition is well-defined comes up in advanced mathematics time and time again.

How can we go about demonstrating that h is in fact well defined? We can do this by showing that no matter how we represent q as a quotient of integers, we get the same answer for our proposed h, namely $\dfrac{g(m)}{g(n)}$. To do this, suppose q can also be represented by $q = \dfrac{a}{b}$ with $a, b \in I_1$ and $b \neq 0$. The goal is to show that $\dfrac{g(m)}{g(n)} = \dfrac{g(a)}{g(b)}$. Because q can be expressed as $\dfrac{m}{n}$ and as $\dfrac{a}{b}$, we must have $\dfrac{m}{n} = \dfrac{a}{b}$. Hence, $mb = an$, so $g(mb) = g(an)$. Now, because g is a ring isomorphism, the latter equation can be written as $g(m)g(b) = g(a)g(n)$. Hence, dividing both sides by $g(n)$ and by $g(b)$ results in $\dfrac{g(m)}{g(n)} = \dfrac{g(a)}{g(b)}$, which is our desired conclusion. (Make sure you understand that when we say "dividing both sides by $g(n)$ and by $g(b)$," we mean multiplying by multiplicative inverses in the field F_2.) Thus, h, as defined above, is in fact well defined.

Next, it is shown that h is onto. To accomplish this, suppose $w \in F_2$. We must show that there exists a $q \in F_1$ such that $w = h(q)$. Now, there exists

$u, v \in I_2$ with $v \neq 0$ such that $w = \frac{u}{v}$. However, recalling that g is onto, there exists $m, n \in I_1$ such that $u = g(m)$ and $v = g(n)$. Furthermore, $n \neq 0$ because $v \neq 0$. Hence, $w = \frac{u}{v} = \frac{g(m)}{g(n)} = h(\frac{m}{n})$. Hence, setting $q = \frac{m}{n}$ "works."

We now show that h is one-to-one. Suppose $m, n, a, b \in I_1$ with $n \neq 0$ and $b \neq 0$ such that $h(\frac{m}{n}) = h(\frac{a}{b})$. We must show that $\frac{m}{n} = \frac{a}{b}$. Now, by the definition of h, we have $\frac{g(m)}{g(n)} = \frac{g(a)}{g(b)}$. Hence, $g(m)g(b) = g(a)g(n)$. Thus, $g(mb) = g(an)$. Now, recall that g is one-to-one. Hence, we must have $mb = an$. From this, we conclude that $\frac{m}{n} = \frac{a}{b}$, as desired.

Next, let $q, r \in F_1$. We show that $h(qr) = h(q)h(r)$. The reader is asked to show that $h(q + r) = h(q) + h(r)$. There are $m, n, a, b \in I_1$ with $n \neq 0$ and $b \neq 0$ such that $q = \frac{m}{n}$ and $r = \frac{a}{b}$. Hence, $h(qr) = h(\frac{m}{n} \cdot \frac{a}{b}) = h(\frac{ma}{nb}) = \frac{g(ma)}{g(nb)} = \frac{g(m)g(a)}{g(n)g(b)} = \frac{g(m)}{g(n)} \cdot \frac{g(a)}{g(b)} = h(\frac{m}{n})h(\frac{a}{b}) = h(q)h(r)$.

Finally, we show that $\forall q, r \in F_1$, we have $q \leq r$ iff $h(q) \leq h(r)$. Suppose $q, r \in F_1$. By Lemma 5.3, there exists $m, n, a, b \in I_1$ with $n > 0$ and $b > 0$ such that $q = \frac{m}{n}$ and $r = \frac{a}{b}$. Now, suppose that $q \leq r$. Then $\frac{m}{n} \leq \frac{a}{b}$. Because $n > 0$ and $b > 0$, we have $mb \leq na$. Hence, $g(mb) \leq g(na)$, so $g(m)g(b) \leq g(a)g(n)$. Now, $g(n) > 0$ and $g(b) > 0$. (Why?) Hence, $\frac{g(m)}{g(n)} \leq \frac{g(a)}{g(b)}$. Thus, $h(\frac{m}{n}) \leq h(\frac{a}{b})$. That is, $h(q) \leq h(r)$. Conversely, suppose that $h(q) \leq h(r)$. It is easy to see that we can "reverse" the above argument to get $q \leq r$.

The reader is now asked to review the argument made in Section 1.4 that there does not exist a rational number whose square is 2. The reader should easily see that we have established all of the assertions made in the proof except for one. We have not yet established what we called **Informal Fact II** in Section 1.4. We need to show that for each rational number r, r can be written as a quotient of integers such that these integers are not both even.

To do this, we introduce the concept of the **greatest common divisor** of two integers. (We could have actually done this in Section 4.3, but it will turn out to be a little easier having absolute value and the rational numbers available.) The greatest common divisor concept will also be used for other tasks than just showing that there does not exist a rational number whose square is 2.

Consider any integers a and b that are not both 0. Let S denote the set of integer common divisors of a and b. That is, $n \in S$ iff $n|a$ and $n|b$. Note that S is nonempty because $1 \in S$. Furthermore, S is bounded above. To see this, note that because a and b are not both 0, they cannot have a common divisor that exceeds the maximum of $\{|a|, |b|\}$. (Also, S is bounded below because an integer n is a common divisor of a and b iff $-n$ is also.) Therefore, S has a greatest element. Hence, the following definition makes sense.

Definition 5.8

Let I be an integer space. Suppose $a, b \in I$ such that a and b are not both 0. Then the **greatest common divisor** of a and b is defined to be the maximum of the set of common divisors of a and b. We denote the greatest common divisor of a and b by $\gcd(a, b)$.

It should be mentioned that in the mathematics literature, the greatest common divisor of a and b is usually denoted by (a, b). Because this can easily be confused with the ordered pair (a, b), we will use the notation $\gcd(a, b)$.

Example 5.4

To find $\gcd(30, 45)$, we note that the set of positive common divisors is $\{1, 3, 5, 15\}$. Hence, $\gcd(30, 45) = 15$. Note that we also have $15 = \gcd(30, 45) = \gcd(-30, 45) = \gcd(30, -45) = \gcd(-30, -45)$.

There are many interesting and important number theory results about the greatest common divisors. For our purposes, we only need a few results. To motivate Theorem 5.15 (which will aid us in showing that there does not exist a rational number whose square is 2), note that $\gcd(a, b)$ is a divisor of both a and b. Hence, both $\dfrac{a}{\gcd(a, b)}$ and $\dfrac{b}{\gcd(a, b)}$ are also integers. (In terms of a rational space F and the set $I = \cap_{X \in G} X$, where G is the set of subrings of F, we can say that $\dfrac{a}{\gcd(a, b)}$ and $\dfrac{b}{\gcd(a, b)}$ are members of I.)

Theorem 5.15

Suppose F is a rational space. Let $I = \cap_{X \in G} X$, where G is the set of subrings of F. Suppose $a, b \in I$ such that a and b are not both 0. Then
$$\gcd\left(\frac{a}{\gcd(a, b)}, \frac{b}{\gcd(a, b)}\right) = 1.$$

Proof: It suffices to show that the only positive divisor of both $\dfrac{a}{\gcd(a, b)}$ and $\dfrac{b}{\gcd(a, b)}$ is 1. Suppose $c \in I$ is a positive divisor of both $\dfrac{a}{\gcd(a, b)}$ and $\dfrac{b}{\gcd(a, b)}$. Then there are $d, e \in I$ such that $\dfrac{a}{\gcd(a, b)} = cd$ and $\dfrac{b}{\gcd(a, b)} = ce$. Hence, $a = (cd)\gcd(a, b)$ and $b = (ce)\gcd(a, b)$. Hence, $c \cdot \gcd(a, b)$ is a common divisor of a and b. Also, note that $c \cdot \gcd(a, b)$ is positive. Because $\gcd(a, b)$ is the greatest positive common divisor of a and b, we must have $c \cdot \gcd(a, b) \le \gcd(a, b)$. Hence, we must have $c \le 1$. Because c is a positive integer, we must have $c = 1$. This proves the result.

Now, to show Informal Fact II of Section 1.4, consider any rational number r. Then r can be written as $r = \dfrac{a}{b}$, in which a and b are integers with b

nonzero. Now, note that we can also express r as $r = (\frac{a}{\gcd(a, b)})/(\frac{b}{\gcd(a, b)})$. Now, by Theorem 5.15, $\frac{a}{\gcd(a, b)}$ and $\frac{b}{\gcd(a, b)}$ do not have any positive common divisors other than 1. In particular, 2 cannot be a common divisor. Hence, $\frac{a}{\gcd(a, b)}$ and $\frac{b}{\gcd(a, b)}$ are not both even!

Thus, we have shown that every rational number can be expressed as a ratio of integers, in which at least one of them is not even. (By the way, in the mathematical literature, if m and n are integers with $\gcd(m, n) = 1$, m and n are called **relatively prime**.)

Informal Fact II is now an established formal fact. Therefore, we formally conclude that there does not exist a rational number whose square is 2. After constructing the real number system, it will be shown that there does exist a *real number* whose square is 2. Of course, this real number cannot be rational, so it will be called **irrational**.

Next, we change gears and show that any rational space is a countable set. This important result may seem surprising to the reader because it seems that there are many, many more rational numbers than positive integers. This is another demonstration of the "mysteries" of the infinite. The following lemma will be useful for this endeavor.

Lemma 5.4

Suppose P is a Peano space. Then for each n, a, $b \in P$, if $n|ab$ and $\gcd(n, a) = 1$, then $n|b$.

Proof: We use the Second Principle of Mathematical Induction on n. The result is obvious for 1. Now, we make the induction hypothesis that the result holds true for all $k \in P$ with $k \leq n$ for some $n \in P$. The goal is to show that the result holds true for $n + 1$.

To accomplish this, suppose a, $b \in P$ such that $n + 1|ab$ and $\gcd(n + 1, a) = 1$. We need to show that $n + 1|b$. By the Fundamental Theorem of Arithmetic, there are primes $p_1, p_2, \ldots, p_r$ (not necessarily distinct) such that $n + 1 = p_1 p_2 \cdots p_r$. Also, by the Fundamental Theorem of Arithmetic, there are primes $q_1, q_2, \ldots, q_s$ such that $b = q_1 q_2 \cdots q_s$. Because $n + 1|ab$, there is a $w \in P$ so that $ab = (n + 1)w$. Hence, we have $a(q_1 q_2 \cdots q_s) = (p_1 p_2 \cdots p_r)w$.

Note that it is not possible for any p_i to divide a, for if $p_i|a$, then p_i is a common divisor of a and $n + 1$, which is not possible because $\gcd(n + 1, a) = 1$. Hence, applying the Fundamental Theorem of Arithmetic to $a(q_1 q_2 \cdots q_s)$, we see that $p_1 = q_k$ for some $k \in \{1, 2, \ldots, s\}$. Now, if $n + 1 = p_1$, we are done. Otherwise, we divide both sides of the equation $a(q_1 q_2 \cdots q_s) = (p_1 p_2 \cdots p_r)w$ by this prime to get $a(q_1 q_2 \cdots q_{k-1} q_{k+1} \cdots q_s) = (p_2 \cdots p_r)w$. Note that $p_2 \cdots p_r < n + 1$. Also, because we know that it is not possible for any p_i to divide a, we must have $\gcd(p_2 \cdots p_r, a) = 1$. Hence, we apply the induction hypothesis to $p_2 \cdots p_r$ to conclude that $(p_2 \cdots p_s)|(q_1 q_2 \cdots q_{k-1} q_k \cdots q_s)$. From this, we see immediately that $n + 1|b$.

Theorem 5.16   Suppose F is a rational space. Then F is countable.

Proof: Define the sets F^+ and F^- by $F^+ = \{r \in F | r > 0\}$ and $F^- = \{r \in F | r < 0\}$. Let $I = \cap_{X \in G} X$, where G is the set of subrings of F. Let $I^+ = \{n \in I | n \geq 1\}$. Now, we know that I^+ is a Peano space. Finally, let $K = \{(m, n) \in I^+ \times I^+ \mid \mathbf{gcd}(m, n) = 1\}$. By Theorem 4.34, $I^+ \times I^+$ is countable. Also, it is clear that K is an infinite subset of $I^+ \times I^+$. Hence, by Theorem 4.30, K is countable.

Define $f: K \to F^+$ by $f(m, n) = \dfrac{m}{n}$ for each $(m, n) \in K$. Theorem 5.15 can be used to quickly show that f is onto. (Why?) We will now show that f is one-to-one. Suppose that $f(m, n) = f(a, b)$ in which (m, n), $(a, b) \in K$. Hence, $\dfrac{m}{n} = \dfrac{a}{b}$, so $mb = na$. Note that $m | na$. Because $\mathbf{gcd}(m, n) = 1$, we have $m | a$ by Lemma 5.4. We also have $a | mb$ and $\mathbf{gcd}(a, b) = 1$, so we have $a | m$ by Lemma 5.4. Because we have both $m | a$ and $a | m$, we must have $a = m$. Plugging this in to the equation $mb = na$ gives us $b = n$. Thus, $(a, b) = (m, n)$, so we can conclude that f is one-to-one. Because f is a bijection and K is countable, F^+ is countable.

From the result that F^+ is countable, it is immediate that F^- is also countable. Hence, by Theorem 4.32, $F^+ \cup F^-$ is countable. By Theorem 4.29, $(F^+ \cup F^-) \cup \{0\}$ is countable. Now, note that $F = (F^+ \cup F^-) \cup \{0\}$. Therefore, F is countable.

Finally for this subsection, recall that if n is any integer, there does not exist an integer that is strictly between n and $n + 1$. However, Theorem 5.17 below states that there is always a rational number strictly between any two distinct rational numbers.

Theorem 5.17   Suppose F is a rational space. Suppose $r, s \in F$ with $r < s$. Then there is a $q \in F$ with $r < q < s$.

Proof: Starting with the inequality $r < s$, it is easy to see that $2r < r + s < 2s$. Hence, $r < \dfrac{r + s}{2} < s$. Because F is a field, $\dfrac{r + s}{2} \in F$. Hence, setting q to $\dfrac{r + s}{2}$ "works."

Field of Quotients

Now it is all well and good to talk about rational spaces. However, we have not yet proved that a rational space exists. We turn our attention to this now. There is a general method in abstract algebra in which any integral domain can be used to construct a field. This field, referred to as the **field**

of quotients, turns out to be a rational space in the case in which the given integral domain is an integer space. Instead of presenting the general method, we construct the field of quotients using an integer space.

Hence, throughout this subsection, we let I be an integer space, with the "usual" notation used for the operations and inequalities. We proved that an integer space exits in Section 3.7. The procedure employed to demonstrate the existence of a rational space is similar to the procedure used in Section 3.7 to demonstrate the existence of an integer space. However, Section 3.7 is optional, so the reader may not have seen the relevant techniques. Consequently, we discuss these techniques again here. Some of the steps of the constructive proof will be left to the student. Before reading this subsection, the student is advised to review the portion of Section 2.4 that is concerned with equivalence classes and equivalence relations.

The basic idea of this construction is quite simple. Consider m, $n \in I$ with $n \neq 0$. How can we "represent" the rational number $\frac{m}{n}$? One very intuitive approach is to use the ordered pair (m, n) to represent $\frac{m}{n}$. The first component of the ordered pair represents the numerator, and the second component represents the denominator. However, there is a problem with this. Regardless of the method that we use to represent $\frac{m}{n}$, we want to have $\frac{m}{n} = \frac{km}{kn}$ for any k, m, $n \in I$ with n and k nonzero. However, if we use ordered pairs for the representation, this does not work because the ordered pair (m, n) does not equal the ordered pair (km, kn) (except for the case in which k is 1.)

Define the set S by $S = \{(m, n) | m, n \in I \text{ and } n \neq 0\}$. The solution is to define a relation $\cong$ on S, and then show that this relation is an equivalence relation. We then show that we can define binary operations and an order on the set of equivalence classes to get a rational space.

Consider any (m, n), $(p, q) \in S$. What criteria should we use to declare that these ordered pairs are "related" [i.e., when do we "want" to have $(m, n) \cong (p, q)$]? Intuitively, we "want" these ordered pairs to be related when they represent the same rational number. That is, we want them to be related when we have $\frac{m}{n} = \frac{p}{q}$. One way to achieve this is to declare that these ordered pairs are related whenever $mq = np$. This will turn out to work perfectly. Hence, we formally define the relation $\cong$ on S by $(m, n) \cong (p, q)$ iff $mq = np$, for each (m, n), $(p, q) \in S$. It is now shown that the relation $\cong$ is an equivalence relation.

Reflexive Property Suppose $(m, n) \in S$. We must show that $(m, n) \cong (m, n)$. That is, we must show that $mn = nm$. Of course, we know that this holds because multiplication is commutative on an integer space.

Symmetric Property Suppose (m, n), $(p, q) \in S$ with $(m, n) \cong (p, q)$. The goal is to show that $(p, q) \cong (m, n)$. That is, the goal is to show that $pn = qm$. Now, because $(m, n) \cong (p, q)$, we have $mq = np$. Of course, the desired result follows immediately.

Transitive Property Suppose (m, n), (p, q), $(r, s) \in S$ with $(m, n) \cong (p, q)$ and $(p, q) \cong (r, s)$. Hence, we have the following two equations:

$$mq = np \tag{1}$$

$$ps = qr. \tag{2}$$

The goal is to show that $(m, n) \cong (r, s)$. That is, we want to show that $ms = nr$.

Case 1: $p = 0$. Hence, $mq = 0$ by Equation 1. Because $q \neq 0$, we must have $m = 0$. Hence, $ms = 0$. We also must have $r = 0$ by Equation 2. Hence, $nr = 0$. Thus, $ms = 0 = nr$.

Case 2: $p \neq 0$. By multiplying Equation 1 and Equation 2, we get $mqps = npqr$. By canceling p and q, we get $ms = nr$.

This demonstrates that $\cong$ is indeed an equivalence relation. In accordance with the notation of Definition 2.19, the equivalence class of each $(m, n) \in S$ shall be denoted by $[(m, n)]$. Now, let F be the set of all of these equivalence classes. That is, $F = \{[(m, n)] | (m, n) \in S\}$. It will be demonstrated that F is a rational space (with appropriately defined operations).

How should we define "addition" on F? Recall that however we define rational numbers, if we have quotients of integers $\frac{m}{n}$ and $\frac{p}{q}$, we "want" their sum to be $\frac{mq + np}{nq}$. With this intuitive motivation, we define a binary operation $\oplus$ on F by $[(m, n)] \oplus [(p, q)] = [(mq + np, nq)]$ for each $[(m, n)]$, $[(p, q)] \in F$. However, the astute reader will see that there is a potential problem with this. We have to demonstrate that this binary operation is well defined. Why? Because we have defined the sum of two equivalence classes in terms of a particular representative of each of these equivalence classes. For example, the equivalence class $[(m, n)]$ is identical to the equivalence class $[(2m, 2n)]$, due to the fact that $(m, n) \cong (2m, 2n)$. (See Theorem 2.7.) Therefore, it is just as legitimate to represent the equivalence class $[(m, n)]$ by the ordered pair $(2m, 2n)$ as it is to represent the equivalence class by (m, n).

To show that this operation is well defined, we must show that if $(a, b) \cong (m, n)$ and $(c, d) \cong (p, q)$, then $(ad + bc, bd) \cong (mq + np, nq)$. From the two former relations, we have the following two equations:

$$an = bm \tag{3}$$

$$cq = dp. \tag{4}$$

The goal is to show that $(ad + bc)nq = bd(mq + np)$. That is, we want to show that $adnq + bcnq = bdmq + bdnp$. Now, multiplying both sides of Equation 3 by dq results in

$$andq = bmdq. \tag{5}$$

Multiplying both sides of Equation 4 by bn results in

$$cqbn = dpbn. \tag{6}$$

Adding Equation 5 and Equation 6 gives the desired result.

The reader is now asked to show that the operation $\oplus$ is commutative and associative. We now define $0'$ by $0' = [(0, 1)]$. Now, for any $[(m, n)] \in F$, we have $[(m, n)] \oplus 0' = [(m, n)] \oplus [(0, 1)] = [(m \cdot 1 + n \cdot 0, n \cdot 1)] = [(m, n)]$. Hence, $0'$ "works" as the zero element of F. We also propose $[(-m, n)]$ to fill the role of the additive inverse of any $[(m, n)] \in F$. Now, $[(m, n)] \oplus [(-m, n)] = [(mn - nm, n^2)] = [(0, n^2)]$. For $[(-m, n)]$ to "work" as the additive inverse, we must have $[(0, n^2)] = [(0, 1)]$. Do we have this? Of course we do, because $(0, n^2) \cong (0, 1)$.

Next, we define "multiplication" on F. We denote this binary operation by $\otimes$. It probably does not come as a surprise that we define $[(m, n)] \otimes [(p, q)]$ to be $[(mp, nq)]$ for each $[(m, n)], [(p, q)] \in F$. As with $\oplus$, we need to show that the binary operation $\otimes$ is well defined, and the reader is asked to do this. The reader is also asked to show that $\otimes$ is commutative and associative. We now define $1'$ by $1' = [(1, 1)]$. Now, for any $[(m, n)] \in F$, we have $[(m, n)] \otimes 1' = [(m, n)] \otimes [(1, 1)] = [(m \cdot 1, n \cdot 1)] = [(m, n)]$. Hence, $1'$ is the identity element. We propose $[(n, m)]$ to fill the role of the multiplicative inverse of any $[(m, n)] \in F$ with $[(m, n)] \neq 0'$. Now, $[(m, n)] \otimes [(n, m)] = [(mn, mn)]$. (Note that this makes sense because both m and n are nonzero.) Because $(mn, mn) \cong (1, 1)$, we have $[(mn, mn)] = [(1, 1)] = 1'$. Hence, $[(m, n)] \otimes [(n, m)] = 1'$. Hence, we can define $[(m, n)]^{-1}$ to be $[(n, m)]$.

It will now be shown that the distributive law holds. Suppose $[(m, n)]$, $[(p, q)]$, $[(r, s)] \in F$. We will show that $[(m, n)] \otimes ([(p, q)] \oplus [(r, s)]) = ([(m, n)] \otimes [(p, q)]) \oplus ([(m, n)] \otimes [(r, s)])$. Now,

$$[(m, n)] \otimes ([(p, q)] \oplus [(r, s)]) = [(m, n)] \otimes [(ps + qr, qs)]$$
$$= [(m(ps + qr), nqs)] = [(mps + mqr, nqs)]$$

Now,

$$([(m, n)] \otimes [(p, q)]) \oplus ([(m, n)] \otimes [(r, s)])$$
$$= [(mp, nq)] \oplus [(mr, ns)] = [(mpns + nqmr, nqns)].$$

The result now follows by noting that $(mps + mqr, nqs) \cong (mpns + nqmr, nqns)$.

Having demonstrated that F is a field, the next step is to show that it is an ordered field. First of all, we note that given an equivalence class $c \in F$, it is always possible to find $m, n \in I$ with $n > 0$ such that $c = [(m, n)]$. (Why?) We define a relation $\leq$ on F as follows: Given $[(m, n)], [(p, q)] \in F$ with $n > 0$ and $q > 0$, we declare that $[(m, n)] \leq [(p, q)]$ iff $mq \leq np$. Again, we have the issue of showing that this definition is well defined. To show this, suppose $(a, b) \cong (m, n)$ and $(c, d) \cong (p, q)$ such that b and d are both positive. The goal is to show that $ad \leq bc$. The statements $(a, b) \cong (m, n)$ and $(c, d) \cong (p, q)$ can be written as the two equations:

$$an = bm \tag{7}$$

$$cq = dp. \tag{8}$$

We also have

$$mq \leq np. \tag{9}$$

Multiplying Equation 7 by q gives

$$anq = bmq. \tag{10}$$

Now, multiplying Equation 9 by b gives

$$bmq \leq bnp. \tag{11}$$

Hence, from Equation 10 and Equation 11, we have

$$anq \leq bnp. \tag{12}$$

Canceling n from this gives

$$aq \leq bp. \tag{13}$$

Multiplying both sides of Equation 13 by d gives

$$aqd \leq bpd. \tag{14}$$

Using Equation 8 and Equation 14 results in

$$aqd \leq bcq. \tag{15}$$

Finally, canceling q from Equation 15 gives us the desired result of $ad \leq bc$.

We will start the proof of transitivity of $\leq$. It is left to the reader to show that $\leq$ satisfies the remaining conditions of Definition 2.20 and that Definition 3.19 is satisfied. Suppose $x, y, z \in F$ with $x \leq y$ and $y \leq z$. The goal is to show that $x \leq z$. Now, there exist $a, b, c, d, e, f \in I$ with b, d, and f positive such that $x = [(a, b)]$, $y = [(c, d)]$, and $z = [(e, f)]$. Hence, we have $ad \leq bc$ and $cf \leq de$. The reader is asked to show that $af \leq be$ (which proves that $x \leq z$).

To complete the proof that F is a rational space, define I_1 by $I_1 = \{[(m, 1)] \mid m \in I\}$. It will be shown that I_1 is an integer space that is a subring of F. Suppose $[(m, 1)], [(n, 1)] \in I_1$. Then $[(m, 1)] \oplus [(n, 1)] = [(m + n, 1)]$, and $[(m, 1)] \otimes [(n, 1)] = [(mn, 1)]$. Hence, $\oplus$ and $\otimes$ are closed on I_1. Also, the additive inverse of any $[(m, 1)] \in I_1$ is $[(-m, 1)]$, which is also in I_1. Hence, I_1 is a subring of F. Now, clearly, for any $[(m, 1)]$, $[(n, 1)] \in I_1$, we have $[(m, 1)] \leq [(n, 1)]$ iff $m \leq n$. Hence, it is easy to show that any subset of I_1 bounded below has a minimum. (Do it!) Hence, I_1 is an integer space. Furthermore, it is now clear that $I_1 = \cap_{X \in G} X$, in which G is the set of subrings of F. (Why?)

Are we done? We still need to show that any $x \in F$ can be expressed as $x = \frac{r}{s}$, in which $r, s \in I_1$. To do this, suppose $x \in F$. Then there exist m, $n \in I$ such that $x = [(m, n)]$ with n nonzero. Now, note that $[(m, n)] = [(m, 1)] \otimes [(1, n)]$. Now, recall that $[(1, n)] = [(n, 1)]^{-1}$. Hence, $[(m, n)] = \frac{[(m, 1)]}{[(n, 1)]}$. Because $[(m, 1)], [(n, 1)] \in I_1$, we are done.

EXERCISES

1. Complete the proof of Theorem 5.14 by showing $h(q + r) = h(q) + h(r)$ for each $q, r \in F_1$.

2. Suppose that a, b, and c are integers such that a and b are not both 0. Show that $\gcd(a + cb, b) = \gcd(a, b)$. (Hint: Use Theorem 4.14.)

3. Fill in all missing pieces of the proof that a rational space exists.

Chapter 6

The Development of the Real Numbers

T he student is now more than ready to delve into the construction of the real number system. In Section 6.1, the postulates obeyed by real numbers are given, and initial properties are derived. In Section 6.2, it is shown that any two fields obeying these postulates are isomorphic. In Section 6.3, it is shown that a set obeying these postulates actually exists, and the ideas presented in demonstrating this will be very useful for the student in preparing to take an undergraduate course in analysis. Finally, some important properties of the real numbers are given in Section 6.4. Upon reading this chapter, the student should be well prepared for an undergraduate course in analysis.

Section 6.1   Postulates for the Real Number System and Some Initial Properties

Here we present the postulates obeyed by the real numbers and then present some elementary properties of any set obeying these postulates. In Section 6.3, we show that we have not done so in vain, by proving that there exists a set that obeys these postulates. Recall that we went through the same type of procedure with the integers and the rational numbers in Chapters 3 and 5, respectively. We need to cover some preliminary material before doing this for the real numbers.

Some More Information on Linear Orders

Before presenting the postulates for the real number system, it is helpful to revisit the topic of linear orders. The student may want to review the discussion of linear orders at the end of Section 2.4, the beginning of Section 3.3, and the beginning of Section 3.6. Key to the development of the real numbers are the concepts of the **least upper bound** and the **greatest lower bound**.

Definition 6.1

Suppose S is a nonempty set with linear order $\leq$. Suppose A is a nonempty subset of S that is bounded above. An element $s \in S$ is called a **least upper bound** of A provided the following two properties hold:

1) s is an upper bound of A.
2) s is the minimum of the set of upper bounds of A.

A least upper bound of A is also called a **supremum** of A. To indicate that s is a supremum of A, we write $s = \mathbf{sup}(A)$. By "abuse of notation," we often leave out the parentheses and write $s = \mathbf{sup}A$.

Definition 6.2

Suppose S is a nonempty set with linear order $\leq$. Suppose A is a nonempty subset of S that is bounded below. An element $s \in S$ is called a **greatest lower bound** of A provided the following two properties hold:

1) s is a lower bound of A.
2) s is the maximum of the set of lower bounds of A.

A greatest lower bound of A is also called an **infimum** of A. To indicate that s is an infimum of A, we write $s = \mathbf{inf}(A)$. By "abuse of notation," we often leave out the parentheses and write $s = \mathbf{inf}A$.

To provide examples intuitively, let S be the set defined by $S = \{x \in \mathbf{R} \mid 1 < x < 2\}$. Then 1 is a greatest lower bound of S because 1 is a lower bound of S, but every smaller number fails to be a lower bound of S. And 2 is a least upper bound of S because 2 is an upper bound of S, but every larger number fails to be an upper bound of S. More formal examples follow after the postulates for the real numbers are specified.

The student should also know that in some texts the authors use the notations **lub** and **glb** to denote **sup** and **inf**, respectively. Due to Theorem 6.1 below, we can refer to a least upper bound and a greatest lower bound of a set as *the* least upper bound and *the* greatest lower bound, respectively.

Theorem 6.1	Suppose S is a nonempty set with linear order $\leq$. Suppose A is a nonempty subset of S. If a least upper bound of A exists, it is unique. If a greatest lower bound of A exists, it is unique.

Proof: Let U be the subset of S consisting of all of the upper bounds of A. If a least upper bound u of A exists, then by definition u is a least element of U. However, Theorem 3.12 states that U can have at most one least element. Let L be the subset of S consisting of all the lower bounds of A. If a greatest lower bound v of A exists, then by definition v is a greatest element of L. However, also by Theorem 3.12, L can have at most one greatest element.

Suppose S is nonempty set with a linear order $\leq$, and suppose A is a nonempty subset of S with $x = \mathbf{sup}A$. Although we have $x \in S$ by definition, the definition does not require that we have $x \in A$. Note that because x is an upper bound of A, we must have $a \leq x$ for each $a \in A$. Now, suppose that u is any upper bound of A. Because x is the minimum of the set of upper bounds of A, we must have $x \leq u$. This means that if we have some $b \in S$ with $b < x$, then b is not an upper bound of A. This means that there must be an element $a \in A$ with $b < a \leq x$. As will be seen, this fact is often used in proofs.

Similarly, suppose the infimum of A exists, with $y = \mathbf{inf}A$. Although we have $y \in S$ by definition, the definition does not require that we have $y \in A$. Note, that because y is a lower bound of A, we must have $y \leq a$ for each $a \in A$. Now, suppose that v is any lower bound of A. Because y is the maximum of the set of lower bounds of A, we must have $v \leq y$. This means that if we have some $c \in S$ with $c > y$, then c is not a lower bound of A. This means that there must be an element $a' \in A$ with $y \leq a' < c$.

Suppose S is a nonempty set with linear order $\leq$. Suppose $A \subseteq S$ with A nonempty. If the maximum of A exists, from now on we will often denote it by $\mathbf{max}(A)$ or by $\mathbf{max}A$. If the minimum of A exists, from now on we will often denote it by **min**(A) or by **min**A. The proof of the next theorem is entirely obvious.

Theorem 6.2 Suppose S is a nonempty set with a linear order $\leq$. Suppose A is a nonempty subset of S. Then the following hold:

1) If **max**A exists, then **sup**A exists and **sup**A = **max**A.
2) If **sup**A exists and **sup**$A \in A$, then **max**A exists and **max**A = **sup**A.
3) If **min**A exists, then **inf**A exists and **inf**A = **min**A.
4) If **inf**A exists and **inf**$A \in A$, then **min**A exists and **min**A = **inf**A.

As will be seen in Example 6.2, simply because the **sup** of a set exists does not necessarily imply that the set has a maximum. Similarly, Example 6.3 shows that simply because the **inf** of a set exists does not necessarily imply that the set has a minimum.

Finally for this subsection, we present a result that will come in handy shortly. (Actually, we could have presented this before.) The proof is left to the reader.

Theorem 6.3 Suppose S is a nonempty set with linear order $\leq$. Suppose A is a nonempty subset of S. If A is finite, then **max**A and **min**A exist.

Complete Ordered Fields

Definition 6.3 below presents the postulates obeyed by the real number system. These postulates are easy to state due to the work that we have already accomplished. Most of the postulates are bundled in the statement that the set of Definition 6.3 is an ordered field. The set obeying the postulates of Definition 6.3 is called a **complete ordered field** in the mathematics literature. Therefore, we use this term rather than calling it a *real space*.

Definition 6.3

Suppose R is an ordered field. Then R is called a **complete ordered field** if it satisfies the condition that each nonempty subset of R that is bounded above has a least upper bound.

Recall that we defined what is meant by a rational space in Section 5.2. We soon will see intuitively that a rational space has "holes" and that these holes are "plugged" in a complete ordered field. First, we need to do some preliminary work. Suppose R is a complete ordered field. Define Z by $Z = \cap_{X \in G} X$, in which G is the set of all subrings of R. Then we know from Theorem 5.13 that Z is an integer space. Now, define Q by $Q = \{\frac{m}{n} | m, n \in Z \text{ and } n \neq 0\}$. We leave it to the reader to show that Q is a rational space. Finally, we let N denote the positive members of Z, so that N is known to be a Peano space. (Note that we have not made use of upper bounds at all yet. We could have showed earlier that *any* ordered field contains a subfield that is a rational space.) Later, we will select a specific complete ordered field that we will call **R** (at last). Then, we will automatically have the sets **Q**, **Z**, and **N**. We shall refer to the sets Q, Z, and N (i.e., using the italicized, unbolded version of these letters) as defined above for a complete ordered field R for the remainder of this section.

The first order of business is to prove the all-important (but very simple) **Principle of Archimedes** (Theorem 6.4). This says that for any positive numbers a and b, if we add b to itself enough times, the result will surpass a. Following this is Theorem 6.5, which is a direct consequence of the Principle of Archimedes. The proof of Theorem 6.5 is left to the reader.

| **Theorem 6.4** | **(Principle of Archimedes)** |

Suppose R is a complete ordered field. Then for any $a, b \in R$ with $a > 0$ and $b > 0$, there exists an $n \in N$ such that $a < nb$.

Proof: First we show that the subset N of R is not bounded above. Suppose (to reach a contradiction) that N is bounded above. Then, by Definition 6.3, N has a least upper bound. Let us call this least upper bound u. Now, because we have $u - 1 < u$, $u - 1$ is not an upper bound of N. Hence, there must be a $k \in N$ with $u - 1 < k$. Adding 1 to this inequality results in $u < k + 1$. However, because N is a Peano space, we have $k + 1 \in N$. But this means that we have found a member of N, namely $k + 1$, that exceeds u. However, this contradicts the statement that u is an upper bound of N. Thus, we conclude that N is not bounded above.

Now, because N is not bounded above, the number (i.e., member of R) $\frac{a}{b}$ is not an upper bound of N. Hence, there must exist an $n \in N$ such that $\frac{a}{b} < n$. Thus, $a < nb$.

| **Theorem 6.5** | Suppose R is a complete ordered field. Then for each $\varepsilon \in R$ with $\varepsilon > 0$, there exists an $n \in N$ such that $\frac{1}{n} < \varepsilon$. |

Next, it is shown that there exists a rational number between any two distinct real numbers.

Theorem 6.6   Suppose R is a complete ordered field. Suppose $a, b \in R$ with $a < b$. Then there exists an $r \in Q$ such that $a < r < b$.

Proof: First, we demonstrate the result holds in the case in which $b > 0$. We then use this result to show that the result holds for any $a, b \in R$. So, suppose $b > 0$. By Theorem 6.5, there exists an $n \in N$ such that $b - a > \frac{1}{n}$. Now, by the Principle of Archimedes, there exists a $k \in N$ such that $b < k \cdot \frac{1}{n}$ (i.e., $b < \frac{k}{n}$). Hence, the set $A = \{m \in N \mid b \le \frac{m}{n}\}$ is nonempty. Thus, A has a least element p. Now, if p is 1, then clearly we have $\frac{p-1}{n} < b$. Otherwise, $p > 1$, so $p - 1$ is a positive integer that is less than p. Hence, $p - 1 \notin A$, so again we have $\frac{p-1}{n} < b$. Now, from the inequality $b - a > \frac{1}{n}$, we have $a - b < -\frac{1}{n}$. Because $p \in A$, we also have $b \le \frac{p}{n}$. Thus, $a = b + (a - b) < \frac{p}{n} - \frac{1}{n} = \frac{p-1}{n} < b$. That is, $a < \frac{p-1}{n} < b$. Because $\frac{p-1}{n} \in Q$, we have shown that the desired result holds in the case in which $b > 0$.

Now, we consider the general case in which $a, b \in R$ with $a < b$. First, we show that there exists an $M \in N$ such that $M > -b$. Now, if $b \ge 0$, we can choose M to be 1. Otherwise, we can invoke the Principle of Archimedes to find an $M \in N$ such that $M \cdot 1 > -b$. In either case, we have found an $M \in N$ such that $M > -b$. Hence, $b + M > 0$. Now, note that $a + M < b + M$. Hence, we can apply the first part of the proof to find an $r' \in Q$ with $a + M < r' < b + M$. Hence, $a < r' - M < b$. Letting $r = r' - M$, and noting that $r \in Q$, we are done. ∎

Next, we define open, closed, and half-open intervals of real numbers. The reader is likely to have seen them before. We use them to provide examples of least upper bounds and greatest lower bounds.

Definition 6.4

Suppose R is a complete ordered field. Suppose $a, b \in R$. Then

1) If $a \le b$, we define the **closed interval** $[a, b]$ by
   $[a, b] = \{x \in R \mid a \le x \le b\}$.
2) If $a < b$, we define the **open interval** (a, b) by
   $(a, b) = \{x \in R \mid a < x < b\}$.
3) If $a < b$, we define the **half-open intervals** $[a, b)$ and $(a, b]$ by
   $[a, b) = \{x \in R \mid a \le x < b\}$ and $(a, b] = \{x \in R \mid a < x \le b\}$.

Even though the notation (a, b) can be used to mean both an ordered pair and an open interval in the mathematics literature, the context should always remove any ambiguity. It should also be mentioned that in the mathematics literature, if $a \in R$, then the notations $[a, \infty)$ and (a, ∞) are often used to denote $\{x \in R \mid x \geq a\}$ and $\{x \in R \mid x > a\}$, respectively. Similarly, the notations $(-\infty, a]$ and $(-\infty, a)$ are often used to denote $\{x \in R \mid x \leq a\}$ and $\{x \in R \mid x < a\}$, respectively.

Example 6.1

Suppose R is an ordered field. Suppose $a, b \in R$ with $a \leq b$. Note that $a = \textbf{min}[a, b]$ and $b = \textbf{max}[a, b]$. Hence, it is clear that $a = \textbf{inf}[a, b]$ and $b = \textbf{sup}[a, b]$.

Example 6.2

Suppose R is a complete ordered field. Suppose $a, b \in R$ with $a < b$. We now show that $b = \textbf{sup}(a, b)$. We first have to show that the set (a, b) is nonempty. However, this is immediate from Theorem 6.6. Now, it is clear that b is an upper bound of (a, b). To show that b is the smallest upper bound of (a, b), it is shown that anything smaller than b cannot be an upper bound of (a, b). Suppose $c < b$. By Theorem 6.6 there is a $q \in Q$ such that $\textbf{max}\{a, c\} < q < b$, so that $q \in (a, b)$. Hence, c cannot be an upper bound of (a, b) because it is less than q. Thus, $b = \textbf{sup}(a, b)$. Note that $\textbf{sup}(a, b) \notin (a, b)$. Note that the set (a, b) does not have a maximum.

Example 6.3

Suppose R is a complete ordered field. Suppose $a, b \in R$ with $a < b$. The reader is asked to show that $a = \textbf{inf}(a, b)$. Note that $\textbf{inf}(a, b) \notin (a, b)$. Note that the set (a, b) does not have a minimum.

Now, note that $N \subset Z \subset Q \subseteq R$. However, we have yet to demonstrate that there is a member of R that is not in Q. Recall that we know that the equation $x^2 = 2$ does not have a solution for x among the elements of Q. Next, we will show that this equation does have a solution among the elements of R. This will show that Q is a proper subset of R. Later, we generalize Theorem 6.7 below.

Theorem 6.7

Suppose R is a complete ordered field. Then there exists an $a \in R$ such that $a^2 = 2$.

Proof: Define the set S by $S = \{q \in Q \mid q > 0 \text{ and } q^2 < 2\}$. We show that $\textbf{sup}S$ exists and that $(\textbf{sup}S)^2 = 2$. To show that $\textbf{sup}S$ exists, we show that S is nonempty and bounded above. Now, it is clear that $1 \in S$, so S is indeed nonempty. Now it will be shown that 2 is an upper bound of S. Suppose $q \in S$. Were we to have $q > 2$, then we would also have $q^2 > 4$, which is not possible because $q^2 < 2$. Therefore, $q \leq 2$ for each $q \in S$, so S is bounded above by 2. Because S is bounded above, it must have a least upper bound. Let us denote the least upper bound of S by a. It will be shown that $a^2 = 2$. To accomplish this, it will be shown that it is not possible to have $a^2 > 2$ and that it is not possible to have $a^2 < 2$. First of all, note that for each $x > 0$, if $x^2 > 2$, then x is an upper bound of S. (Why?)

To show that we do not have $a^2 > 2$, assume that indeed we do have $a^2 > 2$, and we will arrive at a contradiction. We will accomplish this by finding a $b > 0$ with $b^2 > 2$ (making b an upper bound of S) with $b < a$. This will be a contradiction because a is the *least* upper bound of S. Toward this end, let $\delta = \dfrac{a^2 - 2}{2a}$, and let $b = a - \delta$. Because we are assuming that $a^2 > 2$, note that $\delta > 0$. Therefore, $b < a$.

The next goal is to show that $b > 0$ and $b^2 > 2$. Now, $b = a - \delta = a - (\dfrac{a^2 - 2}{2a}) = \dfrac{a^2 + 2}{2a} > 0$. Again using the assumption that $a^2 > 2$, we see that $(a^2 - 2)^2 > 0$. Hence, $a^4 - 4a^2 + 4 > 0$. From this we can easily derive $a^4 + 4a^2 + 4 > 8a^2$, and from this we get $\dfrac{a^2 + 4a^2 + 4}{4a^2} > 2$. Hence, $(\dfrac{a^2 + 2}{2a})^2 > 2$. Because $b = \dfrac{a^2 + 2}{2a}$, we have shown that $b^2 > 2$. Thus, we have arrived at our desired contradiction. We therefore conclude that we cannot have $a^2 > 2$.

To show that we do not have $a^2 < 2$, assume that we do have $a^2 < 2$, and we will arrive at a contradiction. To accomplish this, we will show that the assumption $a^2 < 2$ leads to the conclusion that there is an element of S that is bigger than a, which is clearly not possible because a is an upper bound of S. To that end, let $\delta' = \dfrac{2 - a^2}{a + 2}$, and let $b' = a + \delta'$. Because we are assuming $a^2 < 2$, we see that $\delta' > 0$. Hence, $b' > a$. Next, we want to show that $(b')^2 < 2$. Now, $b' = a + \delta' = a + \dfrac{2 - a^2}{a + 2} = \dfrac{2a + 2}{a + 2}$. Now, from $a^2 < 2$, we clearly have $2a^2 < 4$. From this it is not hard to see that $4a^2 + 8a + 4 < 2a^2 + 8a + 8$. Hence, $\dfrac{4a^2 + 8a + 4}{a^2 + 4a + 4} < 2$, so $(\dfrac{2a + 2}{a + 2}) < 2$. Now, recalling that $b' = \dfrac{2a + 2}{a + 2}$, we have established that $(b')^2 < 2$. Note that we have not yet arrived at our desired contradiction. True, we have found an element b' with $b' > a$ and $(b')^2 < 2$. However, we cannot assert that $b' \in S$ because we do not know if $b' \in Q$. However, by Theorem 6.6, there exists an $r \in Q$ such that $a < r < b'$. Hence $r^2 < (b')^2$, so $r \in S$. It is the existence of r that allows us to assert a contradiction because r is an element of S that is bigger than a. Therefore, it is not possible to have $a^2 < 2$.

Having shown that we do not have $a^2 > 2$ and we also do not have $a^2 < 2$, we can assert that $a^2 = 2$.

In Exercise 13, the reader is asked to show that there is only one positive number satisfying the condition of Theorem 6.7. Of course, the number a of Theorem 6.7 is referred to as $\sqrt{2}$ (or as $2^{\frac{1}{2}}$).

There is something very important to be learned from the proof of Theorem 6.7. Recall that we defined the set S in the proof as

$S = \{q \in Q \mid q > 0 \text{ and } q^2 < 2\}$. Instead of using this set, we could just as well have shown that $\sqrt{2} = \sup\{x \in R \mid x > 0 \textbf{ and } x^2 \leq 2\}$. However, we wanted to make a point concerning the "holes" of a rational space that we promised we would discuss. Note that although S is a set of rational numbers (i.e., members of a rational space), $\sup S$ is not a member of S. This is because we have already shown that there is no member of a rational space whose square is 2. In this sense, with respect to Q, $\sqrt{2}$ is a "hole" in the number line. This hole is "plugged" in R because there is a real number whose square is 2. Similarly, any other such hole in Q will be plugged in R because every bounded subset of Q must have a least upper bound and greatest lower bound in R.

It has now been shown that $N \subset Z \subset Q \subset R$. There is a special name for members of Q and for members of the set $R - Q$.

Definition 6.5

Suppose R is a complete ordered field. Then members of Q are called **rational**, and members of $R - Q$ are called **irrational**. Q is referred to as the **set of rational numbers**, and $R - Q$ is referred to as the **set of irrational numbers**.

So far, we have only identified two irrational numbers: $\sqrt{2}$ and $-\sqrt{2}$. In Chapter 7, it is shown that there are many more irrational numbers than rational numbers. In fact, we show that the set of irrational numbers is uncountable. In the meantime, Theorem 6.8 will increase the number of numbers identified as irrational numbers. The proof is left to the reader. Following this, we show that there is an irrational number between any two distinct real numbers.

Theorem 6.8

Suppose R is a complete ordered field. Suppose $a, b \in R$ with a rational and b irrational. Then $a + b$ is irrational. If $a \neq 0$, then ab is irrational.

Now, we show that there is an irrational number between any two distinct real numbers.

Theorem 6.9

Suppose R is a complete ordered field. Suppose $a, b \in R$ with $a < b$. Then there exists an irrational number s such that $a < s < b$.

Proof: By Theorem 6.6, there is a rational number r such that $a - \sqrt{2} < r < b - \sqrt{2}$. Let $s = r + \sqrt{2}$. Now, we know that $\sqrt{2}$ is irrational. Hence, s is irrational by Theorem 6.8. Because $a < s < b$, the proof is complete.

Finally for this section, we prove a result about greatest lower bounds.

Theorem 6.10 Suppose R is a complete ordered field. Suppose A is a nonempty subset of R that is bounded below. Then **inf**A exists.

Proof: Let B be the set of lower bounds of A. Because A is bounded below, B is nonempty. Let $a \in A$. Then for any $b \in B$, b is a lower bound of A so that $b \leq a$. Thus, a is an upper bound of B. Hence, **sup**B exists.

It will now be shown that **inf**A = **sup**B. To do this, we first show that **sup**B is a lower bound of A. To do this, suppose $a \in A$. We have already noted that a is an upper bound of B. Hence, a cannot be less than the least upper bound of B. That is, **sup**$B \leq a$. We have shown that **sup**$B \leq a$ for each $a \in A$, so **sup**B is indeed a lower bound of A. To show that **sup**B is the greatest lower bound of A, we will show that if g is any lower bound of A, then we must have $g \leq$ **sup**B. However, if g is any lower bound of A, then $g \in B$, so clearly we must have $g \leq$ **sup**B because $b \leq$ **sup**B for any $b \in B$. Thus **inf**A exists and is equal to **sup**B.

Note how much information we have "squeezed" out of the definition of a complete ordered field in this section. It is clear that the concept of the least upper bound is very important. In the next section, we show that any two complete ordered fields are isomorphic.

EXERCISES

1. Prove Theorem 6.3. [Hint: Perform an induction on **card**(A).]

2. Suppose R is a complete ordered field. Let $Z = \cap_{X \in G} X$, in which G is the set of all subrings of R. Let $Q = \{\frac{m}{n} | m, n \in Z$ and $n \neq 0\}$. Show that Q is a rational space. (Hint: Use Theorems 5.5 and 5.6.)

3. Prove Theorem 6.5.

4. Show the result of Example 6.3.

5. Prove Theorem 6.8.

6. Show that there are an infinite number of rational numbers and an infinite number of irrational numbers between any two distinct real numbers. Did you need to use the Axiom of Choice in your proof?

7. Either prove or find a counterexample to the statement "The sum of two irrational numbers must be an irrational number."

8. Either prove or find a counterexample to the statement "The product of two irrational numbers must be an irrational number."

9. Suppose that A is a nonempty bounded subset of a complete ordered field R. Show that **inf**$A \leq$ **sup**A. When does equality occur? That is, under what condition[s] does **inf**A = **sup**A?

10. Suppose A and B are subsets of a complete ordered field R. Furthermore, suppose A is nonempty, B is bounded above, and $A \subseteq B$. Show that $\sup A$ exists and that $\sup A \leq \sup B$.

11. Suppose A and B are subsets of a complete ordered field R. Furthermore, suppose A is nonempty, B is bounded below, and $A \subseteq B$. Show that $\inf A$ exists and that $\inf A \geq \inf B$.

12. Suppose A is a nonempty subset of a complete ordered field R, and suppose A is bounded above. Define the set B by $B = \{-a | a \in A\}$. Show that $\inf B$ exists and that $\inf B = -\sup A$.

13. Show that in a complete ordered field, there is only one positive number whose square is 2.

14. Suppose R is a complete ordered field and $a \in R$ with $a > 0$. Show that there is a unique positive $b \in R$ such that $b^2 = a$. This number b is written as $\sqrt{a}$.

Section 6.2   The Uniqueness of the Real Number System

Let R be a complete ordered field. In Section 6.1, we specified a rational space Q, an integer space Z, and a Peano space N such that $N \subset Z \subset Q \subset R$. Now, suppose we also have another complete ordered field R'. Applying the same method to R' that was applied to R, we can specify a rational space Q', an integer space Z', and a Peano space N' such that $N' \subset Z' \subset Q' \subset R'$. For notational ease, we use the same symbols for the operations and the "usual" linear order for both R and R'. Also, the notation **sup** and **inf** are used for both sets. The context will always resolve any potential ambiguity. We indicate members of R' by using the "prime" notation. That is, the zero of R' shall be denoted by $0'$, the unit element shall be denoted by $1'$, and so on.

The goal of this section is to show that it does not matter whether we use R or R' to be the set of real numbers. We show this by demonstrating the existence of a field isomorphism $\phi: R \to R'$ such that $\phi(x) \leq \phi(y)$ iff $x \leq y$ for each $x, y \in R$. This is what mathematicians mean when they refer to the *uniqueness of the real number system*. Some parts of the proof are left to the reader.

To demonstrate the existence of such an isomorphism, the following steps suffice:

1) Define the mapping $\phi: R \to R'$.

2) Show that ϕ is onto.

3) Show that ϕ is one-to-one.

4) Show that for each $x, y \in R$, we have $\phi(x) \leq \phi(y)$ iff $x \leq y$.

5) Show that $\phi(x + y) = \phi(x) + \phi(y)$ for each $x, y \in R$.

6) Show that $\phi(xy) = \phi(x)\phi(y)$ for each $x, y \in R$.

To begin, recall from Theorem 5.14 that there is a field isomorphism $h\colon Q \to Q'$ such that $h(q_1) \leq h(q_2)$ iff $q_1 \leq q_2$ for each $q_1, q_2 \in Q$. Because h is one-to-one, it is easy to see that we also have $h(q_1) < h(q_2)$ iff $q_1 < q_2$ for each $q_1, q_2 \in Q$. The function h will be used to define ϕ.

For each $x \in R$, define the set A_x by $A_x = \{h(q) | q \in Q \text{ and } q \leq x\}$. We "want" to define $\phi(x)$ to be $\sup(A_x)$, but we first have to show that each A_x is nonempty and bounded above. Now, let $x \in R$. Then there is a $q \in Q$ such that $x - 1 < q < x$ (by Theorem 6.6). Hence, $h(q) \in A_x$, so A_x is nonempty. Also by Theorem 6.6, there is an $r \in Q$ such that $x < r < x + 1$. Now, suppose $q' \in A_x$. Because h is onto, there is a $q_1 \in Q$ such that $q' = h(q_1)$. Note that by the definition of A_x (and using the fact that h is one-to-one), we have $q_1 \leq x$. Hence, $q_1 < r$, so $h(q_1) < h(r)$. That is, $q' < h(r)$. This demonstrates that the set A_x is bounded above by $h(r)$. Hence, we can conclude that $\sup(A_x)$ exists for each $x \in R$. We now formally define $\phi\colon R \to R'$ by $\phi(x) = \sup(A_x)$ for each $x \in R$.

Next, it will be shown that ϕ is onto. Let $z' \in R'$. The goal is to demonstrate that there exists a $z \in R$ such that $\phi(z) = z'$. Define the set B by $B = \{q \in Q \mid h(q) \leq z'\}$. The reader is asked to show that B is nonempty and bounded above (in R, of course). It will be shown that $z' = \phi(\sup B)$. That is, we will show that $z' = \sup\{h(q) | q \in Q \text{ and } q \leq \sup B\}$.

First, it will be shown that z' is an upper bound of the set $\{h(q) | q \in Q \text{ and } q \leq \sup B\}$. To do so, suppose $q_1' \in \{h(q) | q \in Q \text{ and } q \leq \sup B\}$. Then $q_1' = h(q_1)$ for some $q_1 \in Q$ with

$$q_1 \leq \sup B. \tag{1}$$

We want to show that $q_1' \leq z'$. That is, we want to show that $h(q_1) \leq z'$.

Suppose (to reach a contradiction) that $z' < h(q_1)$. Then there exists a $q_2' \in Q'$ such that $z' < q_2' < h(q_1)$. Now, $q_2' = h(q_2)$ for some $q_2 \in Q$. Hence, we have

$$z' < h(q_2) < h(q_1). \tag{2}$$

Therefore,

$$q_2 < q_1. \tag{3}$$

Hence, by Inequality 1 and Inequality 3, we have $q_2 < \sup B$. This says that q_2 is not an upper bound of B. Consequently, there exists a $q_3 \in Q$ with $q_2 < q_3$ and $h(q_3) \leq z'$. This leads to the conclusion that $h(q_2) < h(q_3) \leq z'$. However, this contradicts Inequality 2. Hence, we conclude that $q_1' \leq z'$.

To complete the proof that ϕ is onto, we must show that for any $w' \in R'$, if $w' < z'$, then w' is not an upper bound of $\{h(q) | q \in Q \text{ and } q \leq \sup B\}$. So, consider such a $w' \in R'$. Then there exists a $q_4' \in Q'$ with $w' < q_4' < z'$. Now, there exists a $q_4 \in Q$ such that $q_4' = h(q_4)$, so $w' < h(q_4) < z'$. Therefore, $q_4 \in B$ (by the definition of B), so that $q_4 \leq \sup B$. Thus, $h(q_4) \in \{h(q) | q \in Q \text{ and } q \leq \sup B\}$. Because $w' < h(q_4)$, we see that w' is not an upper bound of $\{h(q) | q \in Q \text{ and } q \leq \sup B\}$ because it is less than a member of that set. This concludes the proof that $z' = \phi(\sup B)$, so ϕ is indeed onto.

Next, it is shown that for any $x, y \in R$, if $x < y$, then $\phi(x) < \phi(y)$. After this is done, the reader is asked to show that ϕ is one-to-one. The reader is also asked to show that for any $x, y \in R$, $x \le y$ iff $\phi(x) \le \phi(y)$.

Suppose $x, y \in R$ with $x < y$. Then there exists an $s_1 \in Q$ such that $x < s_1 < y$. Now consider any $q \in Q$ with $q \le x$. Then $q < s_1$, so $h(q) < h(s_1)$. This tells us that $h(s_1)$ is an upper bound of the set $\{h(q) | q \in Q$ and $q \le x\}$. Thus $\phi(x) \le h(s_1)$. Now, because $s_1 < y$, there exists an $s_2 \in Q$ so that $s_1 < s_2 < y$. Hence, $h(s_1) < h(s_2)$. Now, $h(s_2) \le \phi(y)$. (Why?) Thus, we conclude that $\phi(x) < \phi(y)$.

Let $x, y \in R$. We are now going to show that $\phi(x + y) = \phi(x) + \phi(y)$. Let $A_x = \{h(q) | q \in Q$ and $q \le x\}$, $A_y = \{h(q) | q \in Q$ and $q \le y\}$, and $A_{x+y} = \{h(q) | q \in Q$ and $q \le x + y\}$. Note that $\phi(x) = \sup(A_x)$, $\phi(y) = \sup(A_y)$, and $\phi(x + y) = \sup(A_{x+y})$. Our method of attack is to show first that $\phi(x) + \phi(y) \le \phi(x + y)$. Then, we will assume that strict inequality holds and arrive at a contradiction.

Let $q'_x \in A_x$ and $q'_y \in A_y$. To show that $\phi(x) + \phi(y) \le \phi(x + y)$, we will first show that $q'_x + q'_y \le \phi(x + y)$. Now there exist $q_x, q_y \in Q$ with $q_x \le x$ and $q_y \le y$ such that $q'_x = h(q_x)$ and $q'_y = h(q_y)$. Now, clearly $h(q_x) \le \phi(x)$ and $h(q_y) \le \phi(y)$. Because $q_x + q_y \le x + y$, we see that $h(q_x + q_y) \in A_{x+y}$, so $h(q_x + q_y) \le \phi(x + y)$. However, $h(q_x) + h(q_y) = h(q_x + q_y)$. That is, $q'_x + q'_y \le \phi(x + y)$.

We now know that $q'_x + q'_y \le \phi(x + y)$ for any $q'_x \in A_x$ and $q'_y \in A_y$. Therefore, for any $q'_y \in A_y$, we have $q'_y \le \phi(x + y) - q'_x$ for any $q'_x \in A_x$. That is, $\phi(x + y) - q'_x$ is an upper bound of A_y for any $q'_x \in A_x$. Hence, $\phi(y) \le \phi(x + y) - q'_x$ for any $q'_x \in A_x$ (because $\phi(y)$ is the least upper bound of A_y). From this, we see that $q'_x \le \phi(x + y) - \phi(y)$ for any $q'_x \in A_x$. Thus, $\phi(x + y) - \phi(y)$ is an upper bound of A_x, so $\phi(x) \le \phi(x + y) - \phi(y)$. That is, $\phi(x) + \phi(y) \le \phi(x + y)$.

Now, suppose (to reach a contradiction) that $\phi(x) + \phi(y) < \phi(x + y)$. Then $\phi(x) + \phi(y)$ is not an upper bound of A_{x+y}. Hence, there exists an $r_1 \in Q$ such that

$$\phi(x) + \phi(y) < h(r_1) \le \phi(x + y) \qquad (4)$$

with

$$r_1 \le x + y. \qquad (5)$$

Now, by Inequality 4, there is an $r'_2 \in Q'$ with $\phi(x) + \phi(y) < r'_2 < h(r_1)$. Now, there is an $r_2 \in Q$ such that $r'_2 = h(r_2)$. Hence, we have

$$\phi(x) + \phi(y) < h(r_2) < h(r_1) \qquad (6)$$

Hence, we also have

$$r_2 < r_1. \qquad (7)$$

By Inequality 5 and Inequality 7, we have $r_2 < x + y$. Now, let $\varepsilon = x + y - r_2$, so that $\varepsilon > 0$. There exists $r_3, r_4 \in Q$ with $x - \frac{\varepsilon}{2} < r_3 < x$

and $y - \frac{\varepsilon}{2} < r_4 < y$. Note that $x - r_3 < \frac{\varepsilon}{2}$ and $y - r_4 < \frac{\varepsilon}{2}$. Adding these two inequalities results in

$$x + y - (r_3 + r_4) < \varepsilon. \tag{8}$$

Now, because $r_3 < x$ and $r_4 < y$, we have $h(r_3) \in A_x$ and $h(r_4) \in A_y$. Thus, $h(r_3) \le \phi(x)$ and $h(r_4) \le \phi(y)$. Hence, $h(r_3 + r_4) = h(r_3) + h(r_4) \le \phi(x) + \phi(y)$. Using this and Inequality 6, we see that $h(r_3 + r_4) < h(r_2)$. So, we have

$$r_3 + r_4 < r_2. \tag{9}$$

Now, recall that $\varepsilon = x + y - r_2$. Plugging this into Inequality 8 gives $x + y - (r_3 + r_4) < x + y - r_2$. From this, we get $r_3 + r_4 > r_2$, which contradicts Inequality 9. Therefore, we can (finally) conclude that $\phi(x) + \phi(y) = \phi(x + y)$.

We will now show that $\phi(x)\phi(y) = \phi(xy)$ in the case in which $x, y \in R$ with $x > 0$ and $y > 0$. The reader is asked to show that $\phi(x)\phi(y) = \phi(xy)$ for the remaining cases. Let $B_x = \{h(q) | q \in Q$ and $0 < q \le x\}$, $B_y = \{h(q) | q \in Q$ and $0 < q \le y\}$, and $B_{xy} = \{h(q) | q \in Q$ and $0 < q \le xy\}$. The reader should note that each of these sets is nonempty, and that $\phi(x) = \sup(B_x)$, $\phi(y) = \sup(B_y)$, and $\phi(xy) = \sup(B_{xy})$. Our method of attack is to first show that $\phi(x)\phi(y) \le \phi(xy)$. Then, we will assume that strict inequality holds, and arrive at a contradiction.

Let $u'_x \in B_x$ and $u'_y \in B_y$. Now there exist $u_x, u_y \in Q$ with $0 < u_x \le x$ and $0 < u_y \le y$ such that $u'_x = h(u_x)$ and $u'_y = h(u_y)$. Hence, $0 < u_x u_y \le xy$, so $h(u_x u_y) \in B_{xy}$. This tells us that $h(u_x u_y) \le \phi(xy)$. Hence, $h(u_x)h(u_y) \le \phi(xy)$. That is, $u'_x u'_y \le \phi(xy)$. Hence, $u'_x \le \frac{\phi(xy)}{u'_y}$. From this it can be seen that $\frac{\phi(xy)}{u'_y}$ is an upper bound of B_x for each $u'_y \in B_y$. Hence, $\varphi(x) \le \frac{\phi(xy)}{u'_y}$. Now, from this it can be seen that $u'_y \le \frac{\phi(xy)}{\phi(x)}$ for each $u'_y \in B_y$. Therefore, $\frac{\phi(xy)}{\phi(x)}$ is an upper bound of B_y, so $\varphi(y) \le \frac{\phi(xy)}{\phi(x)}$. Thus, $\phi(x)\phi(y) \le \phi(xy)$.

Assume (to reach a contradiction) that $\phi(x)\phi(y) < \phi(xy)$. Hence, $\phi(x)\phi(y)$ is not an upper bound of B_{xy}, so there exists a $v'_1 \in B_{xy}$ so that $\phi(x)\phi(y) < v'_1 \le \phi(xy)$. Hence, there is a $v_1 \in Q$ with $v'_1 = h(v_1)$ that satisfies

$$0 < v_1 \le xy. \tag{10}$$

Note that $\phi(x)\phi(y) < h(v_1) < \phi(xy)$. Now, there exists a $v'_2 \in Q'$ such that $\phi(x)\phi(y) < v'_2 < h(v_1)$. Of course, there is a $v_2 \in Q$ with $v'_2 = h(v_2)$. Hence, we have

$$\phi(x)\phi(y) < h(v_2) < h(v_1). \tag{11}$$

Thus, we have

$$0 < v_2 < v_1. \tag{12}$$

By Inequality 10 and Inequality 12, we have

$$0 < v_2 < xy. \tag{13}$$

Now, let $\delta = \frac{v_2}{xy}$. Then, by Inequality 13, we have $0 < \delta < 1$. Now, see Exercise 14 of Section 6.1. Note that $0 < \sqrt{\delta} < 1$. (Why?) Therefore, $0 < x\sqrt{\delta} < x$ and $0 < y\sqrt{\delta} < y$. Now, there are v_3, $v_4 \in Q$ so that

$$x\sqrt{\delta} < v_3 < x \tag{14}$$

and

$$y\sqrt{\delta} < v_4 < y. \tag{15}$$

Multiplying these two inequalities gives us $xy\delta < v_3 v_4$. Recalling that $\delta = \frac{v_2}{xy}$, this says that

$$v_2 < v_3 v_4. \tag{16}$$

Also, from Inequality 14 and Inequality 15, we see that $h(v_3) \in B_x$ and $h(v_4) \in B_y$. This tells us that $h(v_3) \leq \phi(x)$ and $h(v_4) \leq \phi(y)$. Hence, $h(v_3 v_4) = h(v_3)h(v_4) \leq \phi(x)\phi(y)$. Hence, by Inequality 11 we have $h(v_3 v_4) < h(v_2)$. From this we can conclude that $v_3 v_4 < v_2$. However, this contradicts Inequality 16. Thus, we (finally) conclude that $\phi(x)\phi(y) = \phi(xy)$. This concludes the proof.

The result that we have proven is very important. Therefore, it deserves to be listed as a theorem. The student should also bask in the glory of having arrived at this point.

Theorem 6.11 Suppose R and R' are complete ordered fields. Then there is a field isomorphism $\phi: R \to R'$ such that for each x, $y \in R$, we have $\phi(x) \leq \phi(y)$ iff $x \leq y$.

EXERCISES

1. Fill in all of the missing pieces of the proof of this section.

2. Show that the mapping $\phi(x)$ of the proof is an extension of the mapping h. That is, show that $\phi(q) = h(q)$ for each $q \in Q$.

Section 6.3 Construction of a Set Obeying the Real Number Postulates

The promise made in the text that the real number system will be constructed based on some simple concepts of set theory is finally fulfilled in this section. It will be demonstrated that a complete ordered field actually exists by specifying one. The construction is based on the construction used in Burril (1967) and Lang (1967). After this section, we shall assume that a complete ordered field has been chosen, and refer to it as **R**. Of course, by virtue of the last section, it does not matter which complete ordered field is chosen; one is as good as any other. The construction of the complete ordered field in this section uses what is known as the method of **Cauchy sequences**. Another well-known method often used in the mathematics literature to construct a complete ordered field is the method of **Dedekind cuts**, but this will not be used in this text. Recall that once **R** has been selected, we automatically have the sets **Q**, **Z**, and **N**.

The author considered labeling this section "Optional" because remaining sections of the book do not make explicit use of the results of the section. Nonetheless, students should see the construction performed in this section at least once in their career. Also, the methods employed in this construction will reappear in analysis courses, so studying this section will make the transition to analysis easier. However, to make this section easier to digest, part of the construction has been placed in Appendix C. The student will be able to get the gist of what is going on without reading Appendix C. The student is encouraged to read Appendix C for the sake of completeness.

Before jumping into the construction, let us briefly review where we are currently, with the help of Figure 6.1. In Section 3.1, we used basic set theory and the Axiom of Infinity (Section 2.7) to prove the existence of a Peano space P. This is illustrated in the left portion of Figure 6.1. In Section 3.7, P was used to prove the existence of an integer space that is denoted I in the figure. This is illustrated by the arrow and box to the right of the box with P in it. Note that I itself contains a Peano space (denoted P' in the figure). We continue moving to the right in the figure, and we recall that we proved the existence of a rational space (denoted Q in the figure) in Section 5.2, using I as an input. We also showed that Q contains an

Figure 6.1
The development of the real number system

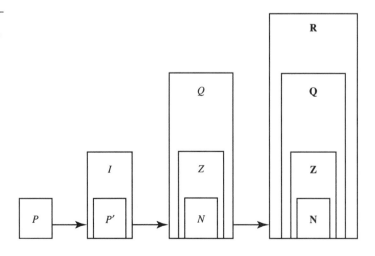

integer space (denoted Z in the figure). Of course, we know that Z contains a Peano space (denoted N in the figure). The right portion of the figure displays what we are about to do. We are about to prove the existence of a complete ordered field, **R**, using Q as in input. Now, we know that **R** contains a rational space, **Q**. We also know that **Q** contains an integer space, **Z**. Finally, we know that **Z** contains a Peano space, **N**. (Obviously, the diagram is not drawn to scale.)

For this section and Appendix C, we take as our starting point, a rational space Q. This rational space contains an integer space, I, and I contains a Peano space, N. We use the "usual" symbols for addition and multiplication of members of Q and for the "normal" order defined on Q. We shall denote the set of positive members of Q as Q^+. That is, $Q^+ = \{q \in Q | q > 0\}$. All the intermediate results that we derive in this section and Appendix C will be labeled lemmas. These lemmas will not formally be used outside of this section and Appendix C. There will not be any examples in the later part of this section because our main goal is to establish the existence of a complete ordered field as quickly as possible.

Preliminary Concepts

Definition 6.6

A sequence $\{q_n\}_{n=1}^{\infty}$ of elements of Q is called a **Cauchy sequence of rational numbers** provided that the following holds: For each $\varepsilon \in Q^+$, there exists an $M \in N$ such that $|q_n - q_m| < \varepsilon$ whenever $m, n \in N$ with $m \geq M$ and $n \geq N$.

We can usually leave out the terms "of rational numbers." The reason why we have included these words is that there are also Cauchy sequences of real numbers and other structures (as the reader will find out in an analysis course).

To understand the meaning of a Cauchy sequence intuitively, recall from Section 5.1 that $|q_n - q_m|$ should be interpreted as the distance between the numbers q_n and q_m. With this in mind, a sequence of rational numbers is a Cauchy sequence if the terms of the sequence get as close to each other as you like forever more (although they might not ever completely reach each other), provided you choose the subscripts of the sequence to be sufficiently large.

Example 6.4   Suppose $r \in Q$ and consider the sequence $\{q_n\}_{n=1}^{\infty}$ in which $q_n = r$ for each $n \in N$. That is, $\{q_n\}_{n=1}^{\infty}$ is a **constant sequence**. Then $\{q_n\}_{n=1}^{\infty}$ is a Cauchy sequence because $|q_n - q_m| = |r - r| = 0$. This is the simplest type of Cauchy sequence.

Example 6.5	We now show that the sequence $\{\frac{1}{n}\}_{n=1}^{\infty}$ is a Cauchy sequence. Let $\varepsilon \in Q^+$. We must show the existence of an $M \in N$ such that $|\frac{1}{n} - \frac{1}{m}| < \varepsilon$ whenever $m, n \in N$ with $m \geq M$ and $n \geq M$. Our strategy will be to find an $M \in N$ such that $\frac{1}{M} < \frac{\varepsilon}{2}$. Why? Because suppose that we succeed in doing this. (In a minute, we shall do so.) Now, consider any $m, n \in N$ with $m \geq M$ and $n \geq M$. Then, $\frac{1}{m} \leq \frac{1}{M} < \frac{\varepsilon}{2}$ and $\frac{1}{n} \leq \frac{1}{M} < \frac{\varepsilon}{2}$. Hence, from Part 3 of Theorem 5.9, we will have $|\frac{1}{n} - \frac{1}{m}| \leq |\frac{1}{n}| + |-\frac{1}{m}| = \frac{1}{n} + \frac{1}{m} < \frac{\varepsilon}{2} + \frac{\varepsilon}{2} = \varepsilon$. That is, we will have $|\frac{1}{n} - \frac{1}{m}| < \varepsilon$. Therefore, to complete the demonstration, we need only show that there exists an $M \in N$ such that $\frac{1}{M} < \frac{\varepsilon}{2}$. Now, because ε is a positive rational number, there exists $a, b \in N$ such that $\varepsilon = \frac{a}{b}$. Hence, our goal is to find an $M \in N$ such that $\frac{1}{M} < \frac{a}{2b}$. That is, we want to find an $M \in N$ so that $M > \frac{2b}{a}$. Now it is clear that the positive integer $2b + 1$ satisfies the inequality $2b + 1 > \frac{2b}{a}$. Hence, setting M to be $2b + 1$ clearly "does the trick."

For the remainder of the section, as well as for Appendix C, let us denote the set of Cauchy sequences (with respect to Q, of course) as S. The method of our proof is to define an equivalence relation on S. We then show that the set of equivalence classes of this equivalence relation, with suitably defined binary operations and a suitably defined order, is a complete ordered field. This work will commence in the following subsection. For the remainder of this subsection, we derive some results that will be needed for this upcoming work.

Definition 6.7

Suppose that $\{q_n\}_{n=1}^{\infty}$ is a sequence of elements of Q, and suppose $L \in Q$. Then $\{q_n\}_{n=1}^{\infty}$ is said to **converge** to L provided that for each $\varepsilon \in Q^+$, there exists an $M \in N$ such that $|q_n - L| < \varepsilon$ whenever $n \in N$ with $n \geq M$. We can also express this by saying that L is the **limit** of the sequence $\{q_n\}_{n=1}^{\infty}$. This is also expressed by writing $\lim_{n \to \infty} q_n = L$. If $\{q_n\}_{n=1}^{\infty}$ converges to some $L \in Q$, we say that $\{q_n\}_{n=1}^{\infty}$ is **convergent**.

In Chapter 7, we define the limit of a sequence of real numbers. (The reader is likely to have encountered this before.) Intuitively, saying that $\lim_{n \to \infty} q_n = L$ means that the terms of the sequence get as close to L as you like forever more (although they might not ever completely reach L), provided you choose the subscript of the sequence to be sufficiently large. If $r \in Q$, the reader is asked to show that $\lim_{n \to \infty} r = r$. (By the expression $\lim_{n \to \infty} r$, we mean the limit of a constant sequence, in which each term of the sequence is r.) The reader is also asked to show that $\lim_{n \to \infty} \frac{1}{n} = 0$.

We next show that if $\{q_n\}_{n=1}^{\infty}$ is a sequence of elements of Q, and if $\{q_n\}_{n=1}^{\infty}$ has limit L_1 and limit L_2, then $L_2 = L_1$. That is, the limit of a convergent sequence is unique.

Lemma 6.1

Suppose that $\{q_n\}_{n=1}^{\infty}$ is a sequence of elements of Q, and suppose $L_1, L_2 \in Q$. If $\{q_n\}_{n=1}^{\infty}$ has limit L_1 and limit L_2, then $L_2 = L_1$.

Proof: Assume (to reach a contradiction) that $L_2 \neq L_1$. Let $\varepsilon = \frac{|L_2 - L_1|}{2}$. Note that $\varepsilon \in Q^+$. Because $\{q_n\}_{n=1}^{\infty}$ has limit L_1, there exists an $M_1 \in N$ such that $|q_n - L_1| < \varepsilon$ whenever $n \in N$ with $n \geq M_1$. Also, because $\{q_n\}_{n=1}^{\infty}$ has limit L_2, there exists an $M_2 \in N$ such that $|q_n - L_2| < \varepsilon$ whenever $n \in N$ with $n \geq M_2$. Now, let $M = \max\{M_1, M_2\}$. Then, whenever $n \in N$ with $n \geq M$ we have both $|q_n - L_1| < \varepsilon$ and $|q_n - L_2| < \varepsilon$. Hence, for any $n \in N$ with $n \geq M$, we have (using the triangle inequality) $|L_2 - L_1| \leq |L_2 - q_n| + |q_n - L_1| = |q_n - L_2| + |q_n - L_1| < \varepsilon + \varepsilon = 2\varepsilon$. Recalling that $\varepsilon = \frac{|L_2 - L_1|}{2}$, the previous statement says that $|L_2 - L_1| < |L_2 - L_1|$. Because this is a contradiction, we have $L_2 = L_1$.

Next, we show that every convergent sequence of rational numbers is a Cauchy sequence.

Lemma 6.2

Suppose that $\{q_n\}_{n=1}^{\infty}$ is a convergent sequence of elements of Q. Then $\{q_n\}_{n=1}^{\infty}$ is a Cauchy sequence.

Proof: By definition, there is an $L \in Q$ such that $\lim_{n \to \infty} q_n = L$. Now, let $\varepsilon \in Q^+$. To show that $\{q_n\}_{n=1}^{\infty}$ is a Cauchy sequence, we must show that there is an $M \in N$ such that $|q_n - q_m| < \varepsilon$ whenever $m, n \in N$ with $m \geq M$ and $n \geq M$. However, using the fact that $\lim_{n \to \infty} q_n = L$ and noting that $\frac{\varepsilon}{2} \in Q^+$, there is an $M \in N$ such that $|q_n - L| < \frac{\varepsilon}{2}$ whenever $n \in N$ with $n \geq M$. Now consider any $m, n \in N$ with $m \geq M$ and $n \geq M$. Then (using the triangle inequality), we have $|q_n - q_m| \leq |q_n - L| + |L - q_m| < \frac{\varepsilon}{2} + \frac{\varepsilon}{2} < \varepsilon$, so we are done.

We have shown that any sequence of rational numbers that converges to a rational number must be a Cauchy sequence of rational numbers. It is natural to wonder if the converse is true. That is, does every Cauchy sequence of rational numbers converge to a rational number? It turns out that the answer to this is "no." In fact, this is what distinguishes a rational space from a complete ordered field. As you will see in your analysis class, any Cauchy sequence of *real numbers* does converge to a *real number*.

Example 6.6

We define the sequence $\{q_n\}_{n=1}^{\infty}$ of rational numbers via recursion as follows: $q_1 = 1$ and $q_{n+1} = \frac{(2/q_n) + q_n}{2}$. Although we omit the proof, the sequence $\{q_n\}_{n=1}^{\infty}$ turns out to be a Cauchy sequence that does not converge to a rational number. In fact, it turns out that this sequence converges to $\sqrt{2}$. (We formally define the limit of a sequence of *real* numbers in Chapter 7.) You will be able to formally prove this after you have started your analysis course. In the meantime, you may want to play with the above sequence and observe by experimentation that the terms of the sequence are getting closer and closer to $\sqrt{2}$.

Next, we show that every Cauchy sequence is bounded. In fact, if $\{q_n\}_{n=1}^{\infty}$ is a Cauchy sequence, we show that there exists a $B \in Q$ with $B \geq 0$ such that $|q_n| \leq B$ for each $n \in N$. (This constant depends on $\{q_n\}_{n=1}^{\infty}$; we are not claiming that the same constant B works for all Cauchy sequences.) This shows that $\{q_n\}_{n=1}^{\infty}$ is bounded because if we have $|q_n| \leq B$ for each $n \in N$, then we have $-B \leq q_n \leq B$ for each $n \in N$. This shows that $\{q_n\}_{n=1}^{\infty}$ is bounded below and bounded above. (Technically, when we say that $\{q_n\}_{n=1}^{\infty}$ is bounded, we mean that the set $\{q_n | n \in N\}$ is bounded.)

Lemma 6.3

Suppose that $\{q_n\}_{n=1}^{\infty}$ is a Cauchy sequence of rational numbers. Then there exists a $B \in Q$ with $B \geq 0$ such that $|q_n| \leq B$ for each $n \in N$.

Proof: Using $\varepsilon = 1$ in Definition 6.6, we see that there exists an $M \in N$ such that $|q_n - q_m| < 1$ whenever $m, n \in N$ with $m \geq M$ and $n \geq M$. We can also select M so that $M > 1$. Hence, for any $n \in N$ with $n \geq M$, we have $|q_n| = |(q_n - q_M) + q_M| \leq |q_n - q_M| + |q_M| < 1 + |q_M|$. Thus, we have shown that $|q_n| < 1 + |q_M|$ for each $n \in N$ with $n \geq M$. Does this mean that we can set B to be $1 + |q_M|$? No, not necessarily. We cannot establish from the above that $|q_n| \leq 1 + |q_M|$ for $n \in \{1, 2, \ldots, M\}$. (We selected $M > 1$ so that the set $\{1, 2, \ldots, M - 1\}$ would be guaranteed to be nonempty.) That is, we have not found a bound that works for *all* $n \in N$. However, if we define B by $B = \mathbf{max}\{|q_1|, |q_2|, \ldots, |q_{M-1}|, 1 + |q_M|\}$, then we do have $|q_n| \leq B$ for all $n \in N$. (We are entitled to take the maximum of a finite set in any ordered set. If this bothers you, then write out a formal proof of this fact.)

Suppose $\{q_n\}_{n=1}^{\infty} \in S$. (Recall that S is the set of Cauchy sequences of rational numbers.) Then we can form a sequence $\{-q_n\}_{n=1}^{\infty}$, obtained by simply taking the additive inverse of each term of $\{q_n\}_{n=1}^{\infty}$. Suppose that we also have a Cauchy sequence $\{r_n\}_{n=1}^{\infty} \in S$. Then, we can form the sequences $\{q_n + r_n\}_{n=1}^{\infty}, \{q_n - r_n\}_{n=1}^{\infty}$, and $\{q_n r_n\}_{n=1}^{\infty}$. The next two corollaries state that all of these sequences are Cauchy sequences. The proof of Lemma 6.5 is left to the reader.

Lemma 6.4

Suppose that $\{q_n\}_{n=1}^{\infty}, \{r_n\}_{n=1}^{\infty} \in S$. Then $\{q_n + r_n\}_{n=1}^{\infty} \in S$ and $\{q_n r_n\}_{n=1}^{\infty} \in S$.

Proof: We first prove that $\{q_n + r_n\}_{n=1}^{\infty}$ is a Cauchy sequence. Let $\varepsilon \in Q^+$. We must show that there exists an $M \in N$ such that $|(q_n + r_n) - (q_m + r_m)| < \varepsilon$ whenever $m, n \in N$ with $m \geq M$ and $n \geq M$. By Lemma 6.3, there exists a $B_1 \in Q$ with $B_1 \geq 0$ such that $|q_n| \leq B_1$ for each $n \in N$. In fact, we can choose B_1 so that $B_1 \geq 1$. Also there exists a $B_2 \in Q$ with $B_2 \geq 1$ such that $|r_n| \leq B_2$ for each $n \in N$. Let $B = \mathbf{max}\{B_1, B_2\}$. Hence, for each $n \in N$, we have $|q_n| \leq B$ and $|r_n| \leq B$. Note that $B \geq 1$.

Because $\{q_n\}_{n=1}^{\infty}$ is a Cauchy sequence, there is an $M_1 \in N$ such that $|q_n - q_m| < \frac{\varepsilon}{2B}$ whenever $m, n \in N$ with $m \geq M_1$ and $n \geq M_1$. Because $\{r_n\}_{n=1}^{\infty}$ is also a Cauchy sequence, there is an $M_2 \in N$ such that $|r_n - r_m| < \frac{\varepsilon}{2B}$ whenever $m, n \in N$ with $m \geq M_2$ and $n \geq M_2$. Let $M = \mathbf{max}\{M_1, M_2\}$. Then we have both $|q_n - q_m| < \frac{\varepsilon}{2B}$ and $|r_n - r_m| < \frac{\varepsilon}{2B}$ whenever $m, n \in N$ with $m \geq M$ and $n \geq M$. Hence, for any such $m, n \in N$ with $m \geq M$ and $n \geq M$, we have $|(q_n + r_n) - (q_m + r_m)| = |(q_n - q_m) + (r_n - r_m)| \leq |q_n - q_m| + |r_n - r_m| < 2(\frac{\varepsilon}{2B}) = \frac{\varepsilon}{B} \leq \varepsilon$. (Note that we used the fact that $B \geq 1$ in the inequality $\frac{\varepsilon}{B} \leq \varepsilon$.) This proves that $\{q_n + r_n\}_{n=1}^{\infty} \in S$. (Note: We will make more use of B in the proof that $\{q_n r_n\}_{n=1}^{\infty} \in S$. We could have gotten away without using it in the above portion of the proof.)

Next, we prove that $\{q_n r_n\}_{n=1}^{\infty}$ is a Cauchy sequence. To do so, again let $\varepsilon \in Q^+$. Also, let B and M be as defined above in the proof that $\{q_n + r_n\}_{n=1}^{\infty} \in S$. We show that $|q_n r_n - q_m r_m| < \varepsilon$ for all $m, n \in N$ such that $m \geq M$ and $n \geq M$. Hence, suppose $m, n \in N$ such that $m \geq M$ and $n \geq M$. Then $|q_n r_n - q_m r_m| = |q_n(r_n - r_m) + r_m(q_n - q_m)| \leq |q_n||r_n - r_m| + |r_m||q_n - q_m|$

$$< B\left(\frac{\varepsilon}{2B}\right) + B\left(\frac{\varepsilon}{2B}\right) = 2\left(\frac{\varepsilon}{2}\right) = \varepsilon.$$

Hence, $\{q_n r_n\}_{n=1}^{\infty} \in S$.

Lemma 6.5

Suppose that $\{q_n\}_{n=1}^{\infty} \in S$. Then $\{-q_n\}_{n=1}^{\infty} \in S$. If we also have $\{r_n\}_{n=1}^{\infty} \in S$, then $\{q_n - r_n\}_{n=1}^{\infty} \in S$.

We are now ready to construct our desired complete ordered field.

The Construction

We begin by defining a relation $\cong$ on S. We will then show that it is an equivalence relation. Consider any $\{q_n\}_{n=1}^{\infty}, \{r_n\}_{n=1}^{\infty} \in S$. We specify that

$\{q_n\}_{n=1}^{\infty} \cong \{r_n\}_{n=1}^{\infty}$ provided that $\lim_{n\to\infty} (q_n - r_n) = 0$; that is, $\{q_n\}_{n=1}^{\infty} \cong \{r_n\}_{n=1}^{\infty}$ iff the sequence $\{q_n - r_n\}_{n=1}^{\infty}$ converges to 0. The reader is asked to supply the easy proofs that the relation $\cong$ is reflexive and symmetric.

We will now show that it is transitive. To do so, suppose $\{q_n\}_{n=1}^{\infty}$, $\{r_n\}_{n=1}^{\infty}$, $\{s_n\}_{n=1}^{\infty} \in S$ with $\{q_n\}_{n=1}^{\infty} \cong \{r_n\}_{n=1}^{\infty}$ and $\{r_n\}_{n=1}^{\infty} \cong \{s_n\}_{n=1}^{\infty}$. The goal is to show that $\{q_n\}_{n=1}^{\infty} \cong \{s_n\}_{n=1}^{\infty}$. That is, we want to show that $\lim_{n\to\infty} (q_n - s_n) = 0$. To do this, suppose $\varepsilon \in Q^{+}$. We will be done provided we can show the existence of an $M \in N$ such that $|q_n - s_n| < \varepsilon$ whenever $n \in N$ with $n \geq M$. Because $\{q_n\}_{n=1}^{\infty} \cong \{r_n\}_{n=1}^{\infty}$, we have $\lim_{n\to\infty} (q_n - r_n) = 0$. Hence, there exists an $M_1 \in N$ such that $|q_n - r_n| < \frac{\varepsilon}{2}$ whenever $n \in N$ with $n \geq M_1$. Similarly, because $\{r_n\}_{n=1}^{\infty} \cong \{s_n\}_{n=1}^{\infty}$, we have $\lim_{n\to\infty} (r_n - s_n) = 0$. Hence, there exists an $M_2 \in N$ such that $|r_n - s_n| < \frac{\varepsilon}{2}$ whenever $n \in N$ with $n \geq M_2$. Now, let $M = \mathbf{max}\{M_1, M_2\}$. Then for each $n \in N$ with $n \geq M$, we have $|q_n - s_n| = |(q_n - r_n) + (r_n - s_n)| \leq |q_n - r_n| + |r_n - s_n| < \frac{\varepsilon}{2} + \frac{\varepsilon}{2} = \varepsilon$. Thus, transitivity is established.

Having demonstrated that $\cong$ is indeed an equivalence relation on S, let us denote the equivalence class of any $\{q_n\}_{n=1}^{\infty} \in S$ by $[\{q_n\}_{n=1}^{\infty}]$. We now proceed to denote the set of equivalence classes of $\cong$ by $\mathbf{R}$. That is, $\mathbf{R} = \{[\{q_n\}_{n=1}^{\infty}] | \{q_n\}_{n=1}^{\infty} \in S\}$. We are now going to define an "addition" on $\mathbf{R}$. Consider any $[\{q_n\}_{n=1}^{\infty}]$, $[\{s_n\}_{n=1}^{\infty}] \in \mathbf{R}$. We define $[\{q_n\}_{n=1}^{\infty}] + [\{s_n\}_{n=1}^{\infty}]$ to be $[\{q_n + s_n\}_{n=1}^{\infty}]$.

But, wait a minute. Recall our discussions in earlier chapters about our definitions being well defined? We need to show that this binary operation is well defined. First of all, we know from Lemma 6.4 that $\{q_n + r_n\}_{n=1}^{\infty} \in S$. Hence, the term $[\{q_n + r_n\}_{n=1}^{\infty}]$ makes sense; it is simply the equivalence class of the Cauchy sequence $\{q_n + r_n\}_{n=1}^{\infty}$. To complete the demonstration that $\cong$ is well defined, we need to show that this binary operation does not depend on any particular representatives of our equivalence classes. (Does this sound familiar?) That is, we need to show that if we have $\{a_n\}_{n=1}^{\infty}$, $\{b_n\}_{n=1}^{\infty} \in S$ such that $\{a_n\}_{n=1}^{\infty} \cong \{q_n\}_{n=1}^{\infty}$ and $\{b_n\}_{n=1}^{\infty} \cong \{r_n\}_{n=1}^{\infty}$, then we have $\{a_n + b_n\}_{n=1}^{\infty} \cong \{q_n + r_n\}_{n=1}^{\infty}$. The reader is asked to complete this argument.

Now, it is easy to see how to define the "zero" of $\mathbf{R}$ and the additive inverse of each $[\{q_n\}_{n=1}^{\infty}] \in \mathbf{R}$. We define the "zero" of $\mathbf{R}$ to be the equivalence class of the constant sequence $\{0\}_{n=1}^{\infty}$. That is, $0 = [\{0\}_{n=1}^{\infty}]$. We define $-[\{q_n\}_{n=1}^{\infty}]$ to be $[\{-q_n\}_{n=1}^{\infty}]$. We know from Lemma 6.5 that $\{-q_n\}_{n=1}^{\infty} \in S$. However, we still need to show that our definition of additive inverses is well defined. Suppose $\{a_n\}_{n=1}^{\infty} \in S$ with $\{a_n\}_{n=1}^{\infty} \cong \{q_n\}_{n=1}^{\infty}$. We must show that $\{-a_n\}_{n=1}^{\infty} \cong \{-q_n\}_{n=1}^{\infty}$. That is, we must show that $\lim_{n\to\infty} (-a_n + q_n) = 0$. (You can tell from the context that we mean the zero element of Q, not the zero element of $\mathbf{R}$.) That is, given any $\varepsilon \in Q^{+}$, we must demonstrate the existence of an $M \in N$ such that $|-a_n + q_n| < \varepsilon$ whenever $n \in N$ with $n \geq M$. Now, note that $|-a_n + q_n| < \varepsilon \Leftrightarrow |a_n - q_n| < \varepsilon$. The result now follows easily because, given that we have $\{a_n\}_{n=1}^{\infty} \cong \{q_n\}_{n=1}^{\infty}$, we know that $\lim_{n\to\infty} (a_n - q_n) = 0$. (Fill in the final step.) Thus, $-[\{q_n\}_{n=1}^{\infty}]$ is well defined.

Next, we define "multiplication" of elements of $\mathbf{R}$. Consider any $[\{q_n\}_{n=1}^{\infty}]$, $[\{r_n\}_{n=1}^{\infty}] \in \mathbf{R}$. We "want" to define $[\{q_n\}_{n=1}^{\infty}] \cdot [\{r_n\}_{n=1}^{\infty}]$ to be

$[\{q_n r_n\}_{n=1}^{\infty}]$. We know from Lemma 6.4 that $\{q_n r_n\}_{n=1}^{\infty} \in S$. However, we still have the problem of showing that this operation is well defined. Specifically, if $\{a_n\}_{n=1}^{\infty} \cong \{q_n\}_{n=1}^{\infty}$ and $\{b_n\}_{n=1}^{\infty} \cong \{r_n\}_{n=1}^{\infty}$, we need to show that $\{a_n b_n\}_{n=1}^{\infty} \cong \{q_n r_n\}_{n=1}^{\infty}$. That is, we must show that $\lim_{n \to \infty} (a_n b_n - q_n r_n) = 0$.

To accomplish this, let $\varepsilon \in Q^+$. We want to show that there exists an $M \in N$ such that $|a_n b_n - q_n r_n| < \varepsilon$ whenever $n \in N$ with $n \geq M$. By Lemma 6.3, we can choose $B_1, B_2 \in Q^+$ such that $|a_n| \leq B_1$ and $|r_n| \leq B_2$ for each $n \in N$. Let $B = \mathbf{max}\{B_1, B_2\}$. Then we have $|a_n| \leq B$ and $|r_n| \leq B$ for each $n \in N$. Now, from the statements $\{a_n\}_{n=1}^{\infty} \cong \{q_n\}_{n=1}^{\infty}$ and $\{b_n\}_{n=1}^{\infty} \cong \{r_n\}_{n=1}^{\infty}$, we have $\lim_{n \to \infty} (a_n - q_n)$ and $\lim_{n \to \infty} (b_n - r_n) = 0$. Hence, there exists M_1, $M_2 \in N$ such that $|a_n - q_n| < \frac{\varepsilon}{2B}$ whenever $n \in N$ with $n \geq M_1$, and $|b_n - r_n| < \frac{\varepsilon}{2B}$ whenever $n \in N$ with $n \geq M_2$. Let $M = \mathbf{max}\{M_1, M_2\}$. Then for each $n \in N$ with $n \geq M$, we have $|a_n b_n - q_n r_n| = |a_n(b_n - r_n) + r_n(a_n - q_n)| \leq |a_n| |b_n - r_n| + |r_n| |a_n - q_n|$

$$\leq B\left(\frac{\varepsilon}{2B}\right) + B\left(\frac{\varepsilon}{2B}\right) = \varepsilon.$$

This completes the proof that our multiplication on **R** is well defined.

We now define the unit element for **R**. We define the unit element 1 of **R** to be the equivalence class of the constant sequence $\{1\}_{n=1}^{\infty}$. That is, $1 = [\{1\}_{n=1}^{\infty}]$.

Next, we want to define multiplicative inverses for nonzero elements of **R**. Before doing this, we need some additional lemmas to accomplish this goal.

| **Lemma 6.6** | Suppose that $\{q_n\}_{n=1}^{\infty} \in S$ has the property that for any $\varepsilon \in Q^+$ and any $n \in N$, there exists a $k \in N$ with $k \geq n$ such that $|q_k| < \varepsilon$. Then $\{q_n\}_{n=1}^{\infty} \in [\{0\}_{n=1}^{\infty}]$. |

Proof: Recall from the definition of equivalence classes that saying $\{q_n\}_{n=1}^{\infty} \in [\{0\}_{n=1}^{\infty}]$ is equivalent to saying $\{q_n\}_{n=1}^{\infty} \cong \{0\}_{n=1}^{\infty}$. That is, we want to show that $\lim_{n \to \infty} q_n = 0$. Hence, it suffices to show that for each $\varepsilon \in Q^+$, there exists an $M \in N$ such that $|q_n| < \varepsilon$ whenever $n \in N$ with $n \geq M$.

Let $\varepsilon \in Q^+$. Because $\{q_n\}_{n=1}^{\infty}$ is a Cauchy sequence, there is an $M_1 \in N$ such that $|q_n - q_m| < \frac{\varepsilon}{2}$ whenever $n, m \in N$ with $n \geq M_1$ and $m \geq M_1$. Now, by the property expressed in the statement of the lemma, there is an $M \in N$ with $M \geq M_1$ such that $|q_M| < \frac{\varepsilon}{2}$. Hence, for each $n \in N$ with $n \geq M$ we have $|q_n| = |(q_n - q_M) + q_M| \leq |q_n - q_M| + |q_M| < \frac{\varepsilon}{2} + \frac{\varepsilon}{2} = \varepsilon$.

Suppose we have $\{q_n\}_{n=1}^{\infty} \in S$ with $\{q_n\}_{n=1}^{\infty} \notin [\{0\}_{n=1}^{\infty}]$. Then, by applying the converse to Lemma 6.6, it is not the case that for every $\varepsilon \in Q^+$ and any $n \in N$ there exists a $k \in N$ with $k \geq n$ such that $|q_k| < \varepsilon$. Restating this is a good example of using the method of proof that was referred to as "the negation of quantifiers" in Chapter 1. Using this, the reader should be able to see that there exists an $\varepsilon \in Q^+$ and an $M \in N$ such that for each $n \in N$ with $n \geq M$, we have $|q_n| \geq \varepsilon$. Also, we can write $[\{q_n\}_{n=1}^{\infty}] \neq 0$ (in which we refer to the zero element of $\mathbf{R}$) instead of writing $\{q_n\}_{n=1}^{\infty} \notin [\{0\}_{n=1}^{\infty}]$. Thus, we have already proved the next lemma.

Lemma 6.7   Suppose that $\{q_n\}_{n=1}^{\infty} \in S$ with $[\{q_n\}_{n=1}^{\infty}] \neq 0$. Then there exists an $\varepsilon \in Q^+$ and an $M \in N$ such that for each $n \in N$ with $n \geq M$, we have $|q_n| \geq \varepsilon$.

Now, consider any $\{q_n\}_{n=1}^{\infty} \in S$ with $[\{q_n\}_{n=1}^{\infty}] \neq 0$. We know from Lemma 6.7 that $q_n = 0$ for at most finitely many $n \in N$. (Here we mean the zero element of Q.) Now define the sequence $\{r_n\}_{n=1}^{\infty}$ as follows: For each $n \in N$, we specify that

$$r_n = \begin{cases} 1 & \text{if} \quad q_n = 0 \\ q_n & \text{if} \quad q_n \neq 0. \end{cases}$$

Because $r_n \neq q_n$ for at most finitely many $n \in N$, we have $\{r_n\}_{n=1}^{\infty} \cong \{q_n\}_{n=1}^{\infty}$. Hence, $[\{r_n\}_{n=1}^{\infty}] = [\{q_n\}_{n=1}^{\infty}]$. What have we accomplished by this? We now have a representative of the equivalence class $[\{q_n\}_{n=1}^{\infty}]$, namely $\{r_n\}_{n=1}^{\infty}$, that is nonzero for each $n \in N$. Therefore, the sequence $\{r_n^{-1}\}_{n=1}^{\infty}$ exists. In the next lemma, we show that $\{r_n^{-1}\}_{n=1}^{\infty}$ is a Cauchy sequence (i.e., that $\{r_n^{-1}\}_{n=1}^{\infty} \in S$).

Lemma 6.8   Suppose that $x \in \mathbf{R}$ with $x \neq 0$, and suppose $\{r_n\}_{n=1}^{\infty} \in x$ such that $r_n \neq 0$ for each $n \in N$. Then $\{r_n^{-1}\}_{n=1}^{\infty} \in S$.

Proof: (Note: Do not be perturbed by the statement $\{r_n\}_{n=1}^{\infty} \in x$. Because $x \in \mathbf{R}$, x is an equivalence class of Cauchy sequences.) Let $\varepsilon \in Q^+$. We want to show that there exists an $M \in N$ such that $|r_n^{-1} - r_m^{-1}| < \varepsilon$ whenever n, $m \in N$ with $n \geq M$ and $m \geq M$. Now, by Lemma 6.7, there exists a $B \in Q^+$ and an $M_1 \in N$ such that $|r_n| \geq B$ for all $n \in N$ with $n \geq M_1$. Also, because $\{r_n\}_{n=1}^{\infty}$ is a Cauchy sequence, there exists an $M_2 \in N$ such that $|r_n - r_m| < B^2\varepsilon$ whenever m, $n \in N$ with $m \geq M_2$ and $n \geq M_2$. Let $M = \max\{M_1, M_2\}$. Then, for any m, $n \in N$ with $m \geq M$ and $m \geq M$, we have $|r_n^{-1} - r_m^{-1}| = |\frac{1}{r_n} - \frac{1}{r_m}| = |\frac{r_m - r_n}{r_n r_m}| = \frac{|r_m - r_n|}{|r_n| \cdot |r_m|} < \frac{B^2\varepsilon}{BB} = \varepsilon$. Hence, $\{r_n^{-1}\}_{n=1}^{\infty} \in S$.

We are ready to define the multiplicative inverse of an $x \in \mathbf{R}$ with $x \neq 0$; that is, for $x \in \mathbf{R}$ with $x \neq [\{0\}_{n=1}^{\infty}]$. Choose an $\{r_n\}_{n=1}^{\infty} \in x$ such that $r_n \neq 0$ for each $n \in N$. We have already discussed how to do this. We then define

x^{-1} to be $[\{r_n^{-1}\}_{n=1}^{\infty}]$, the equivalence class of $\{r_n^{-1}\}_{n=1}^{\infty}$. We know that $\{r_n^{-1}\}_{n=1}^{\infty}$ is a Cauchy sequence by Lemma 6.8.

However, we still have a slight problem. We have to show that this is well defined. In other words, suppose we have $\{s_n\}_{n=1}^{\infty} \cong \{r_n\}_{n=1}^{\infty}$, in which $s_n \neq 0$ for each $n \in N$. We must show that $\{s_n^{-1}\}_{n=1}^{\infty} \cong \{r_n^{-1}\}_{n=1}^{\infty}$. That is, we need to show that the definition of x^{-1} does not depend on which Cauchy sequence is chosen to represent x^{-1}.

Hence, our goal is to show that $\lim\limits_{n \to \infty} (s_n^{-1} - r_n^{-1}) = 0$. Let $\varepsilon \in Q^+$. We want to show that there exists an $M \in N$ such that $|s_n^{-1} - r_n^{-1}| < \varepsilon$ whenever $n \in N$ with $n \geq M$. By Lemma 6.7 there exists a $B_1 \in Q^+$ and an $M_1 \in N$ such that for each $n \in N$ with $n \geq M_1$, we have $|s_n^{-1}| \geq B_1$. Also by Lemma 6.7 there exists a $B_2 \in Q^+$ and an $M_2 \in N$ such that for each $n \in N$ with $n \geq M_2$, we have $|r_n^{-1}| \geq B_2$. Now, let $B = \mathbf{min}\{B_1, B_2\}$, and let $M_3 = \mathbf{max}\{M_1, M_2\}$. Then for any $n \in N$ with $n \geq M_3$, we have $|s_n^{-1}| \geq B$ and $|r_n^{-1}| \geq B$. (Why did we take the **min** of the B's and the **max** of the M's?)

Because $\{s_n\}_{n=1}^{\infty} \cong \{r_n\}_{n=1}^{\infty}$, there exists an $M_4 \in N$ such that for each $n \in N$ with $n \geq M_4$, we have $|s_n - r_n| < B^2 \varepsilon$. Now, let $M = \mathbf{max}\{M_3, M_4\}$. Then, for each $n \in N$ with $n \geq M$, we have $|s_n^{-1} - r_n^{-1}| = |\frac{1}{s} - \frac{1}{r_n}| = |\frac{s_n - r_n}{s_n s_n}| = \frac{|s_n - r_n|}{|s_n| \cdot |r_n|} < \frac{B_2 \varepsilon}{BB} = \varepsilon$. Hence, we have $\lim\limits_{n \to \infty} (s_n^{-1} - r_n^{-1}) = 0$.

We now ask the reader to fill in all of the missing pieces to show that **R** with the above-defined binary operations forms a commutative ring. To complete the demonstration of the existence of a complete ordered field, we must now define an order on **R** and then show that this ordered field is a complete ordered field. We leave the remaining details to Appendix C. In the meantime, we record the result that a complete ordered field exists as a theorem because it is so important.

Theorem 6.12 $\qquad$ A complete ordered field exists.

Exercises

1. Show that $\lim\limits_{n \to \infty} r = r$ for each $r \in Q$.

2. Show that $\lim\limits_{n \to \infty} \frac{1}{n} = 0$.

3. Prove that $\{n\}_{n=1}^{\infty}$ is not a Cauchy sequence.

4. Prove Lemma 6.5.

5. Prove that the relation $\cong$ on S is reflexive and symmetric.

6. Complete the proof that the sum defined on **R** is well defined.

7. Fill in the remaining pieces to show that **R** is a commutative ring.

Section 6.4   Initial Properties of Real Numbers

In this section some additional properties of real numbers are derived. This section contains additional information about the real numbers that every student should know before taking an analysis course. You are likely to find this section much easier than Section 6.3 (assuming that you read Section 6.3). If you did not read Section 6.3, then just take it on faith that a complete ordered field exists. Resolve to read that section someday. For the remainder of this text, we assume that we have chosen a complete ordered field, $\mathbf{R}$. We shall refer to $\mathbf{R}$ as the set of real numbers. Similarly, we shall refer to the set $\mathbf{Q}$ as the set of rational numbers, the set $\mathbf{Z}$ as the set of integers, and the set $\mathbf{N}$ as the set of positive integers. In the first part of this section, we state some identities that are useful and important to know. In the second part we define what is meant by raising a real number to a rational exponent, and we discuss basic properties of this.

Some Basic Identities

We begin with an identity that we have already proved. Recall Example 2.33, in which we showed via induction that we have $1 + 2 + \cdots n = \frac{n(n + 1)}{2}$ for each $n \in \mathbf{N}$. Of course, we "proved" this result before the formal construction of the integers. The reader can take a quick glance at Example 2.33 to see that the argument there is complete now that we have the set $\mathbf{N}$ at our disposal.

Theorem 6.13 $\sum_{i=1}^{n} i = \frac{n(n + 1)}{2}$ for each $n \in \mathbf{N}$.

Suppose $a, r \in \mathbf{R}$. Recall that we defined positive integer exponents in Section 4.2. We now want to present an identity for the sum of terms of a **geometric progression**. A geometric progression is a sequence (finite or infinite) of the numbers, a, ar, ar^2, ar^3, and so on. Note that each term is obtained from the previous term by multiplying by r. The proof of the following theorem is left to the reader. It is not hard to prove it using mathematical induction.

Theorem 6.14   Suppose $a, r \in \mathbf{R}$ with $r \neq 1$. Then $a + ar + ar^2 + \cdots + ar^n = \frac{ar^{n+1} - a}{r - 1}$ for each $n \in \mathbf{N}$.

Now, if we define r^0 to be 1 for each $r \in \mathbf{R}$, then we can write the equation of the theorem as $\sum_{i=0}^{n} ar^i = \frac{ar^{n+1} - a}{r - 1}$.

Example 6.7

Note that for $a = 1$ and $r = 2$, Theorem 6.14 says that $1 + 2 + 2^2 + \cdots + 2^n = 2^{n+1} - 1$ for each $n \in \mathbf{N}$. In other words, the sum of consecutive nonnegative powers of two is one less than the next power of two. Of course, this can also be written as $\sum_{i=0}^{n} 2^i = 2^{n+1} - 1$ for each $n \in \mathbf{N}$.

Next, we begin moving in the direction of proving an important result called the **binomial theorem**. This theorem indicates how to express $(a + b)^n$ in terms of powers of a and b, in which $a, b \in \mathbf{R}$ and $n \in \mathbf{N}$. You have probably seen it before. We begin the discussion by talking about factorials. Recall from Definition 4.6 that $0!$ is defined to be 1 and that $n!$ is defined to be $\prod_{k=1}^{n} k$ for each $n \in \mathbf{N}$. In other words, $n! = 1 \cdot 2 \cdot \cdots \cdot (n-1) \cdot n$ for each $n \in \mathbf{N}$. The reader should have no trouble seeing that $(n + 1)! = (n + 1) n!$ for each $n \in \mathbf{N} \cup \{0\}$. The binomial theorem will be much easier to state after **binomial coefficients** are defined below.

Definition 6.8

Suppose n and k are nonnegative integers with $k \leq n$. The **binomial coefficient** $\binom{n}{k}$ is defined by $\binom{n}{k} = \frac{n!}{k!(n-k)!}$.

Were we just to use Definition 6.8 to calculate binomial coefficients, the calculations could very quickly become quite cumbersome. For example, consider $\binom{8}{3}$. This is defined to be $\frac{8!}{3! \cdot 5!}$. As can be seen, the numerator will be quite large. Fortunately, there is a shortcut in calculating $\binom{n}{k}$. Note that we can write this as $\binom{n}{k} = \frac{n!}{k!(n-k)!} = \frac{1 \cdot 2 \cdot 3 \cdot \cdots \cdot (n-k) \cdot (n-(k-1)) \cdot \cdots \cdot (n-1) \cdot n}{k! \cdot (1 \cdot 2 \cdot 3 \cdot \cdots \cdot (n-k))}$.

Note that the quantity $1 \cdot 2 \cdot \cdots \cdot (n-k)$ can be cancelled from both the numerator and the denominator. We are left with $\binom{n}{k} = \frac{(n-(k-1)) \cdot \cdots \cdot (n-1) \cdot n}{k! \cdot}$. Of course, this can be written as $\binom{n}{k} = \frac{n \cdot (n-1) \cdot \cdots \cdot (n-k+1)}{k! \cdot}$. This is important enough to be recorded as a theorem.

Theorem 6.15

Suppose n and k are nonnegative integers with $k \leq n$. Then $\binom{n}{k} = \frac{n \cdot (n-1) \cdot \cdots \cdot (n-k+1)}{k! \cdot}$.

What is the advantage of using Theorem 6.15? Note that the numerator has only k terms. Consider our example of $\binom{8}{3}$. In this case we have $n = 8$ and $k = 3$. Using Theorem 6.15, we can write it as $\frac{8 \cdot 7 \cdot 6}{3!} = \frac{336}{6} = 56$. We have been spared the necessity of calculating $8!$, and this is a big help. Theorem 6.16 next provides additional help with the calculation of $\binom{n}{k}$.

Theorem 6.16 Suppose n and k are nonnegative integers with $k \leq n$. Then $\binom{n}{n-k} = \binom{n}{k}$.

Proof: By definition, $\binom{n}{n-k} = \dfrac{n!}{(n-k)! \cdot (n-(n-k))!}$. Now, note that $(n-(n-k)) = k$. Hence, we have $\binom{n}{n-k} = \dfrac{n!}{(n-k)! \cdot (n-(n-k))!} = \dfrac{n!}{(n-k) \cdot k!} = \binom{n}{k}$.

Example 6.8 To calculate $\binom{8}{5}$, we could use Theorem 6.15 to get $\binom{8}{5} = \dfrac{8 \cdot 7 \cdot 6 \cdot 5 \cdot 4}{5!}$. However, if we use Theorem 6.16, we get $\binom{8}{5} = \binom{8}{8-5} = \binom{8}{3}$. We can now apply Theorem 6.15 to $\binom{8}{3}$ to get $\binom{8}{5} = \binom{8}{3} = \dfrac{8 \cdot 7 \cdot 6}{3!} = 56$. Note that it is easier to reduce the expression $\dfrac{8 \cdot 7 \cdot 6}{3!}$ than the expression $\dfrac{8 \cdot 7 \cdot 6 \cdot 5 \cdot 4}{5!}$.

The reader should have an easy time showing that for any nonnegative integer n, both $\binom{n}{0}$ and $\binom{n}{n}$ are equal to 1. Also, it is easy to see that both $\binom{n}{1}$ and $\binom{n}{n-1}$ are equal to n for each $n \in \mathbf{N}$. The next theorem expresses an important identity about binomial coefficients.

Theorem 6.17 Suppose $n, k \in \mathbf{N}$ with $k \leq n$. Then $\binom{n}{k} + \binom{n}{k-1} = \binom{n+1}{k}$.

Proof: By definition, $\binom{n}{k} + \binom{n}{k-1} = \dfrac{n!}{k!(n-k)!} + \dfrac{n!}{(k-1)!(n-k+1)!}$. We combine these terms by using a common denominator of $k!(n-k+1)!$. Doing this results in $\dfrac{n!(n-k+1)}{k!(n-k+1)!} + \dfrac{n!k}{k!(n-k+1)!}$. (We used the facts that $(n-k+1)! = (n-k+1)(n-k)!$ and $k! = k(k-1)!$). Simplifying gives us $\dfrac{n!(n-k+1)}{k!(n-k+1)!} + \dfrac{n!k}{k!(n-k+1)!} = \dfrac{n!(n+1)}{k!(n-k+1)} = \dfrac{(n+1)!}{k!(n-k+1)!} = \binom{n+1}{k}$.

Theorem 6.17 gives us a graphical way to calculate the binomial coefficients by drawing what is called **Pascal's triangle**. In this triangle, the upper tip is the row containing only the binomial coefficient $\binom{0}{0}$ (which is 1, of course). The next row contains the binomial coefficients $\binom{1}{0}$ and $\binom{1}{1}$. The next row contains the binomial coefficients $\binom{2}{0}$, $\binom{2}{1}$, and $\binom{2}{2}$. This process continues as long as you like.

Here is the "trick." We know that both nonhorizontal sides of the triangle have to have all 1's on them. This is because each element of the left side is of the form $\binom{n}{0}$, and each element on the right side is of the form $\binom{n}{n}$. Now consider the element $\binom{n+1}{k}$. Vertically, it is situated in the row below the row containing $\binom{n}{k-1}$ and $\binom{n}{k}$. Horizontally, $\binom{n+1}{k}$ is situated between the elements $\binom{n}{k-1}$ and $\binom{n}{k}$. Therefore, to find $\binom{n+1}{k}$, we simply add the element above it and to the left with the element above it and to the right, courtesy of Theorem 6.17. This is illustrated in Figure 6.2.

Figure 6.2
Pascal's triangle

```
         1
       1   1
     1   2   1
   1   3   3   1
 1   4   6   4   1
1  5  10  10  5  1
```

For example, consider the third entry of the sixth row. This is $\binom{5}{2}$. Suppose we had not yet filled this in, and we wanted to use the above row (which we had already filled in) to calculate it. We see from the figure that this entry is horizontally between the entries with values 4 and 6 in the previous row. These are the entries $\binom{4}{1}$ and $\binom{4}{2}$. Because Theorem 6.17 assures us that $\binom{5}{2} = \binom{4}{1} + \binom{4}{2}$, we simply add 4 and 6 to get 10.

We are now ready to state and prove the binomial theorem. For the statement and proof of the theorem, we invoke the convention that $a^0 = 1$ for each $a \in \mathbf{R}$. The theorem states that for each $x, y \in \mathbf{R}$ and each $n \in \mathbf{N}$, we have $(x + y)^n = \binom{n}{0}x^n + \binom{n}{1}x^{n-1}y + \binom{n}{2}x^{n-2}y^2 + \cdots + \binom{n}{n-1}xy^{n-1} + \binom{n}{n}y^n$. This can be written more compactly using summation notation as $(x + y)^n = \sum_{j=0}^{n}\binom{n}{j}x^{n-j}y^j$. The ability to expand $(x + y)^n$ compactly like this is one good reason for defining $0!$ to be 1.

Theorem 6.18

(Binomial Theorem)

Suppose $x, y \in \mathbf{R}$ and $n \in \mathbf{N}$. Then $(x + y)^n = \sum_{j=0}^{n}\binom{n}{j}x^{n-j}y^j$.

Proof: We perform a proof by induction. For 1, the theorem claims that $(x + y)^1 = \binom{1}{0}x^1y^0 + \binom{1}{1}x^0y^1$, which is clearly true. We now make the induction hypothesis and assume that the result holds for some $n \in \mathbf{N}$. That is, we assume that $(x + y)^n = \sum_{j=0}^{n}\binom{n}{j}x^{n-j}y^j$. The goal is to show that $(x + y)^{n+1} = \sum_{j=0}^{n+1}\binom{n+1}{j}x^{n+1-j}y^j$. Now, $(x + y)^{n+1} = (x + y)^n(x + y)$. Hence, by the induction hypothesis, we have $(x + y)^{n+1} = \left(\sum_{j=0}^{n}\binom{n}{j}x^{n-j}y^j\right)(x + y) = \sum_{j=0}^{n}\binom{n}{j}x^{n+1-j}y^j + \sum_{j=0}^{n}\binom{n}{j}x^{n-j}y^{j+1}$. Now, $\sum_{j=0}^{n}\binom{n}{j}x^{n+1-j}y^j$ can be written as $x^{n+1} + \sum_{j=1}^{n}\binom{n}{j}x^{n+1-j}y^j$. Also, $\sum_{j=0}^{n}\binom{n}{j}x^{n-j}y^{j+1}$ can be written as $\left(\sum_{j=0}^{n-1}\binom{n}{j}x^{n-j}y^{j+1}\right) + y^{n+1}$. This expression can be written as

$$\left(\sum_{j=1}^{n}\binom{n}{j-1}x^{n-(j-1)}y^{(j-1)+1}\right) + y^{n+1} = \left(\sum_{j=1}^{n}\binom{n}{j-1}x^{n+1-j}y^j\right) + y^{n+1}.$$

Putting these "pieces" together, we get

$$(x + y)^{n+1} = x^{n+1} + \left(\sum_{j=1}^{n}\binom{n}{j}x^{n+1-j}y^j\right) + \left(\sum_{j=1}^{n}\binom{n}{j-1}x^{n+1-j}y^j\right) + y^{n+1}$$

$$= x^{n+1} + \left(\sum_{j=1}^{n}\left(\binom{n}{j} + \binom{n}{j-1}\right)x^{n+1-j}y^j\right) + y^{n+1}.$$

Now, from Theorem 6.17, we have $\binom{n}{j} + \binom{n}{j-1} = \binom{n+1}{j}$. Thus, $(x+y)^{n+1} = x^{n+1} + \left(\sum_{j=1}^{n}\binom{n+1}{j}x^{n+1-j}y^{j}\right) + y^{n+1} = \sum_{j=0}^{n+1}\binom{n+1}{j}x^{n+1-j}y^{j}$.

It should be pointed out that there are many applications of the binomial theorem. For example, there are applications to physics. We shall make use of the binomial theorem in the proof of Theorem 6.20, where we prove the existence of nth roots.

Rational Exponents

In this subsection we define what is meant by x^r, in which $x \in \mathbf{R}$ with $x > 0$ and $r \in \mathbf{Q}$. We begin with numbers of the form x^n, in which $n \in \mathbf{N}$. We have already defined positive integer exponents (in Definition 4.4) for general binary operations, and we have made extensive use of them (e.g., they were used in the statement and proof of the binomial theorem). Also, by virtue of Theorems 4.4, 4.5, and 4.6, we have already established the following lemma, simply because multiplication of elements of $\mathbf{R}$ is associative and commutative.

Lemma 6.9

1) $x^m x^n = x^{m+n}$ for all $x \in \mathbf{R}$ and $m, n \in \mathbf{N}$.
2) $(xy)^n = x^n y^n$ for all $x, y \in \mathbf{R}$ and $n \in \mathbf{N}$.
3) $(x^m)^n = x^{mn}$ for all $x \in \mathbf{R}$ and $m, n \in \mathbf{N}$.

Next, we want to extend the definition to include exponents of negative integers and an exponent of 0. Note that we have already defined what is meant by x^{-1} for any nonzero x for any field: x^{-1} is the multiplicative inverse of x. (See Definition 5.1 and the comments following it.) By virtue of Definition 5.2, we can also write x^{-1} as $\frac{1}{x}$.

Definition 6.9

Suppose $x \in \mathbf{R}$ with $x \neq 0$. Then x^0 is defined to be 1. For any negative integer n with $n \leq -2$, x^n is defined to be $\frac{1}{x^{-n}}$. That is, $x^n = (x^{-n})^{-1}$.

Note that 0^0 has not been defined by Definition 6.9. Depending on the situation, 0^0 is sometimes defined as 1. For an example of this, consider the binomial theorem applied to $(1 + 0)^2$. Using the summation formula provided by the theorem statement, we get $(1 + 0)^2 = \binom{2}{0} \cdot 1^2 \cdot 0^0 + \binom{2}{1} \cdot 1^1 \cdot 0^1 + \binom{2}{2} \cdot 1^0 \cdot 0^2$.

In this case the success of the formula depends on having $0^0 = 1$. In other situations of advanced mathematics 0^0 is regarded as undefined. We briefly discuss this further in Chapter 7. If n is a negative integer, then 0^n is undefined.

The "standard" laws of integer exponents are presented in Theorem 6.19. The proofs of most of them are left to the reader. To assist the reader with this task, we present some additional lemmas.

Lemma 6.10

Suppose $x \in \mathbf{R}$ with $x \neq 0$, and suppose m and n are negative integers. Then $x^m x^n = x^{m+n}$.

Proof: Note that $-m, -n \in \mathbf{N}$.

$$x^{m+n} = \frac{1}{x^{(-m+-n)}} \qquad \text{(by Definition 6.9)}$$

$$= \frac{1}{x^{-m}x^{-n}} \qquad \text{(by Part 1 of Lemma 6.9)}$$

$$= \left(\frac{1}{x^{-m}}\right)\left(\frac{1}{x^{-n}}\right) \quad \text{(by Part 2 of Theorem 5.5)}$$

$$= x^m x^n. \qquad \text{(by Definition 6.9)}$$

Lemma 6.11

Suppose $x \in \mathbf{R}$ with $x \neq 0$, and suppose $m, n \in \mathbf{N}$. Then $\dfrac{x^m}{x^n} = x^{m-n}$.

Proof:

Case 1: $m = n$. Then $\dfrac{x^m}{x^n} = \dfrac{x^m}{x^m} = 1 = x^0 = x^{m-n}$.

Case 2: $m > n$. Then $\dfrac{x^m}{x^n} = \dfrac{x^{((m-n)+n)}}{x^n} = \dfrac{x^{m-n}x^n}{x^n} = x^{m-n}$.

Case 3: $n > m$. Then $\dfrac{x^m}{x^n} = \dfrac{x^m}{x^{((n-m)+m)}} = \dfrac{x^m}{x^{n-m}x^m} = \dfrac{1}{x^{n-m}} = x^{m-n}$.

Lemma 6.12

Suppose $x \in \mathbf{R}$ with $x \neq 0$, and suppose $m \in \mathbf{N}$ and $-n \in \mathbf{N}$. Then $x^m x^n = x^{m+n}$.

Proof: $x^m x^n = \dfrac{x^m}{x^{-n}} = x^{(m-(-n))} = x^{m+n}$. (We made use of Lemma 6.11.)

We now present the major laws of integer exponents (Theorem 6.19). Only the proof of Part 1 is provided. The student is asked to prove the rest.

Theorem 6.19   Suppose $x, y \in \mathbf{R}$ with $x \neq 0$ and $y \neq 0$, and $m, n \in \mathbf{Z}$. Then

1) $x^m x^n = x^{m+n}$.

2) $\dfrac{x^m}{x^n} = x^{m-n}$.

3) $x^m = \dfrac{1}{x^{-m}}$.

4) $(xy)^m = x^m x^m$.

5) $(x^m)^n = x^{mn}$.

6) $\left(\dfrac{x}{y}\right)^m = \dfrac{x^m}{y^m}$.

7) If $0 < x < y$ and $m \in \mathbf{N}$, then $x^m < y^m$.

8) If $x > 1$ and $m < n$, then $x^m < x^n$.

Proof of Part 1: If $m = 0$ or $n = 0$, the result is obvious. If m and n are both positive, apply Part 1 of Lemma 6.9. If m and n are both negative, apply Lemma 6.10. If one of these numbers is positive and the other is negative, apply Lemma 6.12.

Now, suppose $a \in \mathbf{R}$ with $a \geq 0$. We want to define $a^{\frac{m}{n}}$, where $m, n \in \mathbf{Z}$. Our method is to first define $a^{\frac{1}{n}}$, in which $n \in \mathbf{N}$. To do so, we must prove that there exists a nonnegative real number b such that $b^n = a$. We can then define b to be $a^{\frac{1}{n}}$. That is, we want to show that the nth root of a exists. (We have already shown this in the case in which $a = 2$ and $n = 2$; i.e., $2^{\frac{1}{2}}$. See Theorem 6.7.) The basic idea of the proof is to show that the number $\sup\{x \in \mathbf{R} \mid x \geq 0 \text{ and } x^n \leq a\}$ exists and "works."

Theorem 6.20   Suppose $a \in \mathbf{R}$ with $a \geq 0$, and suppose $n \in \mathbf{N}$. Then there exists a $b \in \mathbf{R}$ with $b \geq 0$ such that $b^n = a$.

Proof: Define the set X by $X = \{x \in \mathbf{R} \mid x \geq 0 \text{ and } x^n \leq a\}$. X is nonempty because $0 \in X$. Our first goal is to show that X is bounded above by $1 + a$. Using the binomial theorem, note that

$$(1 + a)^n = \sum_{k=0}^{n} \binom{n}{k} 1^{n-k} a^k = \sum_{k=0}^{n} \binom{n}{k} a^k \geq na. \tag{1}$$

(The last inequality was obtained by plugging in 1 for k and recalling that $\binom{n}{1} = n$.) Now, assume that X is not bounded above by $1 + a$ (to reach a contradiction). Then there must exist an $x \in X$ with $x > 1 + a$. Note that this implies that $x^n > (a + 1)^n$ (by Part 7 of Theorem 6.19). Also, by the definition of X, we have $x^n \leq a$. Hence, using Inequality 1, we see that $a \geq x^n > (1 + a)^n \geq na$. That is, $a > na$. Because this is impossible, we have our contradiction. Hence, X is bounded above. Thus, X has a least upper bound, which we will denote by b. That is, $b = \sup X$. We will now show that it is not possible to have $b^n > a$. We ask the reader to show that it is not possible to have $b^n < a$. Therefore, the only possible conclusion is that $b^n = a$.

Assume (to reach a contradiction) that $b^n > a$, and let $\delta = b^n - a$. Then $\delta > 0$. By Theorem 6.5, for each $k \in \{1, 2, \ldots, n\}$ there exists an $m_k \in \mathbf{N}$ such that $\frac{1}{m^k} < \frac{\delta}{nb^{n-k}\binom{n}{k}}$. Now, it is clear that $\frac{1}{(m_k)^k} \leq \frac{1}{m^k}$ for each $k \in$

$\{1, 2, \ldots, n\}$. Hence, $\frac{1}{(m_k)^k} < \frac{\delta}{nb^{n-k}\binom{n}{k}}$. We can rewrite this as $\frac{\binom{n}{k}b^{n-k}}{(m_k)^k} < \frac{\delta}{n}$.

Now, let $m = \mathbf{max}\{m_1, m_2, \ldots, m_n\}$. Then, for each $k \in \{1, 2, \ldots, n\}$, we have

$$\frac{\binom{n}{k}b^{n-k}}{(m^k)} < \frac{\delta}{n}. \tag{2}$$

Multiplying Inequality 2 by -1 results in

$$\binom{n}{k}b^{n-k}\frac{(-1)}{m^k} > \frac{-\delta}{n}. \tag{3}$$

It is clear that for each $k \in \{1, 2, \ldots, n\}$ we have

$$\binom{n}{k}b^{n-k}\left(\frac{-1}{m}\right)^k \geq \binom{n}{k}b^{n-k}\frac{(-1)}{m^k}. \tag{4}$$

Hence, from Inequality 3 and Inequality 4, for each $k \in \{1, 2, \ldots, n\}$ we have

$$\binom{n}{k}b^{n-k}\left(\frac{-1}{m}\right)^k > \frac{-\delta}{n}. \tag{5}$$

Hence, we have

$$\sum_{k=1}^{n}\binom{n}{k}b^{n-k}\left(\frac{-1}{m}\right)^k > \sum_{k=1}^{n}\left(\frac{-\delta}{n}\right). \tag{6}$$

However, notice that the right-hand side of Inequality 6 is simply $-\delta$. By the binomial theorem, the left-hand side of Inequality 6 is $(b - \frac{1}{m})^n - b^n$. Hence, we have $(b - \frac{1}{m})^n - b^n > -\delta$. That is,

$$\left(b - \frac{1}{m}\right)^n > b^n - \delta. \tag{7}$$

Now, recall that we defined δ as $\delta = b^n - a$. Hence, Inequality 7 tells us that

$$\left(b - \frac{1}{m}\right)^n > a. \tag{8}$$

This immediately tells us that $b - \frac{1}{m}$ is not an upper bound of X. Hence, there exists an $x \in X$ with

$$b - \frac{1}{m} < x. \tag{9}$$

We now need to show that $b - \frac{1}{m} > 0$. Suppose (to reach a contradiction) that $b - \frac{1}{m} \leq 0$. Then $b \leq \frac{1}{m}$. Because $b \geq 0$, we have $b^n \leq \frac{1}{m^n}$ (using Part 7 of Theorem 6.19). Hence,

$$nb^n \leq \frac{n}{m^n}. \tag{10}$$

Now, by plugging in $k = n$ into Inequality 2 and using $\delta = b^n - a$, we get

$$\frac{1}{m^n} < \frac{b^n - a}{n}. \tag{11}$$

From Inequality 10 and Inequality 11, we have $nb^n \leq \frac{n}{m^n} < b^n - a$. Hence, $nb^n + a < b^n$. However, because $a \geq 0$ and $b \geq 0$, this is clearly not possible, and we have a contradiction. Thus, we have $b - \frac{1}{m} > 0$. Hence, from Inequality 9 and Part 7 of Theorem 6.19, we get $(b - \frac{1}{m})^n < x^n$. However, recall that $x \in X$. Hence, $x^n \leq a$. Thus, $(b - \frac{1}{m})^n < a$. Because this contradicts Inequality 8, we arrived at our contradiction that it is not possible to have $b^n > a$.

Now, suppose $a \geq 0$ and $n \in \mathbf{N}$. Theorem 6.20 does not rule out the possibility that there will be more than one number b such that $b^n = a$; it merely guarantees the existence of one such nonnegative b. Indeed, for example, the reader is well aware that there are two numbers whose square is 4, namely 2 and −2. In fact, it is clear that if n is even, there will be at least two such numbers whose nth power is a: the number b of the proof of Theorem 6.20 and the number −b. The next theorem guarantees that there is exactly one *nonnegative* number whose nth power is a. The easy proof is left to the reader.

Theorem 6.21 Suppose $a \in \mathbf{R}$ with $a \geq 0$, and suppose $n \in \mathbf{N}$. Then there exists a unique nonnegative $b \in \mathbf{R}$ such that $b^n = a$.

Definition 6.10

Suppose $a \in \mathbf{R}$ with $a \geq 0$, and suppose $n \in \mathbf{N}$. Then we define $a^{\frac{1}{n}}$ to be that unique nonnegative real number b such that $b^n = a$. We define $a^{-\frac{1}{n}}$ to be $(a^{\frac{1}{n}})^{-1}$.

Now, suppose $x > 0$ and $q \in \mathbf{Q}$. We know that there exist integers m and n with $n \neq 0$ such that $q = \frac{m}{n}$. It is very tempting to just go ahead and define x^q to be $(x^{\frac{1}{n}})^m$. In fact, we shall end up doing this. But the experienced reader will probably see the potential problem. We have that annoying problem of well-definedness. What if we represent q as a different quotient of integers? The following lemmas will give us what we need. The proof of the first one is left to the reader.

Lemma 6.13

If $x \in \mathbf{R}$ with $x > 0$ and $n \in \mathbf{Z}$ with $n \neq 0$, then

1) $(x^{\frac{1}{n}})^n = x$
2) $(x^n)^{\frac{1}{n}} = x$

Lemma 6.14

Suppose $x \in \mathbf{R}$ with $x > 0$ and suppose $m, n, r, s \in \mathbf{Z}$ with $n \neq 0$ and $s \neq 0$. Suppose that $\frac{m}{n} = \frac{r}{s}$. Then $(x^{\frac{1}{n}})^m = (x^{\frac{1}{s}})^r$.

Proof: If $m = 0$ or $r = 0$, the desired result reduces to $1 = 1$, which is clearly true. For the remainder of the proof, we consider the case in which $m \neq 0$ and $r \neq 0$. Now, we have $ms = nr$. Hence,

$$((x^{\frac{1}{n}})^m)^s = (x^{\frac{1}{n}})^{ms} = (x^{\frac{1}{n}})^{nr} = ((x^{\frac{1}{n}})^n)^r = x^r.$$

(The last equality follows by Part 1 of Lemma 6.13.) Hence, by raising both sides of the above equation to the $\frac{1}{s}$ power and invoking Part 2 of Lemma 6.13, we get

$$(x^{\frac{1}{n}})^m = (x^r)^{\frac{1}{s}}.$$

Hence, to complete the proof, it suffices to show that $(x^r)^{\frac{1}{s}} = (x^{\frac{1}{s}})^r$.

Now, $((x^r)^{\frac{1}{s}})^s = x^r$ (by Part 1 of Lemma 6.13). Also,

$$((x^{\frac{1}{s}})^r)^s = (x^{\frac{1}{s}})^{rs} = (x^{\frac{1}{s}})^{sr} = ((x^{\frac{1}{s}})^s)^r = x^r.$$

(The last equality used Part 1 of Lemma 6.13.)

Thus, we equate these two ways of expressing x^r to get $((x^r)^{\frac{1}{s}})^s = ((x^{\frac{1}{s}})^r)^s$. The desired result now comes from raising both sides to the $\frac{1}{s}$ power and invoking Part 2 of Lemma 6.13.

Consequently, Definition 6.11 now makes complete sense.

Definition 6.11

Suppose $x \in \mathbf{R}$ with $x > 0$. Suppose $q = \frac{m}{n}$, in which $m, n \in \mathbf{Z}$ where $n \neq 0$. Then we define x^q to be $(x^{\frac{1}{n}})^m$.

We now present the laws of exponents for rational powers. All of the proofs are left to the reader.

Theorem 6.22

Suppose $x, y \in \mathbf{R}$ with $x > 0$ and $y > 0$, and $q, r \in \mathbf{Q}$. Then

1) $x^q x^r = x^{q+r}$.
2) $\frac{x^q}{x^r} = x^{q-r}$.

3) $x^q = \dfrac{1}{x^{-q}}$.

4) $(xy)^q = x^q y^q$.

5) $(x^q)^r = x^{qr}$.

6) $\left(\dfrac{x}{y}\right)^q = \dfrac{x^q}{y^q}$.

7) If $0 < x < y$ and $q > 0$, then $x^q < y^q$.

8) If $x > 1$ and $q < r$, then $x^q < x^r$.

Finally, we briefly mention the situation in which $x \in \mathbf{R}$ with $x < 0$. If $n \in$ **Z** with n odd, it is clear how we should define $x^{\frac{1}{n}}$. For example, -27 raised to the $\dfrac{1}{3}$ power is -3. If n is even, then $x^{\frac{1}{n}}$ is not defined within the real number system. Consider a $q \in \mathbf{Q}$. There exists an $m \in \mathbf{Z}$ and an $n \in \mathbf{N}$ with $\gcd(m, n) = 1$ such that $q = \dfrac{m}{n}$. (The reader should be able to prove that this can be done uniquely.) If n is odd, then we can define x^q to be $\left(x^{\frac{1}{n}}\right)^m$. Otherwise, we do not define x^q. The details are left to the reader as a class project.

EXERCISES

1. Prove Theorem 6.14.

2. Find the sum $1 - 4 + 4^2 - 4^3 + \cdots + 4^{50} - 4^{51}$.

3. Show that for any nonnegative integer n, both $\binom{n}{0}$ and $\binom{n}{n}$ are equal to 1.

4. Show that that both $\binom{n}{1}$ and $\binom{n}{n-1}$ are equal to n, for each $n \in \mathbf{N}$.

5. Find the binomial coefficients $\binom{7}{4}$, $\binom{7}{5}$, and $\binom{8}{5}$, and verify that indeed $\binom{8}{5} = \binom{7}{4} + \binom{7}{5}$.

6. Show that $\sum_{j=0}^{n}\binom{n}{j} = 2^n$ for each $n \in \mathbf{N}$. (Hint: $1 + 1 = 2$.) Note that each such sum is the sum of all of the elements of a row of Pascal's triangle.

7. Show that $\sum_{j=0}^{n}(-1)^j\binom{n}{j} = 0$ for each $n \in \mathbf{N}$. (Hint: $1 + -1 = 0$.)

8. Prove Parts 2 though 8 of Theorem 6.19.

9. Complete the proof of Theorem 6.20 by showing that it is not possible to have $b^n < a$.

10. Prove Theorem 6.21. (Hint: Use Part 7 of Theorem 6.19.)

11. Suppose $a \in \mathbf{R}$ with $a > 0$ and suppose that n is an even positive integer. Show that there are exactly two real numbers c that satisfy the property that $c^n = a$.

12. Prove Lemma 6.13.

13. Prove Theorem 6.22.

14. **CLASS PROJECT**! Fill in all the details in the definition of x^q in the case in which $x < 0$. Which parts of Theorem 6.22 hold if we remove the restriction that $x > 0$ and $y > 0$?

Chapter 7

Some Additional Properties of Real Numbers

ongratulations! You should now be well prepared to take advanced mathematics courses. You should have a deep understanding of how mathematical arguments work, as well as how the pieces of the real number system fit together.

The purpose of this chapter is to discuss some topics that the student should know about but are often not presented in mathematics courses. Specifically, we will talk about base b expansions of real numbers and real exponents. We will not provide complete details, but we will provide the student with enough knowledge to "fill in all the holes." There is one famous theorem presented that you will see in your analysis course; this is Cantor's "diagonalization" argument that the set of real numbers is not countable.

To accomplish these two goals, we need to familiarize the student with some material on convergent real sequences and series. A base b expansion of a real number is actually an infinite series, and raising a number to a real exponent actually involves taking the limit of a sequence. Our purpose here is not to be comprehensive with respect to convergent real sequences and series; the student will find a comprehensive discussion of those topics in an analysis course. Rather, our focus is on providing the necessary information on these subjects in order to have a meaningful discussion of the base b expansions and real exponents.

Section 7.1   Convergent Real Sequences I

From now on, if we make a statement such as $a \geq 0$, unless otherwise stated, we are assuming that a is a real number. Also, by the term **real sequence** we mean an infinite sequence $\{a_n\}_{n=1}^{\infty}$ in which $a_n \in \mathbf{R}$ for each $n \in \mathbf{N}$. Now that we have the bolded letter $\mathbf{N}$ at our disposal to represent the set of positive integers, we often use the unbolded capital letter N to be a positive integer (as is done in many analysis texts.) It should be mentioned that we largely imitate the methods of proof found in Johnsonbaugh et al. (2002) for this and the following two sections.

Definition 7.1

Suppose that $\{a_n\}_{n=1}^{\infty}$ is a real sequence and suppose $L \in \mathbf{R}$. Then $\{a_n\}_{n=1}^{\infty}$ is said to **converge** to L provided that for each $\varepsilon > 0$, there exists an $N \in \mathbf{N}$ such that $|a_n - L| < \varepsilon$ whenever $n \in \mathbf{N}$ with $n \geq N$. We can also express this by saying that L is the **limit** of the sequence $\{a_n\}_{n=1}^{\infty}$. This is also expressed by writing $\lim_{n \to \infty} a_n = L$. If $\{a_n\}_{n=1}^{\infty}$ converges to some $L \in \mathbf{R}$, we say that $\{a_n\}_{n=1}^{\infty}$ is **convergent**. If $\{a_n\}_{n=1}^{\infty}$ is not convergent it is said to **diverge** and to be divergent.

Now, do not get "hung up" on notation. There is nothing sacred about using the letter ε. There is also nothing sacred about the letter n. For example, we could also write $\lim_{k \to \infty} a_k = L$. Intuitively, if $\lim_{n \to \infty} a_n = L$, this means that for any positive number that you specify, if you go far enough "out" in the sequence, the distance between the terms of the sequence and L will forevermore be less than your specified positive number.

Example 7.1

Consider the sequence $\{\frac{1}{n}\}_{n=1}^{\infty}$. We will show that $\lim_{n \to \infty} \frac{1}{n} = 0$. Let $\varepsilon > 0$. By Theorem 6.5, there exists an integer $N \in \mathbf{N}$ with $\frac{1}{N} < \varepsilon$. Hence, for any $n \in \mathbf{N}$ with $n \geq N$, we have $\frac{1}{n} \leq \frac{1}{N} < \varepsilon$. Hence, $|\frac{1}{n} - 0| < \varepsilon$ for each $n \in \mathbf{N}$ such that $n \geq N$.

Example 7.2

We will show that the sequence $\{n^2\}_{n=1}^{\infty}$ does not have a limit (i.e., is not convergent). Suppose (to reach a contradiction) that there exists some

$L \in \mathbf{R}$ such that $\lim_{n\to\infty} n^2 = L$. Applying the definition of limit with $\varepsilon = 1$, there must exist an $N \in \mathbf{N}$ such that $|n^2 - L| < 1$ for each $n \in \mathbf{N}$ such that $n \geq N$. From Theorem 5.12 we have $-1 < n^2 - L < 1$. That is, $n^2 < L + 1$ for each $n \in \mathbf{N}$ such that $n \geq N$. However, by the Principle of Archimedes (Theorem 6.4), there exists an $M \in \mathbf{N}$ such that $M > L + 1$. Hence, we also have $M^2 \geq M \geq L + 1$. Note that for any $n \in \mathbf{N}$ with $n \geq M$ we have $n^2 \geq M^2 \geq L + 1$. Now let $K = \mathbf{max}\{M, N\}$. Then we have both $K^2 < L + 1$ and $K^2 \geq L + 1$, which is a contradiction. ∎

Theorem 7.1 tells us that if the real sequence $\{a_n\}_{n=1}^{\infty}$ has a limit, this limit is unique.

| **Theorem 7.1** | Suppose that $\{a_n\}_{n=1}^{\infty}$ is a real sequence with limits L_1 and L_2. Then $L_1 = L_2$. |

Proof: Suppose (to reach a contradiction) that $L_2 \neq L_1$. Let $\varepsilon = \dfrac{|L_2 - L_1|}{2}$. Note that $\varepsilon > 0$. Now, because $\{a_n\}_{n=1}^{\infty}$ has limit L_1, there exists an $N_1 \in \mathbf{N}$ such that $|a_n - L_1| < \varepsilon$ whenever $n \in \mathbf{N}$ with $n \geq N_1$. Also, because $\{a_n\}_{n=1}^{\infty}$ has limit L_2, there exists an $N_2 \in \mathbf{N}$ such that $|a_n - L_2| < \varepsilon$ whenever $n \in \mathbf{N}$ with $n \geq N_2$. Now, let $N = \mathbf{max}\{N_1, N_2\}$. Then, whenever $n \in \mathbf{N}$ with $n \geq N$ we have both $|a_n - L_1| < \varepsilon$ and $|a_n - L_2| < \varepsilon$. Hence, for any $n \in \mathbf{N}$ with $n \geq N$, we have (using the triangle inequality) $|L_2 - L_1| \leq |L_2 - a_n| + |a_n - L_1| = |a_n - L_2| + |a_n - L_1| < \varepsilon + \varepsilon = 2\varepsilon$. Now, recalling that $\varepsilon = \dfrac{|L_2 - L_1|}{2}$, the previous statement says that $|L_2 - L_1| < |L_2 - L_1|$. Because this is a contradiction, we have $L_2 = L_1$.

The reader is asked to prove the next theorem. It simply says that if the terms of two real sequences differ for at most a finite number of terms, and if one converges to some $L \in \mathbf{R}$, then so does the other one.

| **Theorem 7.2** | Suppose that $\{a_n\}_{n=1}^{\infty}$ and $\{b_n\}_{n=1}^{\infty}$ are real sequences such that there is an $N \in \mathbf{N}$ such that $a_n = b_n$ for each $n \in \mathbf{N}$ with $n \geq N$. If $\lim_{n\to\infty} a_n = L$, then $\lim_{n\to\infty} b_n = L$. |

Next, we consider how to find the limits of real sequences that are combinations of real sequences. If $c \in \mathbf{R}$, we can form the sequence $\{c\}_{n=1}^{\infty}$. This is a **constant sequence**, in which every term is simply c. Now, if we also have the real sequence $\{a_n\}_{n=1}^{\infty}$, we can form the real sequence $\{ca_n\}_{n=1}^{\infty}$, which is obtained simply by multiplying each term of $\{a_n\}_{n=1}^{\infty}$ by c.

Theorem 7.3

Suppose that $c \in \mathbf{R}$ and suppose $\{a_n\}_{n=1}^{\infty}$ is a real sequence with limit L. Then

1) $\lim_{n \to \infty} c = c$.
2) $\lim_{n \to \infty} ca_n = cL$. That is, $\lim_{n \to \infty} ca_n = c \cdot (\lim_{n \to \infty} a_n)$.

Proof: The proof of Part 1 is left to the reader. We prove Part 2. If $c = 0$, then $\{ca_n\}_{n=1}^{\infty}$ is the constant sequence $\{0\}_{n=1}^{\infty}$, so $\lim_{n \to \infty} ca_n = 0 = 0 \cdot L$. For the remainder of the proof, consider the case in which $c \neq 0$. Let $\varepsilon > 0$. Note that $\frac{\varepsilon}{|c|} > 0$. Because $\lim_{n \to \infty} a_n = L$, there exists an $N \in \mathbf{N}$ such that $|a_n - L| < \frac{\varepsilon}{|c|}$ for each $n \in \mathbf{N}$ with $n \geq N$. Hence, for any such n, we have $|ca_n - cL| = |c| \cdot |a_n - L| < |c|\frac{\varepsilon}{|c|} = \varepsilon$. The result follows.

Example 7.3

Consider the sequence $\{\frac{-8}{n}\}_{n=1}^{\infty}$. It can be written as $\{(-8) \cdot \frac{1}{n}\}_{n=1}^{\infty}$. We know that $\lim_{n \to \infty} \frac{1}{n} = 0$ from Example 7.1. Hence, from Part 2 of Theorem 7.3 we have $\lim_{n \to \infty} \frac{-8}{n} = (-8) \cdot \lim_{n \to \infty} \frac{1}{n} = (-8) \cdot 0 = 0$.

Now, if we have real sequences $\{a_n\}_{n=1}^{\infty}$ and $\{b_n\}_{n=1}^{\infty}$, we can form the sequences $\{a_n + b_n\}_{n=1}^{\infty}$ and $\{a_n - b_n\}_{n=1}^{\infty}$.

Theorem 7.4

Suppose that $\{a_n\}_{n=1}^{\infty}$ is a real sequence with limit L, and $\{b_n\}_{n=1}^{\infty}$ is a real sequence with limit M. Then

1) $\lim_{n \to \infty} (a_n + b_n) = L + M$.
2) $\lim_{n \to \infty} (a_n - b_n) = L - M$.

Proof: We prove Part 1 and leave Part 2 to the reader. Let $\varepsilon > 0$. Because $\lim_{n \to \infty} a_n = L$, there exists an $N_1 \in \mathbf{N}$ such that for each $n \in \mathbf{N}$ with $n \geq N_1$ we have $|a_n - L| < \frac{\varepsilon}{2}$. Also, because $\lim_{n \to \infty} a_n = M$, there exists an $N_2 \in \mathbf{N}$ such that for each $n \in \mathbf{N}$ with $n \geq N_2$ we have $|b_n - M| < \frac{\varepsilon}{2}$. Let $N = \max\{N_1, N_2\}$. Then for each $n \in \mathbf{N}$ with $n \geq N$, we have $|(a_n + b_n) - (L + M)| = |(a_n - L) + (b_n - M)| \leq |a_n - L| + |b_n - M| < \frac{\varepsilon}{2} + \frac{\varepsilon}{2} = \varepsilon$.

It is important to note that before Theorem 7.4 may be used, it must be established that the sequences $\{a_n\}_{n=1}^{\infty}$ and $\{b_n\}_{n=1}^{\infty}$ are convergent.

Example 7.4

Consider the sequence $\{\frac{n^2 - 6}{5n^2}\}_{n=1}^{\infty}$. Note that $\frac{n^2 - 6}{5n^2} = \frac{1}{5} - \frac{6}{5n^2}$. It is easily seen that the sequence $\{\frac{1}{5}\}_{n=1}^{\infty}$ is convergent with limit $\frac{1}{5}$. It is not hard to show that the sequence $\{\frac{6}{5n^2}\}_{n=1}^{\infty}$ has limit 0. Hence, by Part 2 of Theorem 7.4, $\{\frac{n^2-6}{5n^2}\}_{n=1}^{\infty}$ is convergent, with $\lim(\frac{n^2-6}{5n^2}) = \lim\frac{1}{5} - \lim(\frac{6}{5n^2}) = \frac{1}{5} - 0 = \frac{1}{5}$.

Given real sequences $\{a_n\}_{n=1}^{\infty}$ and $\{b_n\}_{n=1}^{\infty}$, we can form the real sequence $\{a_n b_n\}_{n=1}^{\infty}$. The next lemma will be useful in proving a theorem about the convergence of $\{a_n b_n\}_{n=1}^{\infty}$.

Lemma 7.1

Suppose that $\{a_n\}_{n=1}^{\infty}$ and $\{b_n\}_{n=1}^{\infty}$ are real sequences such that both sequences have limit 0. Then $\lim\limits_{n\to\infty} a_n b_n = 0$.

Proof: Let $\varepsilon > 0$. The goal is to show that there exists an $N \in \mathbf{N}$ such that $|a_n b_n| < \varepsilon$ whenever $n \in N$ with $n \geq \mathbf{N}$. Now, because $\lim\limits_{n\to\infty} a_n = 0$, there is an $N_1 \in \mathbf{N}$ such that for each $n \in \mathbf{N}$ with $n \geq N_1$ we have $|a_n| < 1$. Because $\lim\limits_{n\to\infty} b_n = 0$, there is an $N_2 \in \mathbf{N}$ such that for each $n \in \mathbf{N}$ with $n \geq N_2$ we have $|b_n| < \varepsilon$. Let $N = \max\{N_1, N_2\}$. Then for each $n \in \mathbf{N}$ with $n \geq N$, we have $|a_n b_n| = |a_n| \cdot |b_n| < (1) \cdot \varepsilon = \varepsilon$.

Theorem 7.5

Suppose that $\{a_n\}_{n=1}^{\infty}$ is a real sequence with limit L, and $\{b_n\}_{n=1}^{\infty}$ is a real sequence with limit M. Then $\lim\limits_{n\to\infty} a_n b_n = LM$.

Proof: Note that we can write $a_n b_n$ as $a_n b_n = (a_n - L)(b_n - M) + a_n M + Lb_n - LM$. Now, $\lim\limits_{n\to\infty}(a_n - L) = 0 = \lim\limits_{n\to\infty}(b_n - M)$. (Why?) Hence, by Theorems 7.3 and 7.4 and by Lemma 7.1 the result follows. The student is asked to fill in the details.

The next theorem can easily be proved by using mathematical induction and Theorem 7.5. The proof is left to the reader.

Theorem 7.6

Suppose that $\{a_n\}_{n=1}^{\infty}$ is a real sequence with limit L. Then for each $k \in \mathbf{N}$, we have $\lim\limits_{n\to\infty}(a_n{}^k) = L^k$.

Now, again, suppose that we have real sequences $\{a_n\}_{n=1}^{\infty}$ and $\{b_n\}_{n=1}^{\infty}$. The sequence $\{\frac{a_n}{b_n}\}_{n=1}^{\infty}$ makes sense provided that $b_n \neq 0$ for each $k \in \mathbf{N}$. Now, in some cases, we may be able to show that $b_n = 0$ for at most finitely many n. That is, we may be able to show that the set $\{n \in \mathbf{N} | b_n = 0\}$ is a finite set. Note that this is equivalent to saying that there exists an $N \in \mathbf{N}$ such that $b_n \neq 0$ for all $n \in \mathbf{N}$ with $n \geq N$. (We can also word this by saying that $\{b_n\}_{n=1}^{\infty}$ is *eventually* nonzero.) In such a circumstance, mathematicians often say that the sequence $\{\frac{a_n}{b_n}\}_{n=1}^{\infty}$ exists, when what they really mean is that $\{\frac{a_n}{b_n}\}_{n=N}^{\infty}$ exists. This is "justifiable" in the sense that if we are just interested in convergence, we can define the value of the sequence to be anything we like for members of $\{n \in \mathbf{N} | b_n = 0\}$, provided that this set is finite. (See Theorem 7.2.)

Now, suppose $\{b_n\}_{n=1}^{\infty}$ is convergent with a nonzero limit M. Then it turns out that $\{b_n\}_{n=1}^{\infty}$ is eventually nonzero. Also, the sequence $\{\frac{1}{b_n}\}_{n=1}^{\infty}$ is convergent, with $\lim_{n\to\infty} (\frac{1}{b_n}) = \frac{1}{M}$. This is the subject of Lemma 7.2. This lemma can be used to prove Theorem 7.7. Theorem 7.7 says that if $\lim_{n\to\infty} a_n = L$ and $\lim_{n\to\infty} a_n = M$ with $M \neq 0$, then $\lim_{n\to\infty} (\frac{a_n}{b_n}) = \frac{L}{M}$.

Lemma 7.2

Suppose $\{b_n\}_{n=1}^{\infty}$ is a real sequence such that $\lim_{n\to\infty} b_n = M$ with $M \neq 0$. Then there exists a $K \in \mathbf{N}$ such that $b_n \neq 0$ for each $n \in \mathbf{N}$ with $n \geq K$. Also, $\{\frac{1}{b_n}\}_{n=1}^{\infty}$ is convergent, with $\lim_{n\to\infty} (\frac{1}{b_n}) = \frac{1}{M}$.

Proof: Because $M \neq 0$, $|M| > 0$. Hence, because $\lim_{n\to\infty} b_n = M$, there exists a $K \in \mathbf{N}$ such that $|b_n - M| < \frac{|M|}{2}$ for each $n \in \mathbf{N}$ with $n \geq K$. Because $|M - b_n| = |b_n - M|$, we can write this as $|M - b_n| < \frac{|M|}{2}$ for each $n \in \mathbf{N}$ with $n \geq K$. Now, $|M| = |(M - b_n) + b_n| \leq |M - b_n| + |b_n|$. Hence, for each $n \in \mathbf{N}$ with $n \geq K$, we have $|M| < \frac{|M|}{2} + |b_n|$. From this last inequality it is easy to see that $|b_n| > \frac{|M|}{2}$ for each $n \in \mathbf{N}$ with $n \geq K$. From this, we see that $b_n \neq 0$ for each $n \in \mathbf{N}$ with $n \geq K$. This proves the first half of the theorem. We are now justified in writing the sequence $\{\frac{1}{b_n}\}_{n=1}^{\infty}$.

Let $\varepsilon > 0$. Our goal is to find an $N \in \mathbf{N}$ such that $|\frac{1}{b_n} - \frac{1}{M}| < \varepsilon$ for each $n \in \mathbf{N}$ with $n \geq N$. Now, because $\lim_{n\to\infty} b_n = M$, there exists an $N_1 \in \mathbf{N}$ such that $|b_n - M| < \frac{|M^2|\varepsilon}{2}$ for each $n \in \mathbf{N}$ with $n \geq N_1$. Now, let $N = \mathbf{max}\{N_1, K\}$. For any $n \in \mathbf{N}$ with $n \geq N$, note that $b_n \neq 0$. Also, we have

$$\left|\frac{1}{b_n} - \frac{1}{M}\right| = \left|\frac{M - b_n}{b_n M}\right| = \frac{|M - b_n|}{|b_n M_n|} < \frac{\left(\frac{|M^2|\varepsilon}{2}\right)}{|b_n| \cdot |M_n|} = \frac{|M|\varepsilon}{2|b_n|}.$$

The result follows by recalling that $|b_n| > \frac{|M|}{2}$.

The proof of Theorem 7.7 is left to the reader.

Theorem 7.7

Suppose that $\{a_n\}_{n=1}^{\infty}$ and $\{b_n\}_{n=1}^{\infty}$ are real sequences with limits L and M, respectively, and suppose $M \neq 0$. Then $\lim_{n\to\infty} (\frac{a_n}{b_n}) = \frac{L}{M}$.

Finally for this section, we show that if $\{a_n\}_{n=1}^{\infty}$ is a real convergent sequence, then the set $\{a_n | n \in \mathbf{N}\}$ is bounded. We refer to any sequence $\{a_n\}_{n=1}^{\infty}$ as a **bounded sequence** if the set $\{a_n | n \in \mathbf{N}\}$ is bounded.

Theorem 7.8

Suppose that $\{a_n\}_{n=1}^{\infty}$ is a convergent real sequence. Then there exists a $K > 0$ such that $|a_n| \leq K$ for each $n \in \mathbf{N}$.

Proof: By definition, there exists an $L \in \mathbf{R}$ such that $\lim_{n \to \infty} a_n = L$. Hence, using $\varepsilon = 1$, there exists an $N \in \mathbf{N}$ such that $|a_n - L| < 1$ whenever $n \in \mathbf{N}$ with $n \geq N$. Now, for any $n \in \mathbf{N}$ with $n \geq N$, we have

$$|a_n| = |(a_n - L) + L| \leq |a_n - L| + |L| < 1 + |L|.$$

Of course, for any $n \in \mathbf{N}$ with $n < N$, we have $|a_n| \leq \mathbf{max}\{|a_1|, |a_2| \ldots, a_{N-1}|\}$. Finally, let $K = \mathbf{max}\{|a_1|, |a_2| \ldots, a_{N-1}|, 1 + |L|\}$. Then $K > 0$, and we have $|a_n| \leq K$ for each $n \in \mathbf{N}$.

We still need some additional results about convergent sequences. We will put them into the next section simply to avoid letting this section become too large. The division of the material in this section and the next is therefore arbitrary.

EXERCISES

1. Prove Theorem 7.2.
2. Prove Part 1 of Theorem 7.3.
3. Prove Part 2 of Theorem 7.4. [Hint: $a_n - b_n = a_n + (-1) \cdot b_n$.]
4. Fill in the missing details of the proof of Theorem 7.5.
5. Prove Theorem 7.6.
6. Prove Theorem 7.7.
7. Determine if the sequence $\{a_n\}_{n=1}^{\infty}$ is convergent, in which each $a_n = \sqrt{n+1} - \sqrt{n}$. If it is convergent, find its limit. [Hint: What is $(\sqrt{n+1} - \sqrt{n}) \cdot (\sqrt{n+1} + \sqrt{n})$?]
8. Find an example of a real sequence that is bounded but is not convergent.
9. Suppose that the real sequence $\{a_n\}_{n=1}^{\infty}$ is convergent with limit L. Show that $\lim_{n \to \infty} |a_n| = |L|$. (Hint: Use Theorem 5.11.)

Section 7.2 **Convergent Real Sequences II**

We continue our brief discussion of convergent real sequences.

Definition 7.2

Suppose that $\{a_n\}_{n=1}^{\infty}$ is a real sequence. Then $\{a_n\}_{n=1}^{\infty}$ is said to be

1) **strictly increasing** if $a_n < a_{n+1}$ for each $n \in \mathbf{N}$;
2) **increasing** if $a_n \leq a_{n+1}$ for each $n \in \mathbf{N}$;
3) **strictly decreasing** if $a_{n+1} < a_n$ for each $n \in \mathbf{N}$; or
4) **decreasing** if $a_{n+1} \leq a_n$ for each $n \in \mathbf{N}$.

Note that if a real sequence is *strictly increasing*, it is also *increasing*, and if a real sequence is *strictly decreasing*, it is also *decreasing*. If a sequence is an increasing sequence or a decreasing sequence, it is also called **monotone**. Of course, it is possible for a sequence to be neither decreasing nor increasing; an example is the sequence $\{a_n\}_{n=1}^{\infty}$, in which each $a_n = (-1)^n$ for each $n \in \mathbf{N}$. It should be pointed out that in some texts, the word *increasing* is defined to mean what we have called *strictly increasing*, and similarly the word *decreasing* is defined to mean what we have called *strictly decreasing*. Such texts often use the word *nondecreasing* to mean what we have called *increasing*, and similarly use the word *nonincreasing* to mean what we have called *decreasing*.

Example 7.5

It is easy to see that the sequence $\{\frac{1}{n}\}_{n=1}^{\infty}$ is strictly decreasing. Now consider the sequence $\{\frac{3n-1}{n}\}_{n=1}^{\infty}$. Because each term can be written as $3 - \frac{1}{n}$, we see that this sequence is strictly increasing.

A sequence $\{a_n\}_{n=1}^{\infty}$ is said to be **bounded above** if the set $\{a_n | n \in \mathbf{N}\}$ is bounded above. (See Definition 3.18.) That is, $\{a_n\}_{n=1}^{\infty}$ is bounded above provided there is a $B \in \mathbf{R}$ such that $a_n \leq B$ for each $n \in \mathbf{N}$. Similarly, $\{a_n\}_{n=1}^{\infty}$ is said to be **bounded below** if there is an $A \in \mathbf{R}$ such that $A \leq a_n$ for each $n \in \mathbf{N}$. Note that Theorem 7.8 tells us that a convergent sequence is both bounded below and above because $|a_n| \leq K$ iff $-K \leq a_n \leq K$. Theorem 7.9 provides criteria for the convergence of monotone real sequences.

Theorem 7.9

Suppose that $\{a_n\}_{n=1}^{\infty}$ is a real sequence. Then

1) If $\{a_n\}_{n=1}^{\infty}$ is increasing, then this sequence converges iff it is bounded above.
2) If $\{a_n\}_{n=1}^{\infty}$ is decreasing, then this sequence converges iff it is bounded below.

Proof: We prove Part 1 and leave the proof of Part 2 to the reader. Consider the increasing sequence $\{a_n\}_{n=1}^{\infty}$. We know from Theorem 7.8 that ($\{a_n\}_{n=1}^{\infty}$ is convergent) $\Rightarrow$ ($\{a_n\}_{n=1}^{\infty}$ is bounded above). We now need to show that ($\{a_n\}_{n=1}^{\infty}$ is bounded above) $\Rightarrow$ ($\{a_n\}_{n=1}^{\infty}$ is convergent).

To this end, suppose that $\{a_n\}_{n=1}^{\infty}$ is bounded above. Then the set $\{a_n | n \in \mathbf{N}\}$ has a least upper bound; let us call it L. We will show that $\lim_{n \to \infty} a_n = L$. Let $\varepsilon > 0$. Then $L - \varepsilon$ is not an upper bound of $\{a_n | n \in \mathbf{N}\}$. Thus, there exists an $N \in \mathbf{N}$ such that $L - \varepsilon < a_N$. Because $\{a_n\}_{n=1}^{\infty}$ is increasing, it is not hard to show via induction that $a_N \le a_n$ for each $n \in \mathbf{N}$ with $n \ge N$. (Do it!) Because L is an upper bound of $\{a_n | n \in \mathbf{N}\}$, we clearly have $a_n \le L < L + \varepsilon$. Thus, for any $n \in \mathbf{N}$ with $n \ge N$ we have both $L - \varepsilon < a_N \le a_n$ and $a_n \le L < L + \varepsilon$. Hence, we have $L - \varepsilon < a_n < L + \varepsilon$, so $|a_n - L| < \varepsilon$ (by Theorem 5.12). The result follows.

Example 7.6

Consider the sequence $\{n\}_{n=1}^{\infty}$. Note that this is simply the sequence 1, 2, 3, ... We see that this sequence is both increasing and unbounded above. (In fact, it is strictly increasing.) We conclude immediately from Theorem 7.9 that $\{n\}_{n=1}^{\infty}$ is not convergent.

Next, we discuss the concept of a **subsequence**. Suppose $\{a_n\}_{n=1}^{\infty}$ is a real sequence. Intuitively, a subsequence of $\{a_n\}_{n=1}^{\infty}$ is a sequence formed by possibly removing some of the terms of $\{a_n\}_{n=1}^{\infty}$. For example, we could form the sequence $a_1, a_3, a_5, . . .$, obtained by removing the terms of $\{a_n\}_{n=1}^{\infty}$ for which the subscript is even. We need to make this concept more precise. To do so, we first consider sequences of positive integers.

Definition 7.3

A sequence $\{a_n\}_{n=1}^{\infty}$ of positive integers is said to be a **subsequence** of the sequence $\{n\}_{n=1}^{\infty}$ provided that $\{a_n\}_{n=1}^{\infty}$ is strictly increasing.

Example 7.7

It is easy to see that the sequences $\{n^2 + 3\}_{n=1}^{\infty}$, $\{2n + 1\}_{n=1}^{\infty}$, and $\{n + 3\}_{n=1}^{\infty}$ are subsequences of $\{n\}_{n=1}^{\infty}$. Also, note that $\{n\}_{n=1}^{\infty}$ is itself a subsequence of $\{n\}_{n=1}^{\infty}$.

We are now ready to define a subsequence of an arbitrary real sequence.

Definition 7.4

Suppose $\{a_n\}_{n=1}^\infty$ is a real sequence. A subsequence of $\{a_n\}_{n=1}^\infty$ is the composition of $\{a_n\}_{n=1}^\infty$ with a subsequence of $\{n\}_{n=1}^\infty$.

Example 7.8

Consider the real sequence $\{(\frac{1}{2})^{2n+1}\}_{n=1}^\infty$. Note that this sequence is a subsequence of $\{(\frac{1}{2})^n\}_{n=1}^\infty$. Why? Because $\{(\frac{1}{2})^{2n+1}\}_{n=1}^\infty$ is the composition of $\{(\frac{1}{2})^n\}_{n=1}^\infty$ with $\{2n + 1\}_{n=1}^\infty$, and $\{2n + 1\}_{n=1}^\infty$ is a subsequence of $\{n\}_{n=1}^\infty$. ∎

Often, in proofs, it is useful to express subsequences of $\{n\}_{n=1}^\infty$ by using notation such as $\{n_k\}_{k=1}^\infty$, in which n_k denotes the kth term. Why? Because if we have a real sequence $\{a_n\}_{n=1}^\infty$, then it is easy to express the subsequence of $\{a_n\}_{n=1}^\infty$ formed by the composition of $\{a_n\}_{n=1}^\infty$ with $\{n_k\}_{k=1}^\infty$ as $\{a_{n_k}\}_{k=1}^\infty$. By this we mean that the kth term of the sequence $\{a_{n_k}\}_{k=1}^\infty$ is $a(n(k))$. The reader is asked to show that if $\{n_k\}_{k=1}^\infty$ is a subsequence of $\{n\}_{n=1}^\infty$, then $n_k \geq k$ for each $k \in \mathbf{N}$. The next theorem tells us that if $\{a_n\}_{n=1}^\infty$ is a convergent real sequence with limit L, then every subsequence of $\{a_n\}_{n=1}^\infty$ also has limit L.

Theorem 7.10

Suppose that $\{a_n\}_{n=1}^\infty$ is a real sequence with $\lim\limits_{n\to\infty} a_n = L$. Then every subsequence of $\{a_n\}_{n=1}^\infty$ has limit L.

Proof: Consider any subsequence $\{a_{n_k}\}_{k=1}^\infty$ of $\{a_n\}_{n=1}^\infty$, which is the composition of $\{a_n\}_{n=1}^\infty$ with some subsequence $\{n_k\}_{k=1}^\infty$ of $\{n\}_{n=1}^\infty$. Let $\varepsilon > 0$. Then there exists an $N \in \mathbf{N}$ such that $|a_n - L| < \varepsilon$ whenever $n \in \mathbf{N}$ with $n \geq N$. Now consider any $k \in \mathbf{N}$ with $k \geq N$. Now, we have $n_k \geq k \geq N$. Thus, $|a_{n_k} - L| < \varepsilon$. We have shown that $|a_{n_k} - L| < \varepsilon$ for each $k \in \mathbf{N}$ with $k \geq \mathbf{N}$. Thus, $\lim\limits_{k\to\infty} a_{n_k} = L$.

The next theorem makes use of Theorems 7.9 and 7.10.

Theorem 7.11

Suppose that $0 \leq a < 1$. Then $\lim\limits_{n\to\infty} a^n = 0$.

Proof: $\{a_n\}_{n=1}^\infty$ is decreasing and bounded above by 0. (Why?) Hence, by Part 2 of Theorem 7.9, there exists an $L \in \mathbf{R}$ such that $\lim\limits_{n\to\infty} a^n = L$. Note that the sequence $\{a^{n+1}\}_{n=1}^\infty$ is a subsequence of $\{a^n\}_{n=1}^\infty$. Hence, by Theorem 7.10 $\lim\limits_{n\to\infty} a^{n+1} = L$. However, we can also use Part 2 of Theorem 7.3 to get $\lim\limits_{n\to\infty} a^{n+1} = \lim\limits_{n\to\infty}(aa^n) = a \lim\limits_{n\to\infty}(a^n) = aL$. Thus, we must have $L = aL$, so $L(a - 1) = 0$. Because $a \neq 1$, we must have $L = 0$. Hence, $\lim\limits_{n\to\infty} a^n = 0$.

The proof of the next theorem is left to the reader.

Theorem 7.12

Suppose that $\{a_n\}_{n=1}^{\infty}$ and $\{b_n\}_{n=1}^{\infty}$ are real sequences with limits L and M, respectively, and suppose $a_n \le b_n$ for each $n \in \mathbf{N}$. Then $L \le M$.

Next, we have the following important theorem. This theorem will be used later to help "justify" defining a^0 to be 1 for $a > 0$.

Theorem 7.13

Suppose that $a > 0$. Then $\lim\limits_{n \to \infty} a^{\frac{1}{n}} = 1$.

Proof:

Case 1: $a \ge 1$. Note that $\dfrac{1}{n+1} < \dfrac{1}{n}$ for each $n \in \mathbf{N}$. Hence, by Part 8 of Theorem 6.22, we have $a^{\frac{1}{n+1}} \le a^{\frac{1}{n}}$ for each $n \in \mathbf{N}$. Hence, $\{a^{\frac{1}{n}}\}_{n=1}^{\infty}$ is a decreasing sequence. We also see that this sequence is bounded below by 0. Hence, by Part 2 of Theorem 7.9, there is an $L \in \mathbf{R}$ such that $\lim\limits_{n \to \infty} a^{\frac{1}{n}} = L$. Also, Part 8 of Theorem 6.22 shows us that $a^{\frac{1}{n}} \ge 1$ for each $n \in \mathbf{N}$. (Why?) Hence, $L \ge 1$ by Theorem 7.12. (Why?) In particular, $L \ne 0$. Now, by Theorem 7.6 we have $\lim\limits_{n \to \infty} a^{\frac{2}{n}} = L^2$. However, note that the sequence $\{a^{\frac{2}{2n}}\}_{n=1}^{\infty}$ is a subsequence of $\{a^{\frac{2}{n}}\}_{n=1}^{\infty}$. Hence, by Theorem 7.10 we have $\lim\limits_{n \to \infty} a^{\frac{2}{2n}} = L^2$. However, $a^{\frac{2}{2n}} = a^{\frac{1}{n}}$ for each $n \in \mathbf{N}$, and we have already said that $\lim\limits_{n \to \infty} a^{\frac{1}{n}} = L$. Thus, $L^2 = L$. Because $L \ne 0$, we must have $L = 1$.

Case 2: $0 < a < 1$. Then $\dfrac{1}{a} > 1$. Hence, we apply Case 1 to get $\lim\limits_{n \to \infty} ((\frac{1}{a})^{\frac{1}{n}}) = 1$. The result now follows from Theorem 7.7. (Why?)

Next, we present the **squeeze theorem**. This theorem states that if we have a real sequence $\{c_n\}_{n=1}^{\infty}$ "squeezed" between sequences $\{a_n\}_{n=1}^{\infty}$ and $\{b_n\}_{n=1}^{\infty}$ (i.e., $a_n \le c_n \le b_n$ for each $n \in \mathbf{N}$), and if $\{a_n\}_{n=1}^{\infty}$ and $\{b_n\}_{n=1}^{\infty}$ both converge to some number L, then $\{c_n\}_{n=1}^{\infty}$ also must converge to L.

Theorem 7.14

Suppose that $\{a_n\}_{n=1}^{\infty}$ and $\{b_n\}_{n=1}^{\infty}$ are real sequences, both with limit L. Suppose that $\{c_n\}_{n=1}^{\infty}$ is a real sequence such that $a_n \le c_n \le b_n$ for each $n \in \mathbf{N}$. Then $\lim\limits_{n \to \infty} c_n = L$.

Proof: Let $\varepsilon > 0$. Then, because $\lim\limits_{n \to \infty} a_n = L$, there exists an $N_1 \in \mathbf{N}$ such that for each $n \in \mathbf{N}$ with $n \ge N_1$ we have $L - \varepsilon < a_n < L + \varepsilon$. Also, because $\lim\limits_{n \to \infty} b_n = L$, there exists an $N_2 \in \mathbf{N}$ such that for each $n \in \mathbf{N}$ with $n \ge N_2$ we have $L - \varepsilon < b_n < L + \varepsilon$. Now, let $N = \mathbf{max}\{N_1, N_2\}$. Then for any $n \in \mathbf{N}$ with $n \ge N$, we have $L - \varepsilon < a_n \le c_n \le b_n < L + \varepsilon$. That is, $L - \varepsilon < c_n < L + \varepsilon$. Hence, $\lim\limits_{n \to \infty} c_n = L$.

Example 7.9

Consider the sequence $\{\frac{1}{n!+6}\}_{n=1}^{\infty}$. Note that $0 \leq \frac{1}{n!+6} \leq \frac{1}{n}$ for each $n \in \mathbf{N}$. That is, the sequence $\{\frac{1}{n!+6}\}_{n=1}^{\infty}$ is "squeezed" between the sequences $\{0\}_{n=1}^{\infty}$ and $\{\frac{1}{n}\}_{n=1}^{\infty}$. Because these latter two sequences both have limit 0, Theorem 7.14 tells us that $\{\frac{1}{n!+6}\}_{n=1}^{\infty}$ also has limit 0. Note that we could also haveshown that $\{\frac{1}{n!+6}\}_{n=1}^{\infty}$ has limit 0 by noting that this sequence is a subsequence of the sequence $\{\frac{1}{n}\}_{n=1}^{\infty}$ and then using Theorem 7.10.

Finally, we present one more theorem, which can prove very useful in analysis.

Theorem 7.15

Suppose that $a > 0$. Then $\lim\limits_{n \to \infty} (\frac{1}{n^a}) = 0$.

Proof: Let $\varepsilon > 0$. By Theorem 6.4 there exists an $N \in \mathbf{N}$ such that $N > (\frac{1}{\varepsilon})^{(\frac{1}{a})}$. Then for any $n \in \mathbf{N}$ with $n \geq N$, we have $\frac{1}{n^a} < \varepsilon$. The result follows by noting that $|\frac{1}{n^a}| = \frac{1}{n^a}$.

EXERCISES

1. Provide an example of a real sequence that is both decreasing and increasing.
2. Prove Part 2 of Theorem 7.9.
3. Show that if $\{n_k\}_{k=1}^{\infty}$ is a subsequence of $\{n\}_{n=1}^{\infty}$, then $n_k \geq k$ for each $N \in \mathbf{N}$.
4. Prove Theorem 7.12. (Hint: First prove the following: Suppose $\{a_n\}_{n=1}^{\infty}$ is a real sequence with limit L, and suppose $a_n \geq 0$ for each $n \in \mathbf{N}$. Then $L \geq 0$.)
5. Relax the requirement of the statement in Theorem 7.12 that "$a_n \leq b_n$ for each $n \in \mathbf{N}$" to "there exists an $N \in \mathbf{N}$ such that $a_n \leq b_n$ for each $n \in \mathbf{N}$ with $n \geq N$." Show that we still have $L \leq M$.
6. Suppose that $\{a_n\}_{n=1}^{\infty}$ is a real sequence such that both subsequences $\{a_{2n}\}_{n=1}^{\infty}$ and $\{a_{2n-1}\}_{n=1}^{\infty}$ converge to the real number L. Show that $\{a_n\}_{n=1}^{\infty}$ converges to L.

Section 7.3 ## A Brief Introduction to Infinite Series

In this section we discuss those aspects of infinite series that are pertinent to our discussion of base b expansions of real numbers in Section 7.4.

Consider a real sequence $\{a_n\}_{n=1}^{\infty}$. From this sequence we can form the sequence $\{s_n\}_{n=1}^{\infty}$, in which $s_n = a_1 + a_2 + \cdots + a_n$ for each $n \in \mathbf{N}$. That is, $s_n = \sum_{k=1}^{n} a_k$ for each $n \in \mathbf{N}$. The sequence $\{s_n\}_{n=1}^{\infty}$ is called the **sequence of partial sums** of $\{a_n\}_{n=1}^{\infty}$. An **infinite series** is an ordered pair

$(\{a_n\}_{n=1}^\infty, \{s_n\}_{n=1}^\infty)$, in which $\{a_n\}_{n=1}^\infty$ is a real sequence, and $\{s_n\}_{n=1}^\infty$ is its corresponding sequence of partial sums. Let us make this formal.

Definition 7.5

Suppose $\{a_n\}_{n=1}^\infty$ is a real sequence. An **infinite series** is an ordered pair $(\{a_n\}_{n=1}^\infty, \{s_n\}_{n=1}^\infty)$, in which $\{s_n\}_{n=1}^\infty$ is the real sequence in which $s_n = \sum_{k=1}^n a_k$ for each $n \in \mathbf{N}$. The sequence $\{s_n\}_{n=1}^\infty$ is called the **sequence of partial sums** of the infinite series.

Instead of using the "messy" notation $(\{a_n\}_{n=1}^\infty, \{s_n\}_{n=1}^\infty)$, we usually express the infinite series by writing $\sum_{n=1}^\infty a_n$. Of course, there is nothing sacred about using the letter 'n'. We could, for example, express the infinite series as $\sum_{m=1}^\infty a_m$. We also do not need to use the letter 's' to write the sequence of partial sums.

Example 7.10

Consider the sequence $\{1\}_{n=1}^\infty$. This gives rise to the infinite series $\sum_{n=1}^\infty 1$. Let us denote the sequence of partial sums by $\{s_n\}_{n=1}^\infty$. Note that $s_n = \sum_{k=1}^n 1 = n$ for each $n \in \mathbf{N}$. That is, the sequence of partial sums is $\{n\}_{n=1}^\infty$. ∎

Example 7.11

Consider the sequence $\{(\frac{1}{2})^n\}_{n=1}^\infty$. This gives rise to the infinite series $\sum_{n=1}^\infty (\frac{1}{2})^n$. Now, denote the sequence of partial sums by $\{t_n\}_{n=1}^\infty$. Note that we have $t_n = \frac{1}{2} + (\frac{1}{2})^2 + \cdots + (\frac{1}{2})^n$. However, we can simplify $\{t_n\}_{n=1}^\infty$. Recall the identity for the sum of terms of a geometric progression (Theorem 6.14). Using this, we have $1 + \frac{1}{2} + (\frac{1}{2})^2 + \cdots + (\frac{1}{2})^n = \frac{(1/2)^{n+1} - 1}{1/2 - 1}$. This can be simplified to $2 - \frac{1}{2^n}$. Note that we have shown that $t_n + 1 = 2 - \frac{1}{2^n}$. Hence, $t_n = 1 - \frac{1}{2^n}$. That is, our sequence of partial sums is $\{1 - \frac{1}{2^n}\}_{n=1}^\infty$. ∎

A sequence of partial sums $\{s_n\}_{n=1}^\infty$ of an infinite series $\sum_{n=1}^\infty a_n$ may either converge or diverge. It is easy to see that the sequence of partial sums of Example 7.10 diverges. However, the sequence of partial sums of Example 7.11 converges. In fact, $\lim_{n \to \infty} (1 - \frac{1}{2^n}) = 1$. (What limit theorems from Section 7.2 have we used?) These facts motivate the next definition.

Definition 7.6

An infinite series $\sum_{n=1}^\infty a_n$ is said to **converge** (i.e., to be **convergent**), provided its sequence of partial sums converges. Otherwise, it is said to **diverge** (i.e., to be **divergent**). If the sequence of partial sums is convergent and converges to S, we say that S is the **sum** of the series $\sum_{n=1}^\infty a_n$.

Now, recall that in this text we have introduced you to a number of "abuses of notation" that are used in the mathematics community. Here is one more. If the series $\sum_{n=1}^{\infty} a_n$ is convergent with sum S, we write $\sum_{n=1}^{\infty} a_n = S$, and we can use the expression $\sum_{n=1}^{\infty} a_n$ in equations. Hence, for convergent series we use the same symbol to represent the sum of the series as we do for the series itself. So, for Example 7.11, we have $\sum_{n=1}^{\infty} (\frac{1}{2})^n = 1$. We can also write this as $\frac{1}{2} + \frac{1}{4} + \frac{1}{8} + \cdots = 1$.

Sometimes we have a series that "starts" at an integer other than 1. That is, we could have a series like $\sum_{n=k}^{\infty} a_n$ for some $k \in \mathbf{Z}$. Of course, this only makes sense if we have the real sequence $\{a_n\}_{n=k}^{\infty}$. The sequence of partial sums of $\sum_{n=k}^{\infty} a_n$ is the sequence $\{s_n\}_{n=1}^{\infty}$ in which $s_n = a_k + a_{k+1} + \cdots + a_{k+n-1}$ for each $n \in \mathbf{N}$. That is, $s_1 = a_k$, $s_2 = a_k + a_{k+1}$, $s_3 = a_k + a_{k+1} + a_{k+2}$, and so on.

When you take an analysis course, you will find out that it is often much easier to determine if a given series converges or not than it is to simplify the sum of a convergent series. However, there is one very important convergent series whose sum is easy to calculate: a **geometric series**.

Theorem 7.16   **(Sum of a Geometric Series)**

Suppose a and r are real numbers, and $0 \le r < 1$. Then $\sum_{n=0}^{\infty} ar^n$ converges, and $\sum_{n=0}^{\infty} ar^n = \dfrac{a}{1-r}$. (We define 0^0 to be 1 for this series.)

Proof: Let $\{s_n\}_{n=1}^{\infty}$ denote the sequence of partial sums. Then for each $n \in \mathbf{N}$, we have $s_n = a + ar + ar^2 + \cdots ar^{n-1}$. Note that this is simply the sum of a finite number of terms of a geometric progression. Hence, from Theorem 6.14 we have $s_n = \dfrac{a(r^n - 1)}{r-1} = \dfrac{a(1-r^n)}{1-r}$. Now, because $0 \le r < 1$, we can invoke Theorem 7.11, to get $\lim_{n \to \infty} r^n = 0$. Hence, it is now easy to show that $\lim_{n \to \infty} s_n = \dfrac{a}{1-r}$. (Do it!) Hence, $\sum_{n=0}^{\infty} ar^n = \dfrac{a}{1-r}$.

It is not hard to show that the conclusion of Theorem 7.16 holds also if $-1 < r < 0$, but we will not bother with this because it will not be pertinent to the remainder of this text. (In any case, you will see it again in your analysis course.)

Next, we present a few very useful theorems for our purposes.

Theorem 7.17   Suppose that $\sum_{n=1}^{\infty} a_n$ and $\sum_{n=1}^{\infty} b_n$ are convergent series with sums L and M, respectively. Then the series $\sum_{n=1}^{\infty} (a_n + b_n)$ is convergent, and $\sum_{n=1}^{\infty} (a_n + b_n) = L + M$.

Proof: Let $\{s_n\}_{n=1}^{\infty}$, $\{t_n\}_{n=1}^{\infty}$, and $\{u_n\}_{n=1}^{\infty}$ denote the sequences of partial sums of the series $\sum_{n=1}^{\infty} a_n$, $\sum_{n=1}^{\infty} b_n$, and $\sum_{n=1}^{\infty} (a_n + b_n)$, respectively. We have $\lim\limits_{n \to \infty} (\sum_{k=1}^{n} a_k) = L$, and $\lim\limits_{n \to \infty} (\sum_{k=1}^{n} b_k) = M$ by definition. Now, $u_n = \sum_{k=1}^{n}(a_k + b_k) = (\sum_{k=1}^{n} a_k) + (\sum_{k=1}^{n} b_k)$. Hence, by Theorem 7.4 $\{u_n\}_{n=1}^{\infty}$ is convergent, and $\lim\limits_{n \to \infty} u_n = \lim\limits_{n \to \infty} (\sum_{k=1}^{n} a_k) + \lim\limits_{n \to \infty} (\sum_{k=1}^{n} b_k) = L + M$. The result follows immediately.

Note that we can also write the conclusion of Theorem 7.17 as $\sum_{n=1}^{\infty}(a_n + b_n) = (\sum_{n=1}^{\infty} a_n) + (\sum_{n=1}^{\infty} b_n)$.

The proofs of the following three theorems are left to the reader.

Theorem 7.18

Suppose that $\sum_{n=1}^{\infty} a_n$ is convergent, and $c \in \mathbf{R}$. Then the series $\sum_{n=1}^{\infty}(ca_n)$ is convergent, and $\sum_{n=1}^{\infty}(ca_n) = c \sum_{n=1}^{\infty} a_n$.

Theorem 7.19

Suppose that $\sum_{n=1}^{\infty} a_n$ and $\sum_{n=1}^{\infty} b_n$ are convergent series with sums L and M, respectively. Then the series $\sum_{n=1}^{\infty}(a_n - b_n)$ is convergent, and $\sum_{n=1}^{\infty}(a_n - b_n) = L - M$.

Theorem 7.20

Suppose that $\sum_{n=1}^{\infty} a_n$ is a convergent series. Then for each $k \in \mathbf{N}$, $\sum_{n=k+1}^{\infty} a_n$ is convergent, and $\sum_{n=1}^{\infty} a_n = \sum_{n=1}^{k} a_n + \sum_{n=k+1}^{\infty} a_n$.

Finally, we have Theorem 7.21, which will prove useful in the next section.

Theorem 7.21

Suppose we have sequences $\{a_n\}_{n=1}^{\infty}$ and $\{b_n\}_{n=1}^{\infty}$, in which the terms of $\{a_n\}_{n=1}^{\infty}$ are all nonnegative. Furthermore suppose that $\sum_{n=1}^{\infty} b_n$ converges, and that $a_n \leq b_n$ for each $n \in \mathbf{N}$. Then $\sum_{n=1}^{\infty} a_n$ converges, and $\sum_{n=1}^{\infty} a_n \leq \sum_{n=1}^{\infty} b_n$.

Proof: Let $\{s_n\}_{n=1}^{\infty}$ and $\{t_n\}_{n=1}^{\infty}$ denote the partial sums of $\sum_{n=1}^{\infty} a_n$ and $\sum_{n=1}^{\infty} b_n$, respectively. Now $s_n = \sum_{k=1}^{n} a_k \leq \sum_{k=1}^{n} b_k = t_n$ for each $n \in \mathbf{N}$. Because $\{t_n\}_{n=1}^{\infty}$ is convergent, it is bounded above by some number B (by Theorem 7.8). Hence, $s_n \leq t_n \leq B$ for each $n \in \mathbf{N}$. This tells us that $\{s_n\}_{n=1}^{\infty}$ is also bounded above by B. Now, because $a_n \geq 0$, it is easy to see that $\{s_n\}_{n=1}^{\infty}$ is an increasing sequence. (Why?) We have shown that $\{s_n\}_{n=1}^{\infty}$ is an increasing sequence that is bounded above. Thus, $\{s_n\}_{n=1}^{\infty}$ converges by Theorem 7.9. That is, $\sum_{n=1}^{\infty} a_n$ is convergent. Also, because we have $s_n \leq t_n$

for each $n \in \mathbf{N}$, Theorem 7.12 tells us that $\lim\limits_{n \to \infty} s_n \leq \lim\limits_{n \to \infty} t_n$. That is, $\sum_{n=1}^{\infty} a_n \leq \sum_{n=1}^{\infty} b_n$.

EXERCISES

1. Find the sequence of partial sums of the series $\sum_{n=1}^{\infty}(-1)^n$. Is this series convergent or divergent?

2. Prove Theorem 7.18.

3. Prove Theorem 7.19.

4. Prove Theorem 7.20.

5. Simplify the sum of the series $\sum_{n=5}^{\infty} \frac{3}{10^n}$.

6. Use Theorem 7.21 to show that $\sum_{n=1}^{\infty} \frac{1}{n!}$ converges.

7. Suppose the sequences $\{a_n\}_{n=1}^{\infty}$ and $\{b_n\}_{n=1}^{\infty}$ are given by $a_n = (\frac{1}{3})^n$ and $b_n = (\frac{1}{4})^n$ for each $n \in \mathbf{N}$. Find the sum of each one of the series $\sum_{n=1}^{\infty} a_n$, $\sum_{n=1}^{\infty} b_n$, and $\sum_{n=1}^{\infty} a_n b_n$, respectively. Does $\sum_{n=1}^{\infty} a_n b_n = (\sum_{n=1}^{\infty} a_n) \cdot (\sum_{n=1}^{\infty} b_n)$?

Section 7.4   Representation of Real Numbers

Recall that we discussed how to express positive integers using any integer $b \geq 2$ as a "base" in Section 4.3. Specifically, we showed that any $n \in \mathbf{N}$ can be uniquely expressed as $n = a_k b^k + a_{k-1} b^{k-1} + \cdots + a_1 b + a_0$ for which k is a nonnegative integer, each of $a_0, a_1, \ldots, a_k$ are integers satisfying $0 \leq a_i \leq b - 1$ and $a_k > 0$. Expressing integers in this way makes it easier to perform calculations with them.

Similarly, we need a useful way to express real numbers to facilitate calculations with them. If you doubt this, then I hope you are not the one calculating the tip at a restaurant! A very well-known method of representation of real numbers is developed in this section. In the process of doing this, we also present a variant of a famous proof that the set of real numbers is uncountable. (Recall that an uncountable set is an infinite set that cannot be put into a one-to-one correspondence with the set of positive integers.)

We begin with a concept that should be very familiar to you. You have been writing the decimal expansions of real numbers since grade school. For example, you know that the number $\frac{1}{3}$ can be written as $\frac{1}{3} = 0.333 \cdots$. What does this mean? It means that $\frac{1}{3}$ can be written as the infinite series $\frac{1}{3} = \frac{3}{10} + \frac{3}{10^2} + \frac{3}{10^3} + \cdots$. That is, $\frac{1}{3} = \sum_{n=1}^{\infty} \frac{3}{10^n}$. Now, there is nothing mathematically "special" about the number 10. We will show that any integer $b \geq 2$ will do. First, we briefly discuss a topic that will allow us to confine our attention to the interval $[0, 1)$. (Recall from Definition 6.4 that

$[0, 1) = \{x \in R | 0 \le x < 1\}$.) Specifically, it will be useful for us to define what is meant by the **greatest integer in** x, for each real number x. We will use this concept in Theorem 7.23. This theorem will allow us to show that if we can find the base b expansion of each number in $[0, 1)$, then we can do so for any real number.

Let x be any real number. Now, we know from the Principle of Archimedes (Theorem 6.4) that there exists an integer n so that $-x \le n$. Hence, $-n \le x$. Define the set S by $S = \{k \in \mathbf{Z} | k \le x\}$. Then S is nonempty because $-n \in S$. Now, we can apply the Principle of Archimedes to x and assert the existence of an integer m so that $x \le m$. Hence, S is a nonempty set of integers that is bounded above by an integer. Thus, S has a greatest element; that is, a maximum (by Theorem 3.53.) Let us denote this maximum by $\mathbf{gint}(x)$. Note that by definition, $\mathbf{gint}(x)$ is the largest integer that is less than or equal to x. We have justified Definition 7.7 below.

Definition 7.7

Let $x \in \mathbf{R}$. Then we define the **greatest integer in** x, denoted by $\mathbf{gint}(x)$, to be the largest integer that is less than or equal to x.

In many texts, the notation $[x]$ is used to denote the greatest integer in x. Because that notation has already been used enough for other reasons in this text, we will use the notation $\mathbf{gint}(x)$. We now have the following theorem.

Theorem 7.22

Let $x \in \mathbf{R}$. Then

1) $\mathbf{gint}(x) = x$ iff $x \in \mathbf{Z}$.
2) $x - 1 < \mathbf{gint}(x) \le x$.

Proof:

1) If $\mathbf{gint}(x) = x$, then x must be an integer because $\mathbf{gint}(x)$ is an integer. Conversely, suppose $x \in \mathbf{Z}$. Then x is an integer such that $x \le x$, but $x + 1$ is an integer with $x + 1 > x$. Hence, x is the largest integer that is $\le x$ (i.e., $\mathbf{gint}(x) = x$).

2) We have $\mathbf{gint}(x) \le x$ by definition. Suppose (to reach a contradiction) that $x - 1 \ge \mathbf{gint}(x)$. Then $\mathbf{gint}(x) + 1 \le x$. Note that $\mathbf{gint}(x) + 1$ is an integer. Then $\mathbf{gint}(x) + 1$ is an integer that is less than or equal to x and is larger than $\mathbf{gint}(x)$. This contradicts the fact that $\mathbf{gint}(x)$ is the *largest* integer that is less than or equal to x.

Example 7.12   It is easy to see that $\mathbf{gint}(7 + \frac{3}{8}) = 7$ and $\mathbf{gint}(-7 - \frac{3}{8}) = -8$.

How can we use the greatest integer in x to "find a representation for" x? The next theorem paves the way.

Theorem 7.23   Let $x \geq 0$. Then there exists a unique nonnegative integer n and a unique $u \in [0, 1)$ such that $x = n + u$.

Proof: Let $n = \mathbf{gint}(x)$. We know from Theorem 7.22 that $x - 1 < n \leq x$. Because $x \geq 0$, we have $-1 \leq x - 1 < n$. Hence, because n is an integer, we must have $n \geq 0$. (Why?) Note that we can write x as $x = n + (x - n)$. From the inequality $x - 1 < n \leq x$, we can derive the inequality $0 \leq x - n < 1$. Now, we set $u = x - n$. Hence, we have $x = n + u$, in which n is a nonnegative integer and $u \in [0, 1)$.

To prove uniqueness, suppose we also have $x = n_1 + u_1$, in which n_1 is a nonnegative integer and $u \in [0, 1)$. Hence, $n_1 \leq x$ and $n_1 + 1 > n_1 + u_1 = x$. That is, n_1 is the greatest integer that is less than or equal to x. Thus, $n_1 = \mathbf{gint}(x) = n$. Now, we have $n + u = x = n_1 + u_1 = n + u_1$. From this we immediately get $u_1 = u$.

For the remainder of this section let b denote any integer such that $b \geq 2$. Let $x \geq 0$. We know from Theorem 7.23 that we can write x as $x = n + u$, in which n is a nonnegative integer and $u \in [0, 1)$. However, as already noted, we can express n in base b as $n = a_k b^k + a_{k-1} b^{k-1} + \cdots + a_1 b + a_0$. Thus, all that remains to be done is to find a way to express u in base b. (We are not concerned with negative numbers because for any $x < 0$, once we know how to express $-x$ we can express x by simply attaching a negative sign to the expression of $-x$.) We soon make additional use of the greatest integer $\mathbf{gint}$.

Now, consider a sequence $\{a_n\}_{n=1}^{\infty}$ in which $a_n \in \{0, 1, \ldots, b - 1\}$ for each $n \in \mathbf{N}$. We can form the infinite series $\frac{a_1}{b} + \frac{a_2}{b^2} + \frac{a_3}{b^3} + \cdots$; that is, the infinite series $\sum_{n=1}^{\infty} \frac{a_n}{b^n}$. For the remainder of this section we call such a series a **base b series**, and we often denote it as $(0.a_1 a_2 a_3 \cdots)_b$. Of course, when $b = 10$, we have the familiar decimal expansion (and we do not need to write the 'b'). We now show that every base b series converges to a number in the interval $[0, 1]$.

Theorem 7.24   Let $(0.a_1 a_2 a_3 \cdots)_b$ be a base b series. Then $(0.a_1 a_2 a_3 \cdots)_b$ converges to a number in the interval $[0, 1]$.

Proof: We need to show that the series $\sum_{n=1}^{\infty} \frac{a_n}{b^n}$ converges to a number in $[0, 1]$. Because each $a_n \in \{0, 1, \cdots, b - 1\}$, note that $\frac{a_n}{b^n} \leq \frac{b - 1}{b^n}$. Our plan of attack is to show that $\sum_{n=1}^{\infty}\frac{b-1}{b^n}$ converges. Then, Theorem 7.21 tells us that $\sum_{n=1}^{\infty}\frac{a_n}{b^n}$ converges and that $\sum_{n=1}^{\infty}\frac{a_n}{b^n} \leq \sum_{n=1}^{\infty}\frac{b-1}{b^n}$. Now, note that $\sum_{n=1}^{\infty}\frac{b-1}{b^n}$ can be written as $\sum_{n=1}^{\infty}(b-1)(\frac{1}{b})^n$. We know from Theorem 7.16 that $\sum_{n=0}^{\infty}(b-1)(\frac{1}{b})^n$ converges to $\frac{b-1}{1-1/b}$, which can be simplified to $\sum_{n=0}^{\infty}(b-1)(\frac{1}{b})^n = b$. Thus, $\sum_{n=1}^{\infty}(b-1)(\frac{1}{b})^n = (\sum_{n=0}^{\infty}(b-1)(\frac{1}{b})^n) - (b-1) = b - (b-1) = 1$. It is now clear that $0 \leq \sum_{n=1}^{\infty}\frac{a_n}{b^n} \leq 1$. That is, $(0.a_1a_2a_3\cdots)_b \in [0,1]$.

Note that for $b=10$, the statement $\sum_{n=1}^{\infty}\frac{b-1}{b^n}=1$ tells us that $0.999\cdots=1$. For general base b, we have $(0.b-1\,b-1\,b-1\cdots)_b = 1$. In the exercises, the reader is asked to show that if at least one digit is not $b-1$, then $(0.a_1a_2a_3\cdots)_b < 1$. You will no doubt recall that you have often come upon decimals that have a finite sequence of digits that repeat forever. (We often say that the series "terminates" with this repetition.) For example, the decimal 0.513 "terminates" with the repeating digit 0. As another example, by the decimal that we write as $0.1262626\cdots$, we mean that the digits '26' continue to repeat (i.e., it terminates with the repetition of these digits). You may recall from your earlier education that you often used the notation $0.1\overline{26}$ to express this repetition.

We can also do this for base b series. When we write $(0.\overline{r_1r_2r_3\cdots r_m})_b$, this mean that the digits $r_1, r_2, \ldots, r_m$ begin repeating immediately to the right of the decimal point. That is, by $(0.\overline{r_1r_2r_3\cdots r_m})_b$, we mean $(0.r_1r_2r_3\cdots r_m r_1r_2r_3\cdots r_m\cdots)_b$. We also can have base b series such as $(0.a_1a_2\cdots a_n \overline{r_1r_2r_3\cdots r_m})_b$. By this we mean that there may be some leading digits before the beginning of the repetition.

We are not very concerned (in this text) about whether or not we "captured" the earliest possible start of the repetition or its earliest possible length. So, for example, for our purposes it does not matter whether we express the decimal expansion $0.1\overline{26}$ as $0.126\overline{26}$. The point is that we "caught" the repetition eventually. (The exception to this is that there will be situations in which the expansion terminates with the repetition of the digit $b-1$, and we will want to know when this repetition begins.) The reader is asked in the exercises to make precise this concept of repetition.

Now, we know that any base b series converges to a number in the interval $[0,1]$. Next, we show that for every $x \in [0,1]$ there exists a base b series that converges to x. We will use the letter 'q' to represent the digits of this series because it reminds us of the word *quotient*, and thinking of the digits in this way will shortly be useful for us.

Theorem 7.25 If $x \in [0,1]$, then there exists a base b series $(0.q_1q_2q_3\cdots)_b$ that converges to x.

Proof: The idea behind the proof is quite simple. We choose q_1 to be the largest integer in the set $\{0, 1, \ldots, b - 1\}$ such that $\frac{q_1}{b} \leq x$. Next, we invoke the Generalized Recursion Theorem: Having chosen $q_1, q_2, \ldots, q_n$, we choose q_{n+1} to be the largest integer in the set $\{0, 1, \ldots, b - 1\}$ such that $\frac{q_1}{b} + \frac{q_2}{b^2} + \cdots + \frac{q_n}{b^n} + \frac{q_{n+1}}{b^{n+1}} \leq x$. We will be a bit more formal.

If $x = 1$, we already know that $x = (0.b - 1 b - 1 b - 1 \cdots)_b$. For the remainder of the proof, let us consider the case in which $0 \leq x < 1$. Hence,

$$0 \leq bx < b. \tag{1}$$

We now define the sequence $\{q_n\}_{n=1}^{\infty}$ as follows: We define q_1 by

$$q_1 = \mathbf{gint}(bx). \tag{2}$$

We now invoke the Generalized Recursion Theorem. For $n \in \mathbf{N}$, having chosen $q_1, q_2, \ldots, q_n$, we specify

$$q_{n+1} = \mathbf{gint}(b^{n+1}x - q_1 b^n - q_2 b^{n-1} \cdots - q_n b). \tag{3}$$

Note that for $n \in \mathbf{N}$ with $n \geq 2$, we can write Equation 3 as

$$q_n = \mathbf{gint}(b^n x - q_1 b^{n-1} - q_2 b^{n-2} \cdots - q_{n-1} b). \tag{4}$$

It will now be shown that $0 \leq x - \frac{q_1}{b} - \frac{q_2}{b^2} - \cdots - \frac{q_n}{b^n} < \frac{1}{b^n}$ for each $n \in \mathbf{N}$. From Equation 2 and Theorem 7.22 we have $bx - 1 < q_1 \leq bx$. Rearranging this gives us $0 \leq x - \frac{q_1}{b} < \frac{1}{b}$, so the result holds for 1. Now consider $n \in \mathbf{N}$ with $n \geq 2$. From Equation 4 and Theorem 7.22, we have

$$b^n x - q_1 b^{n-1} - q_2 b^{n-2} \cdots - q_{n-1} b - 1 < q_n$$
$$< b^n x - q_1 b^{n-1} - q_2 b^{n-2} \cdots - q_{n-1} b.$$

Dividing this inequality by b^n and doing some simple rearranging gives us the desired result. Hence, for each $n \in \mathbf{N}$, we have

$$0 \leq x - \frac{q_1}{b} - \frac{q_2}{b^2} - \cdots - \frac{q_n}{b^n} < \frac{1}{b^n}. \tag{5}$$

We now have two things left to do. We have to show that $q_n \in \{0, 1, \ldots, b - 1\}$ for each $n \in \mathbf{N}$, and we have to show that $(0.q_1 q_2 q_3 \cdots)_b = x$. To accomplish the former, we first show that $q_n \geq 0$ for each $n \in \mathbf{N}$, and we then show that $q_n < b$ for each $n \in \mathbf{N}$.

We see immediately from Inequality 1 and Equation 2 that $q_1 \geq 0$. Now consider an $n \in \mathbf{N}$ with $n \geq 2$. Using Inequality 5 with $n - 1$, we see that $\frac{q_1}{b} + \frac{q_2}{b^2} + \cdots + \frac{q_{n-1}}{b^{n-1}} \leq x$. Multiplying this by b^n and doing some simple rearranging gives us $b^n x - q_1 b^{n-1} - q_2 b^{n-2} \cdots - q_{n-1} b \geq 0$. Hence, from Equation 4 we have $q_n \geq 0$. (Why?)

We see immediately from Inequality 1 and Equation 2 that $q_1 < b$. Now consider an $m \in \mathbf{N}$ with $m \geq 2$. Then m can be written as $m = n + 1$, with $n \geq 1$. (We are doing this because we want to use Equation 3, which is terms of $n + 1$.) Now (to reach a contradiction), suppose $q_m \geq b$. That is,

$q_{n+1} \geq b$. From Inequality 5 we have $b^{n+1}x - q_1 b^n - q_2 b^{n-1} \cdots - q_n b \geq b$. (Why?) Dividing this inequality by b gives us

$$b^n x - q_1 b^{n-1} - \cdots - q_{n-1} b - q_n \geq 1. \tag{6}$$

(For $n = 1$, this should be written as $bx - q_1 \geq 1$.)

Now, from Inequality 6, q_n is the *greatest integer* that is less than or equal to the quantity $b^n x - q_1 b^{n-1} - q_2 b^{n-2} \cdots - q_{n-1}b$. Therefore, we must have $b^n x - q_1 b^{n-1} - q_2 b^{n-2} \cdots - q_{n-1}b < q_n + 1$. Note that this contradicts Inequality 6. We have completed the demonstration that $q_n \in \{0, 1, \ldots, b - 1\}$ for each $n \in \mathbf{N}$.

We now show that $(0.q_1 q_2 q_3 \cdots)_b = x$. From Inequality 5 we see that the sequence $\{x - \frac{q_1}{b} - \frac{q_2}{b^2} - \cdots - \frac{q_n}{b^n}\}_{n=1}^{\infty}$ is "squeezed" between the sequence $\{0\}_{n=1}^{\infty}$ and the sequence $\{\frac{1}{b^n}\}_{n=1}^{\infty}$. Now, we can write $\frac{1}{b^n}$ as $(\frac{1}{b})^n$. We also note that $0 < \frac{1}{b} < 1$ (because $b \geq 2$). Hence, by Theorem 7.11, $\lim_{n \to \infty} \frac{1}{b^n} = 0$. Thus, from the "squeeze" theorem (Theorem 7.14), we have $\lim_{n \to \infty} (x - \frac{q_1}{b} - \frac{q_2}{b^2} - \cdots - \frac{q_n}{b^n}) = 0$. From this we see that $\lim_{n \to \infty} (\frac{q_1}{b} + \frac{q_2}{b^2} + \cdots + \frac{q_n}{b^n}) = x$. (Why?) Now the sequence $\{\frac{q_1}{b} + \frac{q_2}{b^2} + \cdots + \frac{q_n}{b^n}\}_{n=1}^{\infty}$ is the se-quence of partial sums of the infinite series $\sum_{n=1}^{\infty} \frac{q_n}{b^n}$. Thus, $\sum_{n=1}^{\infty} \frac{q_n}{b^n} = x$. That is, $(0.q_1 q_2 q_3 \cdots)_b = x$.

Next, we show that if the base b series $(0.a_1 a_2 a_3 \cdots)_b \neq 1$, and if this expansion "terminates" with the repetition of the digit $b - 1$, then there is another way to express it as a base b expansion. To show this, we know that there must exist a $k \in \mathbf{N}$ such that $a_k \neq b - 1$. Why? Because if all of the digits are $b - 1$, we know that $(0.a_1 a_2 a_3 \cdots)_b = 1$. But we have stipulated that $(0.a_1 a_2 a_3 \cdots)_b \neq 1$. Because $(0.a_1 a_2 a_3 \cdots)_b$ "terminates" with the repetition of the digit $b - 1$, but not all of the digits are $b - 1$, there must be a largest positive integer n such that $a_n \neq b - 1$. Hence, $(0.a_1 a_2 a_3 \cdots)_b$ can be written as $(0.a_1 a_2 a_3 \cdots a_n \overline{b - 1})_b$, in which $a_n \in \{0, 1, \ldots, b - 2\}$.

Now, by Theorem 7.20, $(0.a_1 a_2 a_3 \cdots a_n \overline{b - 1})_b = (\sum_{i=1}^{n} \frac{a_i}{b^n}) + (\sum_{i=n+1}^{\infty} \frac{b-1}{b^i})$. We now ask the reader to show that $(\sum_{i=n+1}^{\infty} \frac{b-1}{b_i}) = \frac{1}{b^n}$. Thus, $(0.a_1 a_2 a_3 \cdots a_n \overline{b - 1})_b = (0.a_1 a_2 a_3 \cdots a_{n-1} a_n + 1 \overline{0})$. Note that this is a proper base b series because $a_n + 1 \in \{0, 1, \ldots, b - 1\}$.

Example 7.13 The decimal expansion $0.34\overline{9}$ can also be written as $0.35\overline{0}$.

The next theorem says that if the number x has a base b series $(0.a_1 a_2 a_3 \cdots)_b$ that does not terminate with the repetition of the digit $b - 1$, then there is no other way to express it as a base b series. The theorem further says that if $x \neq 1$, and has a base b series $(0.a_1a_2a_3\cdots)_b$ that does "terminate" with the repetition of the digit $b - 1$, there is precisely one more way to express x as a base b series (which is what we just discussed.) The proof is left to the reader.

Theorem 7.26   Suppose $x \in [0, 1)$. If x has a base b series that terminates with the repetition of the digit $b - 1$, then x has precisely two representations as a base b series. Otherwise, x has precisely one representation as a base b series.

Next, we present a slight variant of Cantor's infamous "diagonal proof" that the set **R** of real numbers is uncountable. Recall (from Theorem 4.35) that if a set S has an uncountable subset, then S is itself uncountable. Instead of directly proving that **R** is uncountable, we will show that **R** contains an uncountable subset.

Theorem 7.27   The set **R** of real numbers is uncountable.

Proof: Let S denote the set of real numbers in $[0, 1)$ that are expressable as a base 10 series using only the digits in $\{0, 1, \ldots, 8\}$. Note that by Theorem 7.26, each member of S is expressable as a base 10 series in one and only one way. Now, because $S \subseteq \mathbf{R}$, it suffices to prove that S is uncountable.

Suppose (to reach a contradiction) that S is countable. Then the totality of the member of S can be represented as a sequence $\{s_n\}_{n=1}^{\infty}$ of distinct real numbers. Hence, $S = \{s_n | n \in \mathbf{N}\}$. Now we can write members of the sequence in their base 10 decimal expansions as

$$s_1 = 0.a_{11}a_{12}a_{13}\cdots$$
$$s_2 = 0.a_{21}a_{22}a_{23}\cdots$$
$$s_3 = 0.a_{31}a_{32}a_{33}\cdots$$
$$\vdots$$

in which each digit is in the set $\{0, 1, \ldots, 8\}$; that is, a_{mn} is the nth digit of the base 10 series expansion of the mth member of the sequence. Now, we define the sequence $\{u_n\}_{n=1}^{\infty}$ as follows:

$$u_n = \begin{cases} 7 & \text{if} \quad a_{nn} \neq 7 \\ 1 & \text{if} \quad a_{nn} = 7 \end{cases} \text{ for each } n \in \mathbf{N}.$$

We define u by $u = 0.u_1u_2u_3\cdots$. Note that this real number is an element of S because it has been expressed as a base 10 series with each digit in the set $\{0, 1, \ldots, 8\}$. However, note that $u_n \neq a_{nn}$ for each $n \in \mathbf{N}$. That is, the nth digit of u is not the same as the nth digit of u_n for each $n \in \mathbf{N}$. Now, we know from Theorem 7.26 that each u_n has one and only one base 10 series. Therefore, $u \neq s_n$ for each $n \in \mathbf{N}$. However, this contradicts our earlier statement that u is a member of S. Therefore, it is not possible to list the elements of S in a sequence. That is, S is uncountable.

Why is this called the "diagonal" proof? Because to construct a contradiction, it uses the diagonal of the matrix formed by writing out the base 10 series of each of the elements of S to arrive at a contradiction. That is, it uses the numbers a_{nn}.

Finally for this section, we discuss a little more about base b series that terminate with repeating digits. The reader probably recalls from earlier grades that a number has a base 10 series that terminates with repeating digits iff the number is rational. This is true for any base b as well. We will prove part of this and leave part of it to the reader.

Theorem 7.28

Suppose $x \in [0, 1]$, and suppose x has a base b series that terminates with repeating digits. Then $x \in \mathbf{Q}$ (i.e., x is rational).

Proof:

Case 1: The base b series for x is given by $(0.\overline{r_1 r_2 r_3 \cdots r_m})_b$. Let $K = r_1 b^{m-1} + r_2 b^{m-2} + \cdots + r_{m-1} b + r_m$. Then we have $b^m x = K + (0.\overline{r_1 r_2 r_3 \cdots r_m})_b$. (Which infinite series theorems justify this step?) Hence, $b^m x = K + x$, so $x = \dfrac{K}{b^m - 1} \in \mathbf{Q}$.

Case 2: The base b series for x is given by $(0.a_1 a_2 \cdots a_n \overline{r_1 r_2 r_3 \cdots r_m})_b$ (i.e., the repetition might not begin immediately). Let $L = a_1 b^{n-1} + a_2 b^{n-2} + \cdots + a_{n-1} b + a_n$. Then we have $b^n x = L + (0.\overline{r_1 r_2 r_3 \cdots r_m})_b$. By Case 1, $(0.\overline{r_1 r_2 r_3 \cdots r_m})_b \in \mathbf{Q}$. Because $L \in \mathbf{N}$, we have $b^n x \in \mathbf{Q}$. Thus, $x \in \mathbf{Q}$.

━━━━━━

We provide an outline of the proof for the next theorem, which states that every rational number in $[0, 1]$ has a base b series that terminates with repeating digits.

Theorem 7.29

Suppose $x \in [0, 1] \cap \mathbf{Q}$. Then x has a base b series that terminates with repeating digits.

Outline of Proof: If $x = 1$, then we know that $x = (0.\overline{b-1})_b$. For the remainder of the proof, consider the case in which $x < 1$. Because x is also rational, there must exist a nonnegative integer m and a positive integer n with $m < n$ such that $x = \dfrac{m}{n}$. Now, define a sequence $\{(q_k, r_k)\}_{k=1}^{\infty}$ via the Recursion Theorem as follows: Let $q_1 = \mathbf{gint}(\dfrac{bm}{n})$ and $r_1 = bm - nq_1$. Now having (q_k, r_k) for some $k \in \mathbf{N}$, we define (q_{k+1}, r_{k+1}) by $q_{k+1} = \mathbf{gint}(\dfrac{br_k}{n})$ and $r_{k+1} = r_k b - nq_{k+1}$.

You will need to show that each $q_k \in \{0, 1, \ldots, b-1\}$, and that $(0.q_1 q_2 q_3 \cdots)_b = x$. One "sneaky" way to do this is to show that each q_k is the same as in the proof of Theorem 7.25.

Next, you will need to show that $r_k \in \{0, 1, \ldots, n-1\}$ for each $k \in \mathbf{N}$. From this, you should be able to show that there exist $u, v \in \mathbf{N}$ with $u \neq v$ such that $r_u = r_v$. Now, you can use this to "jump-start" your repetition of digits.

─────

A little reflection will help you see that each r_k in the above proof outline is the kth remainder when you find the base b series of $x = \frac{m}{n}$ via your "normal" long-division algorithm that you learned in grade school. Of course, each q_k is the kth quotient. Try a long division to convince yourself that this is true.

EXERCISES

1. Let $(0.a_1 a_2 a_3 \cdots)_b$ be a base b series. Suppose there exists a $k \in \mathbf{N}$ such that $a_k \neq b - 1$. Show that $(0.a_1 a_2 a_3 \cdots)_b \in [0, 1)$.

2. Come up with a precise mathematical definition of what is meant by a base b series terminating with repeating digits.

3. Show that $(\sum_{i=n+1}^{\infty} \frac{b-1}{b^i}) = \frac{1}{b^n}$ for each $n \in \mathbf{N}$ (provided $b > 1$).

4. Prove Theorem 7.26.

5. Recall the Division Algorithm of Section 4.3, in which we showed that for each $a, b \in \mathbf{Z}$ with $b > 0$, there are unique integers q and r such that $a = bq + r$ with $0 \leq r < b$. Show that $q = \mathbf{gint}(\frac{a}{b})$.

6. Plot the graph of $\mathbf{gint}(x)$ for $x \in [-2, 2]$.

7. **CLASS PROJECT**! Fill in all the details of the proof of Theorem 7.29 that were left out by the lazy author.

─────────────

Section 7.5   Real Exponents

In this section we briefly discuss what is meant by a^x in which $a > 0$ and x is an arbitrary real number. Of course, this has already been done (in Section 6.4) for the case in which x is rational (i.e., $x \in \mathbf{Q}$).

Many introductory analysis books define a^x after first discussing integration. After the exponential and natural logarithm functions have been defined, a^x can be defined by $a^x = e^{(x \ln a)}$. In this section we briefly show the reader how a^x can be defined by using limits. Details are available in Johnsonbaugh et al. (2002).

The basic idea for defining a^x is quite simple. First we define it for $a \geq 1$. To do this, we choose any increasing sequence $\{q_n\}_{n=1}^{\infty}$ of rational numbers that converges to x. We then define a^x to be $\lim_{n \to \infty} (a^{q_n})$. Now, recall that $\{q_n\}_{n=1}^{\infty}$ increasing means that $q_n \leq q_{n+1}$ for each $n \in \mathbf{N}$. (See Definition 7.2.) Why do we want $\{q_n\}_{n=1}^{\infty}$ to be increasing? First of all, note that there exists a rational number r with $x \leq r$. (You can get this from Theorem 6.6.) Then, because each $q_n \leq x$, we have $q_n \leq r$ for each $n \in \mathbf{N}$. (How do we know that each $q_n \leq x$?) Because $a \geq 1$, we see that $a^{q_n} \leq a^r$. That is, the sequence $\{a^{q_n}\}_{n=1}^{\infty}$ is bounded above by a^r. It is also easy to see that $\{a^{q_n}\}_{n=1}^{\infty}$ is increasing because $\{q_n\}_{n=1}^{\infty}$ is increasing and $a \geq 1$. Now, recall from Theorem 7.9 that every increasing sequence that is bounded above converges. This is what allows us to assert that $\lim_{n \to \infty}(a^{q_n})$ exists. After defining a^x for $a \geq 1$, we can define it to be $\left(\frac{1}{a}\right)^{-x}$ for $0 < a < 1$. Of course, if x is rational, then this "new" definition of a^x will turn out to coincide with the "old" definition of raising a number to a rational power that we covered in Section 6.4.

Although the basic idea is not all that hard, there are two problems that have to be dealt with. First, we need to show that there really does exist an increasing sequence $\{q_n\}_{n=1}^{\infty}$ of rational numbers that converges to x.

Theorem 7.30

Suppose $x \in \mathbf{R}$. Then there exists an increasing sequence $\{q_n\}_{n=1}^{\infty}$ of rational numbers such that $\lim_{n \to \infty} q_n = x$.

Proof: Recall (from Theorem 6.6) that there is always a rational number between two distinct real numbers. We define $\{q_n\}_{n=1}^{\infty}$ by using the Recursion Theorem as follows: Choose q_1 to be a rational number such that $x - 1 < q_1 < x$. Now for each $n \in \mathbf{N}$, having chosen q_n, choose $q_{n+1} \in \mathbf{Q}$ such that $\max\{q_n, x - \frac{1}{n+1}\} < q_{n+1} < x$. From this recursive definition, it is not hard to see that $\{q_n\}_{n=1}^{\infty}$ is increasing and that $x - \frac{1}{n} < q_n < x$ for each $n \in \mathbf{N}$. We can now invoke the "squeeze theorem" (Theorem 7.14) to see that $\lim_{n \to \infty} q_n = x$.

We should make a remark for the purist at heart. If you are not a purist, then disregard this paragraph. Recall the discussion in Section 4.5 on the Axiom of Choice. At first blush, it may appear that we need to use this in the proof of 7.30 because for each $n \in \mathbf{N}$, we are "choosing" a rational number without specifying a recipe for indicating which one to choose. The response to this is that we could modify the proof of Theorem 6.6 to indicate a specific rational number between any two given distinct real numbers. To do this, we would first have to modify the proof of the Principle of Archimedes (Theorem 6.4) to specify which integer to choose. It is often easy to do such a specification of integers in terms of least integer or greatest integer. If you are interested, then go do this!

We briefly discuss the second problem. Recall that we "want" to define a^x for $a \geq 1$ to be $\lim_{n \to \infty}(a^{q_n})$ where $\{q_n\}_{n=1}^{\infty}$ is an increasing sequence of rational numbers that converges to x. We know from Theorem 7.30 that there is a such a sequence of rational numbers. However, we still have the problem of showing that our desired definition of a^x is well defined. (Recall the issue of "well defined"?) Suppose we have another increasing sequence $\{r_n\}_{n=1}^{\infty}$ of rational numbers that converges to x. We need to show that $\lim\limits_{n \to \infty} (a^{r_n}) = \lim\limits_{n \to \infty} (a^{q_n})$. That is, we need to show that the definition of a^x does not depend on *which* increasing sequence of rational numbers converging to x is chosen. The author leaves it to the interested student to do this someday. After starting your analysis class, you will be better prepared to do this. After, the reader should be able (with some work) to show that the law of exponents of Theorem 6.22 holds for real exponents.

The author promised (in Section 6.4) a brief discussion of why 0^0 is sometimes regarded as undefined in mathematics. For that matter, why is a^0 defined to be 1 for $a > 0$? Let us discuss the latter issue first. One answer is that it makes the "laws" of exponents work nicely. For example, we can now assert that $a^b a^c = a^{(b+c)}$ for any real numbers b and c; we do not have to worry about cases in which b or c is zero.

There is also a better intuitive answer. Recall from Theorem 7.13 that if $a > 0$, then $\lim\limits_{n \to \infty} a^{\frac{1}{n}} = 1$. Hence, for example, the number $3^{\left(\frac{1}{5000}\right)}$ is indeed very close to 1. As the exponent is getting closer and closer to 0, a raised to this exponent is getting closer and closer to 1. Therefore, it is "natural" to define a^0 to be 1. In fact, it turns out that if $\{a_n\}_{n=1}^{\infty}$ is any sequence of positive real numbers converging to a with $a > 0$, and if $\{b_n\}_{n=1}^{\infty}$ is any real sequence converging to 0, then the sequence $\{a_n^{b_n}\}_{n=1}^{\infty}$ converges to 1.

Now, let us get back to the issue of 0^0. If it were "natural" to define it to be 1, then it would be the case that for every sequence $\{a_n\}_{n=1}^{\infty}$ of positive numbers converging to 0, and for every real sequence $\{b_n\}_{n=1}^{\infty}$ converging to 0, the sequence $\{a_n^{b_n}\}_{n=1}^{\infty}$ would converge to 1. However, consider the sequences $\{a_n\}_{n=1}^{\infty}$ and $\{b_n\}_{n=1}^{\infty}$ in which $a_n = \frac{1}{n^n}$ and $b_n = \frac{1}{n}$. It is clear that both of these sequences converge to 0. However, the sequence $\{a_n^{b_n}\}_{n=1}^{\infty}$ converges to 0, and the reader is asked to show this.

Given the above example, maybe it makes sense to define 0^0 to be 0. To counter this, the reader is asked in the exercises to find sequences $\{a_n\}_{n=1}^{\infty}$ and $\{b_n\}_{n=1}^{\infty}$ that both converge to 0 such that the sequence $\{a_n^{b_n}\}_{n=1}^{\infty}$ converges to 1. (See the hint.) It turns out that there is no "natural" way to define 0^0. This concept of "naturalness" is related to the concept of continuity, which you will study in your analysis course.

We have come to the end of the journey for this text. Hopefully, however, you are just at the beginning of your mathematical journey. One major construction that we did not undertake in this text is the construction of the complex number system; that is, numbers of the form $a + bi$ where i is the square root of -1. However, you will find this construction in many, many texts. You will also most likely find it to be quite simple compared to many of the constructions of this text.

EXERCISES

1. Consider the sequences $\{a_n\}_{n=1}^{\infty}$ and $\{b_n\}_{n=1}^{\infty}$ in which $a_n = \frac{1}{n^n}$ and $b_n = \frac{1}{n}$. Show that $\lim\limits_{n\to\infty} a_n^{b_n} = 0$.

2. Find real sequences $\{a_n\}_{n=1}^{\infty}$ and $\{b_n\}_{n=1}^{\infty}$ that both converge to 0, such that $\lim\limits_{n\to\infty} a_n^{b_n} = 1$. (Hint: You may use the fact that $\lim\limits_{n\to\infty} n^{\frac{1}{n}} = 1$, even though we did not prove this in this text. You will see a proof of this result in your analysis course.)

Appendix A

Proof of the Cantor–Schröder–Bernstein Theorem

Theorem 4.28 **(Cantor–Schröder–Bernstein)**

Suppose that A and B are nonempty sets. Suppose there are functions $f: A \to B$ and $g: B \to A$ that are both one-to-one. Then $A \approx B$.

The reader should be able to verify each step of the proof. However, it is difficult to convey an intuitive "feel" for what is going on.

Proof: If f is onto, then f is itself a bijection, and we are done. Similarly for g. For the remainder of the proof, we examine the case in which f and g are not onto. Hence, $B - \mathbf{im}(f)$ is nonempty. For each $y \in B - \mathbf{im}(f)$, we define a sequence $\{x_{y,n}\}_{n=1}^{\infty}$ as follows via the Recursion Theorem: $x_{y,1}$ is defined to be $g(y)$, and given $x_{y,n}$ for some $n \in \mathbf{N}$, $x_{y,n+1}$ is defined to be $g(f(x_{y,n}))$. Hence, for a given $y \in B - \mathbf{im}(f)$, the terms of the sequence $\{x_{y,n}\}_{n=1}^{\infty}$ are $g(y)$, $g(f(g(y)))$, $g(f(g(f(g(y)))))$, and so on. The sets A, B, $\mathbf{im}(f)$, and $\mathbf{im}(g)$ are displayed in Figure A.1.

With this definition, it is clear that each term of $\{x_{y,n}\}_{n=1}^{\infty}$ is in $\mathbf{im}(g)$ for each $y \in B - \mathbf{im}(f)$. Now define the set E by $\bigcup_{y \in B - \mathbf{im}(f)}(\bigcup_{n=1}^{\infty}\{x_{y,n}\})$. Note that $E \subseteq \mathbf{im}(g) \subseteq A$.

Figure A.1
Initial sets in the
proof of Theorem
4.28

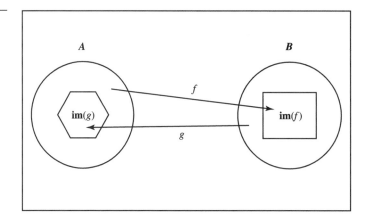

Before presenting our proposed mapping from A to B, we explain how we will "abuse notation" to simplify expressions. Suppose $x \in \mathbf{im}(g)$. Because g is one-to-one, there is a unique $b \in B$ such that $x = g(b)$. Hence, $g^{-1}(\{x\}) = \{b\}$. Now, it is not "legal" to write $g^{-1}(x) = b$ because we are analyzing the case in which g is not onto, so the function g^{-1} does not exist. (Of course, the inverse does exist if we regard g as a function with range $\mathbf{im}(g)$.) For the remainder of the proof, we adopt the convention that when we write $g^{-1}(x)$, we mean that unique element of $g^{-1}(\{x\})$.

Next, we propose a mapping $h\colon A \to B$ that we will show is a bijection. Define h by

$$h(x) = \begin{cases} g^{-1}(x) & if & x \in E \\ f(x) & if & x \in A - E \end{cases}.$$

We must show that h is onto and one-to-one.

Showing h Is Onto

Consider any $y \in B$. We must show that there exists an $a \in A$ such that $y = h(a)$. Now, let $x = g(y)$, so that $y = g^{-1}(x)$. If $x \in E$, then $h(x) = g^{-1}(x) = y$, so setting a to be x "works." For the remainder of the proof that h is onto, we analyze the case in which $x \notin E$. Now, because $x \notin E$, we must have $y \notin B - \mathbf{im}(f)$, for otherwise we would have $x = g(y) = x_{y,1} \in E$. (Recall the definition of $x_{y,1}$.) Because $y \in B$ but $y \notin B - \mathbf{im}(f)$, we must have $y \in \mathbf{im}(f)$. Hence, there must exist an $a \in A$ such that $y = f(a)$.

It will now be shown by proof by contradiction that $a \notin E$. To do so, suppose that $a \in E$. Then $a = x_{z,n}$ for some $z \in B - \mathbf{im}(f)$ and some $n \in \mathbf{N}$. Now, recalling that $x = g(y)$ and $y = f(a)$, we have $x = g(y) = g(f(a)) = g(f(x_{z,n})) = x_{z,n+1} \in E$. However, this is not possible because we are considering the situation in which $x \notin E$. We conclude that we must have $a \notin E$. Hence, $y = f(a) = h(a)$ (by the definition of h). We have met our goal of showing that there exists an $a \in A$ such that $y = h(a)$. Hence, h is onto.

Showing h Is One-to-One

To prove that h is one-to-one, we will show that if $h(u) = h(v)$ for some $u, v \in A$, then $u = v$. Hence, suppose $h(u) = h(v)$ for some $u, v \in A$.

Case 1: Both $u, v \notin E$. Then $h(u) = f(u)$ and $h(v) = f(v)$. Hence, $f(u) = f(v)$, so $u = v$ because f is one-to-one.

Case 2: Both $u, v \in E$. Then $g^{-1}(u) = g^{-1}(v)$. Because g is one-to-one, it is easy to see that $u = v$. (Why?)

Case 3: One of these elements is in E and the other is not. Let us say $u \in E$ and $v \notin E$. (The case in which $v \in E$ and $u \notin E$ is completely analogous.) Note that the statement $h(u) = h(v)$ translates to $g^{-1}(u) = f(v)$, so $u = g(f(v))$. Now, there are two subcases to consider.

Case 3a: $u = x_{z,1}$ for some $z \in B - \mathbf{im}(f)$. Hence, $u = g(z)$. However, this says that $g(z) = g(f(v))$. Because g is one-to-one, we conclude that $z = f(v)$. However, this is not possible because $z \in B - \mathbf{im}(f)$, but $f(v) \in \mathbf{im}(f)$. Hence, this case is not possible.

Case 3b: $u = x_{z,n+1}$ for some $z \in B - \mathbf{im}(f)$ and some $n \in \mathbf{N}$. Then $u = g(f(x_{z,n}))$. Because we also have $u = g(f(v))$, we must have $g(f(x_{z,n})) = g(f(v))$. Because g is one-to-one, we have $f(x_{z,n}) = f(v)$. Because f is one-to-one, we conclude that $x_{z,n} = v$. However, this is not possible because $x_{z,n} \in E$, but $v \notin E$. Hence, this case is also not possible.

Because Cases 3a and 3b are impossible, we see that Case 3 is impossible.

Hence, we have shown that in all possible cases, we have $u = v$. Hence, h is one-to-one.

Conclusion of Proof

Because h is both onto and one-to-one, it is a bijection. Hence, $A \approx B$.

Appendix B

Using the Axiom of Choice to Prove Some Results about Infinite Sets

<table>
<tr>
<td>**Axiom of Choice**</td>
<td>Suppose F is a nonempty family of nonempty sets. Then there exists a function $f: F \to \cup_{A \in F} A$ such that $f(A) \in A$ for each $A \in F$.</td>
</tr>
</table>

Proof of Theorem 4.39

<table>
<tr>
<td>**Theorem 4.39**</td>
<td>Let S be any infinite set. Then there exists a countable subset of S.</td>
</tr>
</table>

Proof: We first explain the intuitive motivation for the proof. The goal is to prove the existence of a set $A = \{a_1, a_2, \ldots\}$ of distinct elements, with $A \subseteq S$. This is accomplished as follows: Because A is infinite, it is nonempty, so we can choose an $a_1 \in S$. Now, we invoke the Generalized Recursion Theorem. Having chosen distinct elements $a_1, a_2, \ldots, a_n$ for some $n \in \mathbf{N}$, we note that the set $S - \{a_1, a_2, \ldots, a_n\}$ must be nonempty because S is infinite. Hence, we choose $a_{n+1} \in S - \{a_1, a_2, \ldots, a_n\}$. The need to invoke the Axiom of Choice is due to the fact that we are choosing an arbitrary member from an infinite collection of sets.

We now formally present the proof that involves the formal invocation of the Axiom of Choice and the Generalized Recursion Theorem. We let F

be the family of sets defined by $F = \mathbf{P}(S) - \{\varnothing\}$ (i.e., F is the collection of all subsets of S, with the exception of the empty set). Now, by the Axiom of Choice, there exists a function $h: F \to \cup_{B \in F} B$ with $h(B) \in B$. Note that $h(B) \in S$ for each $B \in F$ because $B \subseteq S$.

Now, define $g: \cup_{n=1}^{\infty} S^n \to S$ such that for each $n \in \mathbf{N}$ and each $(s_1, s_2, \ldots, s_n) \in S^n$, we specify that $g(s_1, s_2, \ldots, s_n) = h(S - \{s_1, s_2, \ldots, s_n\})$. Note that this makes sense because $S - \{s_1, s_2, \ldots, s_n\}$ is nonempty by Theorems 4.24 and 4.27. Now, by the Generalized Recursion Theorem, there is a unique sequence $\{a_n\}_{n=1}^{\infty}$ with range S such that 1) $a_1 = h(S)$, and 2) $a_{n+1} = g(a_1, a_2, \ldots, a_n)$ for each $n \in \mathbf{N}$. It is now easy to show that the elements of the sequence $\{a_n\}_{n=1}^{\infty}$ are distinct. Define A by $A = \{a_1, a_2, \ldots\}$. Then it is clear that A is a countable subset of S.

Proof of Theorem 4.40

Theorem 4.40 Let S be any infinite set. Then there exists an $A \subset S$ such that $S \approx A$.

We will make use of the following lemma. Note that the proof of the lemma does not use the Axiom of Choice.

Lemma B.1 Let S be any countable set. Then there exists an $A \subset S$ such that $S \approx A$.

Proof: Because S is countable, its elements can be listed as $S = \{s_1, s_2, \ldots\}$, in which each s_i is distinct. Now, define the function $f: S \to S$ by $f(s_n) = s_{2n}$ for each $n \in \mathbf{N}$. It is clear this mapping is one-to-one. Now, define A by $A = \mathbf{im}(f)$. It is clear that $A \subset S$. Finally define $g: S \to A$ by $g(s) = f(s)$ for each $s \in S$. Then g is a bijection. Hence, $S \approx A$.

Now we are ready to prove Theorem 4.40. Note that although the proof does not make direct reference to the Axiom of Choice, it does so indirectly by making use of Theorem 3.39, and the proof of Theorem 3.39 does use the Axiom of Choice.

Proof of Theorem 4.40: If S is countable, we are done by Lemma B.1. For the rest of the proof, we examine the case in which S is not countable. By Theorem 4.39 there exists a countable subset T of S. By Lemma B.1, there exists a $B \subset T$ such that $T \approx B$. Hence, there exists a bijection $g: T \to B$. Now, by Theorem 4.37 $S - T$ is nonempty because we are looking at the case in which S is not countable. Now, define $f: S \to (S - T) \cup B$ by

$$f(s) = \begin{cases} s & \text{if} \quad s \in S - T \\ g(s) & \text{if} \quad s \in T \end{cases}.$$

Figure B.1

Sets in the proof of
Theorem 4.40

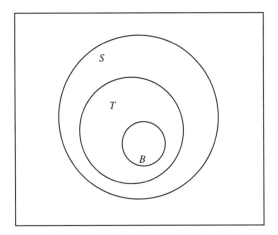

Then it is clear that f is a bijection. Hence, $S \approx (S - T) \cup B$. It is also clear that $(S - T) \cup B \subset S$. (Figure B.1 may help you visualize this. S is everything in the outer circle, T is everything in the outer circle inside S, and B is everything inside the inner circle.)

Proof of Theorem 4.41

Theorem 4.41

Suppose we have the countable family F of countable sets. Then $\cup_{A \in F} A$ is countable.

We need to prove a few lemmas first. It is helpful to write a union of a countable family of sets as a union of a countable family of pairwise disjoint sets. The proof of Lemma B.2 below does not use the Axiom of Choice.

Lemma B.2

Any union of a countable family of sets can be written as a union of a countable family of pairwise disjoint sets.

Proof: Suppose F is a countable family of sets. Then the members of F can be expressed as the members of a sequence $\{A_n\}_{n=1}^{\infty}$, so that $F = \{A_1, A_2, \ldots\}$. Define the sequence $\{B_n\}_{n=1}^{\infty}$ by $B_1 = A_1$ and $B_n = A_n - (A_1 \cup A_2 \cup \cdots \cup A_{n-1})$ for each $n \in \mathbf{N}$ with $n > 1$. It is clear that the sets of $\{B_n\}_{n=1}^{\infty}$ are pairwise disjoint. (Why?) Now it is shown that $\cup_{n=1}^{\infty} B_n = \cup_{n=1}^{\infty} A_n$. It is clear that $\cup_{n=1}^{\infty} B_n \subseteq \cup_{n=1}^{\infty} A_n$, so the proof is completed by showing that $\cup_{n=1}^{\infty} A_n \subseteq \cup_{n=1}^{\infty} B_n$. To show this, suppose $a \in \cup_{n=1}^{\infty} A_n$. Then $a \in A_n$ for some $n \in \mathbf{N}$. Hence, the set $\{m \in \mathbf{N} | a \in A_m\}$ is nonempty. Let k denote the least element of $\{m \in \mathbf{N} | a \in A_m\}$. If $k = 1$, then $a \in B_1$ (because $B_1 = A_1$). Now, consider the case in which $k > 1$.

Then $a \in A_r$ for any $r \in \mathbf{N}$ such that $1 \leqslant r \leqslant k - 1$. Hence, $a \in A_k - (A_1 \cup A_2 \cup \cdots \cup A_{k-1})$ (i.e., $a \in B_k$). Hence, $a \in \cup_{n=1}^{\infty} B_n$. We have shown that $\cup_{n=1}^{\infty} A_n \subseteq \cup_{n=1}^{\infty} B_n$.

The proof of the next lemma does make use of the Axiom of Choice.

Lemma B.3   Suppose we have the countable family F of pairwise disjoint countable sets. Then $\cup_{A \in F} A$ is countable.

Proof: Because F is countable, the members of F can be expressed as the members of a sequence $\{A_n\}_{n=1}^{\infty}$, so that $F = \{A_1, A_2, \ldots\}$, with each A_n distinct. Now, for each $n \in \mathbf{N}$, let F_n denote the set of bijections with domain $\mathbf{N}$ and range A_n. (We know that each F_n is nonempty because each A_n is countable.) By the Axiom of Choice, there exists a function $f:\{F_n | n \in \mathbf{N}\} \to \cup_{n=1}^{\infty} F_n$ such that $f(F_n) \in F_n$ for each $n \in \mathbf{N}$. (Note that we have simply "chosen" a bijection in each F_n.) Let us denote $f(F_n)$ by the sequence $\{a_{n,i}\}_{n=1}^{\infty}$ for each $n \in \mathbf{N}$, in which each $a_{n,i}$ is distinct. Note that $a_{n,i} \in A_n$ for each $n, i \in \mathbf{N}$. Now, define $g: \cup_{A \in F} A \to \mathbf{N}$ by $g(a_{n,i}) = 2^n 3^i$ for each $n, i \in \mathbf{N}$. Then, by the same technique as in the proof of Theorem 4.34 (either Proof 1 or Proof 2), we see that $\cup_{A \in F} A \approx \mathbf{N}$, so $\cup_{A \in F} A$ is countable.

It appears (at first glance) that we are ready to prove Theorem 4.41. We simply use Lemma B.2 to write the union of the sets in the family F as a union of pairwise disjoint sets and then apply Lemma B.3 to this pairwise disjoint family. However, there is one problem with this: Some (or all) of the sets in this pairwise disjoint family could be finite. (In fact, some could be the empty set, but they could not all be the empty set.) Therefore, we will need the following lemma, whose proof also uses the Axiom of Choice. (We need the Axiom of Choice because the *family of sets* of Lemma B.4 is not finite, even though each member of the family is a finite set.)

Lemma B.4   Suppose we have the countable family F of pairwise disjoint nonempty finite sets. Then $\cup_{A \in F} A$ is countable.

Proof: Because F is countable, the members of F can be expressed as the members of a sequence $\{A_n\}_{n=1}^{\infty}$, so that $F = \{A_1, A_2, \ldots\}$, with each A_n distinct. Now, because each A_n is finite but nonempty, there is a positive integer p_n such that A_n is of size p_n for each $n \in \mathbf{N}$. Now, for each $n \in \mathbf{N}$, let F_n denote the set of bijections with domain $\mathbf{I}_{p_n}$ and range A_n. (We know that each F_n is nonempty because each A_n is nonempty.) By the Axiom of

Choice, there exists a function $f: \{F_n | n \in \mathbf{N}\} \to \cup_{n=1}^{\infty} F_n$ such that $f(F_n) \in F_n$ for each $n \in \mathbf{N}$. (Note that we have simply "chosen" a bijection in each F_n.) Let us denote $f(F_n)$ by the finite sequence $\{a_{n,i}\}_{i=1}^{p_n}$ for each $n \in \mathbf{N}$, in which each $a_{n,i}$ is distinct. Note that $a_{n,i} \in A_n$ for each $n \in \mathbf{N}$ and $i \in \mathbf{I}_{p_n}$. Now, define $g: \cup_{A \in F} A \to \mathbf{N}$ by $g(a_{n,i}) = 2^n 3^i$ for each $n \in \mathbf{N}$ and $i \in \mathbf{I}_{p_n}$. Then, by the same technique as in the proof of Theorem 4.34 (either Proof 1 or Proof 2), we see that $\cup_{A \in F} A \approx \mathbf{N}$, so $\cup_{A \in F} A$ is countable.

We are now ready to prove Theorem 4.41. The proof uses the Axiom of Choice via the invocation of Lemmas B.3 and B.4.

Proof of Theorem 4.41: Because F is countable, the members of F can be expressed as the members of a sequence $\{A_n\}_{n=1}^{\infty}$, so that $F = \{A_1, A_2, \ldots\}$. Define the sequence $\{B_n\}_{n=1}^{\infty}$ by $B_1 = A_1$ and $B_n = A_n - (A_1 \cup A_2 \cup \cdots \cup A_{n-1})$ for each $n \in \mathbf{N}$ with $n > 1$. Then by the proof of Lemma B.2, the sets of $\{B_n\}_{n=1}^{\infty}$ are pairwise disjoint, and $\cup_{n=1}^{\infty} B_n = \cup_{n=1}^{\infty} A_n$. Note that $A_1 \subseteq \cup_{n=1}^{\infty} B_n$ with A_1 countable. Hence, we know (by the contrapositive of Part 1 of Theorem 4.19) that $\cup_{n=1}^{\infty} B_n$ is infinite. From the definition of each B_n, we see that $B_n \subseteq A_n$ for each $n \in \mathbf{N}$. Hence, by Theorem 4.30, each B_n is countable or finite.

Let $G = \{B_n | n \in \mathbf{N}\}$. Now define V and W by $V = \{C \in G | C \text{ is countable}\}$ and $W = \{C \in G | C \text{ is finite but nonempty}\}$. Hence, $\cup_{n=1}^{\infty} B_n = (\cup_{C \in V} C) \cup (\cup_{C \in W} C)$. Because $V \subseteq G$ and G is a countable family, V must be either a countable family or a finite family. (V could be empty.) If V is a countable family, then $(\cup_{C \in V} C)$ is countable by Lemma B.3. If V is finite but nonempty, then $(\cup_{C \in V} C)$ is countable by Theorem 4.33. (If V is empty, then of course $(\cup_{C \in V} C)$ is empty.) Similarly, $W \subseteq G$, so W must be either a countable family or a finite family. (W could be empty.) If W is countable, then $(\cup_{C \in W} C)$ is countable by Lemma B.4. Otherwise $(\cup_{C \in W} C)$ is finite, by Theorem 4.23. Now, note that if V is empty, then W must be countable because otherwise $\cup_{n=1}^{\infty} B_n$ would be finite, and we have already noted that $\cup_{n=1}^{\infty} B_n$ is infinite. Therefore, $\cup_{n=1}^{\infty} B_n$ is either the union of two countable sets or the union of a countable set and a finite set. Then, by Theorems 4.29 and 4.32, $\cup_{n=1}^{\infty} B_n$ is countable. Because $\cup_{n=1}^{\infty} B_n = \cup_{n=1}^{\infty} A_n = \cup_{A \in F} A$, $\cup_{A \in F} A$ is countable.

Appendix C

Completion of the Construction of a Set Obeying the Real Number Postulates

R ecall from Section 6.3 that we had shown that the set **R** that we defined is a field. We must now show that **R** is an ordered field and that it is a complete ordered field. We will accomplish this in three subsections. The first subsection will contain results about Cauchy sequences in ordered fields. The next subsection shows that **R** is an ordered field, and the last subsection shows that **R** is a complete ordered field. For the first subsection, the reader is asked to temporarily forget about the notation used in Section 6.3. This subsection will employ its own notation. For the last two subsections, the notation of Section 6.3 shall be in force.

Cauchy Sequences in Ordered Fields

This subsection uses its own notation, so temporarily forget about the notation of Section 6.3 for its duration. For this subsection, we let F denote an ordered field. Now, we know that F contains an integer space I by Theorem 5.13. We also know that I contains a Peano space N. Let us denote the set of positive members of F by F^+. We can define Cauchy sequences and convergent sequences for F, just like we did for rational spaces.

Definition C.1

A sequence $\{q_n\}_{n=1}^{\infty}$ of elements of F is called a **Cauchy sequence** provided that the following holds: For each $\varepsilon \in F^{+}$, there exists an $M \in N$ such that $|q_n - q_m| < \varepsilon$ whenever $m, n \in N$ with $m \geq M$ and $n \geq M$.

Definition C.2

Suppose that $\{q_n\}_{n=1}^{\infty}$ is a sequence of elements of F, and suppose $L \in F$. Then $\{q_n\}_{n=1}^{\infty}$ is said to **converge** to L provided that for each $\varepsilon \in F^{+}$, there exists an $M \in N$ such that $|q_n - L| < \varepsilon$ whenever $n \in N$ with $n \geq M$. We can also express this by saying that L is the **limit** of the sequence $\{q_n\}_{n=1}^{\infty}$. This is also expressed by writing $\lim_{n \to \infty} q_n = L$. If $\{q_n\}_{n=1}^{\infty}$ converges to some $L \in F$, we say that $\{q_n\}_{n=1}^{\infty}$ is **convergent**.

The following lemma indicates that the limit of a convergent sequence in F is unique. The proof is identical to the proof of Lemma 6.1 provided that each occurrence of 'Q^{+}' in the proof of Lemma 6.1 is replaced with 'F^{+}'.

Lemma C.1   Suppose that $\{q_n\}_{n=1}^{\infty}$ is a sequence of elements of F, and suppose $L_1, L_2 \in F$. If $\{q_n\}_{n=1}^{\infty}$ has limit L_1 and limit L_2, then $L_2 = L_1$.

Definition C.3

The ordered field F is said to **satisfy the Cauchy criterion** if each Cauchy sequence of elements of F is convergent.

Definition C.4

The ordered field F is said to be an **Archimedean field** if, for each $x \in F$, there exists an $n \in N$ such that $x < n$.

The reader may recall that we have seen the concept of Definition C.4 in Theorems 6.4 and 6.5.

Suppose that F is an Archimedean field, and let $\varepsilon \in F^{+}$. Note that there exists an $n \in N$ such that $\dfrac{1}{\varepsilon} < n$. Hence, $\dfrac{1}{n} < \varepsilon$.

Now, we state and prove the important result of this subsection.

| Lemma C.2 | Suppose that F is an Archimedean field that satisfies the Cauchy criterion. Then every nonempty subset S of F that is bounded above has a least upper bound. |

Proof: Let S be a nonempty subset of F that is bounded above. For each $n \in N$, we define the set T_n by $T_n = \{y \in I \mid nx \le y \ \forall \ x \in S\}$. Note that we have

$$x \le \frac{y}{n} \tag{1}$$

for each $x \in S$ and each $y \in T_n$. We will show that each T_n is nonempty and bounded below. Now, because S is bounded above, it must have an upper bound $b \in F$. Because F is an Archimedean field, there exists a $y \in I$ with $nb \le y$. (Of course, if $b \le 0$, we do not need to use the Archimedean property here. Also, why is it acceptable to use '$\le$' instead of '$<$'?) Now consider any $x \in S$. Because b is an upper bound of S, we have $x \le b$. Hence, $nx \le nb$. Because $nb \le y$, we have $nx \le y$. That is, we have $nx \le y$ for each $x \in S$. Thus (by the definition of T_n), we conclude that $y \in T_n$. This shows that T_n is nonempty. To show that T_n is bounded below, consider any $x \in S$. Note from the definition of T_n that $nx \le k$ for any $k \in T_n$. Thus, T_n is bounded below by nx.

We have shown for each $n \in N$, T_n is a nonempty set of integers (i.e., members of I) that is bounded below. Therefore, each T_n has a minimum; let us denote this minimum by y_n for each $n \in N$. Note that this means that $y_n - 1$ is not a lower bound of T_n. Hence, for each $n \in N$, there exists an $x_n \in S$ such that $y_n - 1 < nx_n \le y_n$ for each $n \in N$. Hence,

$$\frac{y_n}{n} - \frac{1}{n} < x_n \le \frac{y_n}{n} \tag{2}$$

for each $n \in N$.

Now, let $z_n = \frac{y_n}{n}$. We will prove that $\{z_n\}_{n=1}^{\infty}$ is a Cauchy sequence. To begin to show this, consider $m, n \in N$, and consider the case in which $\frac{y_n}{n} \le \frac{y_m}{m}$. We now claim that $\frac{y_m}{m} - \frac{1}{m} < \frac{y_n}{n} \le \frac{y_m}{m}$. Assume (to reach a contradiction) that $\frac{y_n}{n} \le \frac{y_m}{m} - \frac{1}{m}$. Then Inequality 1 tells us that $\frac{y_m}{m} - \frac{1}{m}$ is an upper bound of S. However, from Inequality 2 we have $\frac{y_m}{m} - \frac{1}{m} < x_m$. Because $x_m \in S$, we have contradicted our statement that $\frac{y_m}{m} - \frac{1}{m}$ is an upper bound of S. Thus, we have $\frac{y_m}{m} - \frac{1}{m} < \frac{y_n}{n} \le \frac{y_m}{m}$. Now, in the case in which $\frac{y_m}{m} \le \frac{y_n}{n}$, we simply reverse the roles of m and n in the above argument to get $\frac{y_n}{n} - \frac{1}{n} < \frac{y_m}{m} \le \frac{y_n}{n}$. Note that in either case, we get

$$\left| \frac{y_n}{n} - \frac{y_m}{m} \right| \le \frac{1}{\min\{m, n\}} \tag{3}$$

for each $m, n \in N$. Now, let $\varepsilon \in F^+$. From the remark following Definition C.4, there is an $M \in N$ such that $\frac{1}{M} < \varepsilon$. Hence, for any $m, n \in N$ with $m \ge M$ and $n \ge M$, we have $\left| \frac{y_n}{n} - \frac{y_m}{m} \right| \le \frac{1}{\min\{m, n\}} \le \frac{1}{M} < \varepsilon$. Thus, $\{z_n\}_{n=1}^{\infty}$ is a Cauchy sequence.

Because F satisfies the Cauchy criterion, there must exist a $w \in F$ such that $\lim_{n \to \infty} z_n = w$. Our goal is to show that w is the least upper bound of S. To do this, we first show that w is an upper bound of S. Suppose (to reach a contradiction) that w is not an upper bound of S. Then, there exists an $x \in S$ such that $w < x$. Hence, $x - w \in F^+$. Hence, there must exist an $n \in N$ such that $|z_n - w| < \frac{x - w}{2}$. Now note that $z_n - w \leq |z_n - w| = |w - z_n|$. Hence, $w - z_n \geq -|w - z_n| > -(\frac{x - w}{2})$. Thus, we have $x - z_n = (x - w) + (w - z_n) \geq (x - w) - |w - z_n| > (x - w) + -(\frac{x - w}{2}) = \frac{x - w}{2} > 0$, so we have $x > z_n$. Because $x \in S$, this tells us that z_n is not an upper bound of S. However, recalling that $z_n = \frac{y_n}{n}$, we see from Inequality 1 that z_n is an upper bound of S. Hence, we have a contradiction. We conclude that w is indeed an upper bound of S.

To complete the proof, we need to show that w is the smallest upper bound of S. To do this, we will show that if $u < w$, then u is not an upper bound of S. Now, from Inequality 2 we see that $z_n - \frac{1}{n} < x_n \leq z_n$ for each $n \in N$. Hence, $|z_n - x_n| \leq \frac{1}{n}$ for each $n \in N$. Because F is an Archimedean field, there exists an $m_1 \in N$ such that $\frac{1}{m_1} < \frac{w - u}{4}$. Hence, we have $|z_{m_1} - x_{m_1}| < \frac{w - u}{4}$. Because $\lim_{n \to \infty} z_n = w$, there must exist an $m_2 \in N$ such that $|z_k - w| < \frac{w - u}{4}$ for all $k \in N$ with $k \geq m_2$. Let $m = \max\{m_1, m_2\}$. Then we have both $|z_m - x_m| < \frac{w - u}{4}$ and $|z_m - w| < \frac{w - u}{4}$. Now, note that $z_m - x_m \leq |z_m - x_m|$. Hence, $x_m - z_m \geq -|z_m - x_m|$. Similarly, $z_m - w \geq -|z_m - w|$. Thus, we have $x_m - u = (w - u) + (x_m - z_m) + (z_m - w) \geq (w - u) - |x_m - z_m| - |z_m - w| \geq (w - u) - (\frac{w - u}{4}) - (\frac{w - u}{4}) = \frac{w - u}{2} > 0$.

Hence, $u < x_m$. Because $x_m \in S$, we see that u is not an upper bound of S. Thus, w is indeed the smallest upper bound of S. This concludes the proof that S has a least upper bound.

Showing That R Is an Ordered Field

From now until the end of this appendix, the notation of Section 6.3 is in force, so the reader should review that section. Recall that we defined a set **R** that we showed to be a field. Also, recall that we used some of the same notation for **R** that we did for Q. The context should always indicate whether we are "working" with elements of **R** or elements of Q. In this subsection, we will define an order on **R**, and we will show that **R** is an ordered field with respect to this order. In the next subsection we will show that **R** is an Archimedean field and has the Cauchy criterion. Note that by Lemma C.2, this will complete the proof that **R** is a complete ordered field.

Consider $a, b \in \mathbf{R}$ with $a = [\{a_n\}_{n=1}^{\infty}]$ and $b = [\{b_n\}_{n=1}^{\infty}]$. We "want" to define $a < b$ to mean that there exists a $c \in Q^+$ and an $M \in N$ such that $b_n - a_n > c$ for each $n \in N$ with $n \geq M$. We could then define $a \leq b$ to mean that $a < b$ or $a = b$. (Of course, to say that $a = b$ is equivalent to saying that $\{a_n\}_{n=1}^{\infty} \cong \{b_n\}_{n=1}^{\infty}$.)

Of course, we have the "usual" problem of ensuring that this definition is well defined. Hence, suppose we also have $\{a_n\}_{n=1}^{\infty} \cong \{q_n\}_{n=1}^{\infty}$ and $\{b_n\}_{n=1}^{\infty} \cong \{r_n\}_{n=1}^{\infty}$. Let us first examine the case in which $a = b$. Then we want to show that $\{q_n\}_{n=1}^{\infty} \cong \{r_n\}_{n=1}^{\infty}$. However, we immediately have this due to the transitivity of equivalence relations.

Let us now consider the case in which $a \neq b$. Recall that we have a $c \in Q^+$ and an $M \in N$ such that

$$b_n - a_n > c \qquad (1)$$

for each $n \in N$ with $n \geq M$. We need to show that there exists a $c_2 \in Q^+$ and an $M_2 \in N$ such that $r_n - q_n > c_2$ for each $n \in N$ with $n \geq M_2$. Now, we know that $\lim_{n \to \infty} (a_n - q_n) = 0$ and $\lim_{n \to \infty} (b_n - r_n) = 0$. Then there exists $K_1, K_2 \in N$ such that $|a_k - q_k| < \frac{c}{4}$ and $|b_m - r_m| < \frac{c}{4}$ for each $k, m \in N$ with $k \geq K_1$ and $m \geq K_2$. Let $K = \max\{K_1, K_2\}$. Then we have both $|a_n - q_n| < \frac{c}{4}$ and $|b_n - r_n| < \frac{c}{4}$ for each $n \in N$ with $n \geq K$. We can express this as $-\frac{c}{4} < a_n - q_n < \frac{c}{4}$ and $-\frac{c}{4} < r_n - b_n < \frac{c}{4}$. Hence, $-q_n > -a_n - \frac{c}{4}$ and $r_n > b_n - \frac{c}{4}$ for each $n \in N$ with $n \geq K$. This tells us that $r_n - q_n > b_n - a_n - \frac{c}{2}$ for each $n \in N$ with $n \geq K$. Hence, using Inequality 1, we see that $r_n - q_n > \frac{c}{2}$ for each $n \in N$ with $n \geq \max\{M, K\}$. This completes the proof that our relation $\leq$ on **R** is well defined.

Now, to show that the relation $\leq$ on **R** is a linear order, it must be shown that $\leq$ is reflexive, antisymmetric, transitive, and that any two elements are comparable. Recall that this last condition means that for each $a, b \in \mathbf{R}$, we have $a \leq b$ or $b \leq a$. The proofs that $\leq$ is reflexive, antisymmetric, and transitive are easy, and we ask the reader to supply them. We shall prove that any two elements of **R** are comparable. The next lemma will help considerably in this effort. Recall from Section 6.3 that S denotes the set of Cauchy sequences of elements of Q.

Lemma C.3

Let $\{a_n\}_{n=1}^{\infty} \in S$. Then exactly one of the following conditions holds:

1) $\{a_n\}_{n=1}^{\infty} \cong \{0\}_{n=1}^{\infty}$.

2) There exists a $c \in Q^+$ and an $M \in N$ such that $a_n \geq c$ for each $n \in N$ with $n \geq M$.

3) There exists a $c \in Q$ with $c < 0$ and an $M \in N$ such that $a_n \leq c$ for each $n \in N$ with $n \geq M$.

Proof: It is easy to see that these three conditions are mutually exclusive. Therefore, it suffices to show that at least one of them holds. If Part 1 holds, then we are done. For the remainder of the proof, consider the case in which Part 1 does not hold. By Lemma 6.7, there exists a $K \in N$ such that $|a_n| \geq c$ for each $n \in N$ with $n \geq K$. Hence, for each $n \in N$ with $n \geq K$,

$a_n \geq c$ provided a_n is positive, and $-a_n \geq c$ provided that a_n is negative. Let $E_1 = \{n \in N | a_n > 0\}$, and let $E_2 = \{m \in N | a_m < 0\}$. Now, note that $|a_n - a_m| = a_n - a_m \geq 2c$ for each $n \in E_1$ and $m \in E_2$. Suppose (to reach a contradiction) that both E_1 and E_2 are infinite sets. Then it is not the case that there exists a $K_1 \in N$ with the property that $|a_n - a_m| < 2c$ for all $n, m \in N$ with $n \geq K_1$ and $m \geq K_1$. However, this contradicts the fact that $\{a_n\}_{n=1}^\infty$ is a Cauchy sequence. Therefore, E_1 is finite or E_2 is finite. If E_2 is finite, then Part 2 holds. Otherwise, if E_1 is finite, then Part 3 holds.

Now we can prove that any two elements of **R** are comparable. Suppose $a, b \in \mathbf{R}$ with $a = [\{a_n\}_{n=1}^\infty]$ and $b = [\{b_n\}_{n=1}^\infty]$. Then $a - b = [\{a_n - b_n\}_{n=1}^\infty]$. If the Cauchy sequence $\{a_n - b_n\}_{n=1}^\infty$ satisfies Part 1 of Lemma C.3, then $a - b = 0$ (i.e., $a = b$). If Part 2 is satisfied, then $a > b$. If Part 3 is satisfied, then $a < b$. This concludes the proof that $\leq$ is a linear order on **R**. In fact, it is now easy to show that Definition 3.19 is satisfied so that **R** is in fact an ordered field. We leave the details to the reader.

The Conclusion of the Proof That R Is a Complete Ordered Field

We have already shown that **R** is an ordered field. Therefore, we can take the absolute value of elements of **R** just as we can for any ordered field. By a typical "abuse of notation," we shall reuse the absolute value sign, $| \ |$, to also indicate the absolute value of elements of **R**. Again, the context will always indicate if we are "working" with elements of Q or elements of **R**. We shall use the symbol $\mathbf{R}^+$ to indicate the positive members of **R**. Because **R** is an ordered field, it also contains a rational space **Q**, an integer space **I**, and a Peano space **N**. With a little reflection, the reader should be able to see that $\mathbf{N} = \{[\{n\}_{k=1}^\infty] | n \in N\}$ and $\mathbf{Q} = \{[\{q\}_{k=1}^\infty] | q \in Q\}$. Now, suppose that $q \in Q$. For purposes of ease of notation, we shall denote $[\{q\}_{n=1}^\infty]$ by $\bar{q}$. Hence $\bar{q} \in \mathbf{R}$ for each $q \in Q$. For the remainder of this appendix, we express the zero element and unit element of **R** as $\bar{0}$ and $\bar{1}$, respectively.

Recall that our method of proof is to show that **R** satisfies the conditions of Lemma C.2 because this will prove that each nonempty subset of **R** that is bounded above has a least upper bound. We can prove that **R** is an Archimedean field immediately.

Lemma C.4

R is an Archimedean field.

Proof: Let $x \in \mathbf{R}$ with $x = [\{a_n\}_{n=1}^\infty]$. We need to show the existence of an $M \in \mathbf{N}$ such that $x \leq M$. Now, by Lemma 6.3 there exists a $b \in Q$ such that $a_n \leq b$ for each $n \in N$. Furthermore, it is obvious that we can choose b so that $b \in Q^+$. Now recalling that Q is a rational space, there exist $k_1, k_2 \in N$

such that $b = \frac{k_1}{k_2}$. Hence, $a_n \leq \frac{k_1}{k_2} \leq k_1$ for each $n \in N$. Now, we know that we must either have $x > \bar{k}_1$ or else $x \leq \bar{k}_1$. It is easily seen that the condition that $a_n \leq k_1$ for each $n \in N$ renders the inequality $x > \bar{k}_1$ impossible.

Next, we begin taking absolute values of members of **R**.

Lemma C.5

Let $\{a_n\}_{n=1}^{\infty} \in S$, and let $c \in Q^+$. Suppose there is an $M \in N$ such that $|a_n| \leq c$ for each $n \in N$ with $n \geq M$. Then $|x| \leq \bar{c}$, where $x = [\{a_n\}_{n=1}^{\infty}]$.

Proof: Note that we have $-c \leq a_n \leq c$ for each $n \in N$ with $n \geq M$. It is now obvious that we have $-\bar{c} \leq x$ and $x \leq \bar{c}$. Hence, $-\bar{c} \leq x \leq \bar{c}$. That is, $|x| \leq \bar{c}$.

Now, because **R** is an ordered field, we can talk of convergent sequences and Cauchy sequences of elements of **R**. That is, we can apply Definitions C.1 and C.2 to the ordered field **R**. If we have a sequence $\{a_n\}_{n=1}^{\infty}$ of elements of Q, by $\{\bar{a}_n\}_{n=1}^{\infty}$, we shall mean the corresponding sequence of elements of **R**. (Technically, we should write this as $\{\overline{a_n}\}_{n=1}^{\infty}$, but such cumbersome notation is unnecessary as long as the reader knows what is meant.)

Lemma C.6

Let $\{a_n\}_{n=1}^{\infty} \in S$, and let $x = [\{a_n\}_{n=1}^{\infty}]$. Then the sequence $\{\bar{a}_n\}_{n=1}^{\infty}$ converges to x.

Proof: Let $\varepsilon \in \mathbf{R}^+$. By Lemma C.4 there exists a $K \in N$ such that $\frac{1}{\varepsilon} < \bar{K}$. Hence, we have $\frac{1}{K} < \varepsilon$. Now, because $\{a_n\}_{n=1}^{\infty}$ is a Cauchy sequence of rational numbers, there exists an $M \in N$ such that $|a_n - a_m| < \frac{1}{K}$ for each $m, n \in N$ with $m \geq M$ and $n \geq M$.

For any $m \in N$, consider the Cauchy sequence $\{a_n - a_m\}_{n=1}^{\infty}$. Define y_m by $y_m = [\{[a_n - a_m]_{n=1}^{\infty}\}]$. Hence, we can form the sequence $\{y_m\}_{m=1}^{\infty}$ of elements of **R**. Because $|a_n - a_m| < \frac{1}{K}$ for each $m, n \in N$ with $m \geq M$ and $n \geq M$, we see from Lemma C.5 that $|y_m| \leq \frac{1}{K} < \varepsilon$ for each $m, n \in N$ with $m \geq M$. Hence, we have shown that $|y_m| < \varepsilon$ for each $m \in N$ with $m \geq M$. Thus, $\lim_{m \to \infty} y_m = 0$. Now, note that $y_m = [\{a_n\}_{n=1}^{\infty}] - [\{a_m\}_{n=1}^{\infty}] = x - \bar{a}_m$. We have shown that $\lim_{m \to \infty} \bar{a}_m = x$.

We are now ready to prove the final result.

Lemma C.7　　The ordered field **R** satisfies the Cauchy criterion.

Proof: Let $\{A_n\}_{n=1}^{\infty}$ be a Cauchy sequence of elements of **R**. The goal is to prove that this sequence converges. Now, recall that each A_n is an equivalence class of a Cauchy sequence $\{x_{n,m}\}_{m=1}^{\infty}$ of rational numbers. Now, by Lemma C.6 we have $\lim_{m\to\infty} \bar{x}_{n,m} = A_n$ for each $n \in \mathbf{N}$. Therefore, for each $n \in$ **N**, there is an element of Q that we shall call a_n such that $|A_n - \bar{a}_n| < \dfrac{1}{n}$.

Let $\varepsilon \in \mathbf{R}^+$. Because **R** is an Archimedean field, there is an $M_1 \in \mathbf{N}$ such that $\dfrac{1}{M_1} < \dfrac{\varepsilon}{3}$. Because $\{A_n\}_{n=1}^{\infty}$ is a Cauchy sequence, there is an $M_2 \in \mathbf{N}$ such that $|A_n - A_m| < \dfrac{\varepsilon}{3}$ for each $m, n \in \mathbf{N}$ with $n \geq M_2$ and $m \geq M_2$. Now let $M = \mathbf{max}\{M_1, M_2\}$. Then for each $m, n \in \mathbf{N}$ with $n \geq M$ and $m \geq M$, we have $|\bar{a}_n - \bar{a}_m| \leq |\bar{a}_n - A_n| + |A_n - A_m| + |A_m - \bar{a}_m| < \dfrac{\varepsilon}{3} + \dfrac{\varepsilon}{3} + \dfrac{\varepsilon}{3} = \varepsilon$. Thus, $\{\bar{a}_n\}_{n=1}^{\infty}$ is a Cauchy sequence. Hence, $\{a_n\}_{n=1}^{\infty}$ must be a Cauchy sequence of elements of Q. (Why?) Hence, by Lemma C.6 $\{a_n\}_{n=1}^{\infty}$ converges to some $A \in \mathbf{R}$. Now, we have $|A_n - A| \leq |A_n - \bar{a}_n| + |\bar{a}_n - A|$ for each $n \in \mathbf{N}$. Now it is clear that both $|A_n - \bar{a}_n|$ and $|a_n - A|$ can each be "made" smaller than $\dfrac{\varepsilon}{2}$ for sufficiently large $n \in \mathbf{N}$. Thus, $\{A_n\}_{n=1}^{\infty}$ converges to A.

This completes the proof that **R** is a complete ordered field.

References

Burrill, Claude W. *Foundations of Real Numbers.* McGraw-Hill, New York, 1967.

Cupillari, Antonella. *The Nuts and Bolts of Proofs.* Academic Press, San Diego, 2001.

D'Angelo, John P., and Douglas B. West. *Mathematical Thinking: Problem-Solving and Proofs.* Prentice Hall, Upper Saddle River, NJ, 1997.

Devlin, Keith J. *Sets, Functions, and Logic: Basic Concepts of University Mathematics.* Chapman and Hall, London, 1981.

Gaughan, Edward. *Introduction to Analysis.* Wadsworth Publishing Company, Belmont, CA, 1968.

Goldrei, Derek. *Classic Set Theory.* Chapman & Hall, London, 1996.

Harrison, John. *Theorem Proving with the Real Numbers.* Springer-Verlag, London, 1998.

Herstein, I. N. *Topics in Algebra.* John Wiley & Sons, New York, 1975.

Jacobson, Nathan. *Basic Algebra I.* W. H. Freeman and Co., San Francisco, 1974.

Johnson, D. L. *Elements of Logic Via Numbers and Sets.* Springer-Verlag, London, 1998.

Johnsonbaugh, Richard, and W. E. Pfaffenberger. *Foundations of Mathematical Analysis.* Dover Publications, Mineola, NY, 2002.

Landau, Edmund. *Foundations of Analysis.* Chelsea Publishing Company, New York, 1966.

Lang, Serg. *Algebraic Structures.* Addison-Wesley, Reading, MA, 1967.

Protter, M. H., and C. B. Morrey. *A First Course in Real Analysis.* Springer-Verlag, New York, 1977.

Rana, Inder K. *From Numbers to Analysis.* World Scientific Publishing Co., Singapore, 1998.

Rodgers, Nancy. *Learning to Reason: An Introduction to Logic, Sets, and Relations.* John Wiley & Sons, New York, 2000.

Rosen, Kenneth, H. *Discrete Mathematics and Its Applications.* McGraw-Hill, New York, 1995.

Rosen, Kenneth, H. *Elementary Number Theory and Its Applications.* Addison-Wesley, Reading, MA, 2000.

Rudin, Walter. *Principles of Mathematical Analysis.* McGraw-Hill, New York, 1976.

Solow, Daniel. *How to Read and Do Proofs.* John Wiley & Sons, New York, 1990.

Wolf, Robert S. *Proof, Logic, and Conjecture: The Mathematician's Toolbox.* W. H. Freeman and Co., New York, 1998.

Index

Outstanding New Titles:

Computer Science Illuminated, Second Edition
Nell Dale and John Lewis
ISBN: 0-7637-2626-5
©2004

Introduction to Programming with Visual Basic .NET
Gary J. Bronson and David Rosenthal
ISBN: 0-7637-2478-5
©2005

Information Security Illuminated
Michael G. Solomon and Mike Chapple
ISBN: 0-7637-2677-X
©2005

Calculus: The Language of Change
David W. Cohen and James Henle
ISBN: 0-7637-2947-7
©2005

Applied Calculus For Scientists & Engineers: A Journey in Dialogues
Frank Blume
ISBN: 0-7637-2877-2
©2005

The Tao of Computing
Henry Walker
ISBN: 0-7637-2552-8
©2005

Databases Illuminated
Catherine Ricardo
ISBN: 0-7637-3314-8
©2004

Foundations of Algorithms Using Java Pseudocode
Richard Neapolitan and Kumarss Naimipour
ISBN: 0-7637-2129-8
©2004

Artificial Intelligence Illuminated
Ben Coppin
ISBN: 0-7637-3230-3
©2004

Programming and Problem Solving with C++, Fourth Edition
Nell Dale and Chip Weems
ISBN: 0-7637-0798-8
©2004

Java 5 Illuminated: An Active Learning Approach
Julie Anderson and Herve Franceschi
ISBN: 0-7637-1667-7
©2005

Programming in C++, Third Edition
Nell Dale and Chip Weems
ISBN: 0-7637-3234-6
©2005

Computer Networking Illuminated
Diane Barrett and Todd King
ISBN: 0-7637-2676-1
©2005

Computer Systems, Third Edition
J. Stanley Warford
ISBN: 0-7637-3239-7
©2005

A Gateway to Higher Mathematics
Jason Goodfriend
ISBN: 0-7637-2733-4
©2006

Linear Algebra with Applications, Fifth Edition
Gareth Williams
ISBN: 0-7637-3235-4
©2005

Readings in CyberEthics, Second Edition
Richard Spinello and Herman Tavani
ISBN: 0-7637-2410-6
©2004

C#.NET Illuminated
Art Gittleman
ISBN: 0-7637-2593-5
©2005

http://www.jbpub.com/

JONES AND BARTLETT
PUBLISHERS
BOSTON   TORONTO   LONDON   SINGAPORE

1.800.832.0034

Take Your Courses to the Next Level

Turn the page to preview new and forthcoming titles in Computer Science and Math from Jones and Bartlett...

Providing solutions for students and educators in the following disciplines:

- Introductory Computer Science
- Java
- C++
- Databases
- C#
- Data Structures

- Algorithms
- Network Security
- Software Engineering
- Discrete Mathematics
- Engineering Mathematics
- Complex Analysis

Please visit http://computerscience.jbpub.com/ and http://math.jbpub.com/ to learn more about our exciting publishing programs in these disciplines.